W9-CDP-053

BETWEEN SCIENCE AND VALUES

BETWEEN SCIENCE AND VALUES

Loren R. Graham

COLUMBIA UNIVERSITY PRESS
NEW YORK 1981

Clothbound editions of Columbia University Press books are Smyth-sewn and printed on permanent and durable acid-free paper.

Library of Congress Cataloging in Publication Data

Graham, Loren R.
Between science and values.

Bibliography: p.
Includes index.
1. Science—Social aspects. 2. Science—History.
3. Science—Philosophy. I. Title.
Q175.5.G72 303.4'83 81-4436
ISBN 0-231-05192-1 AACR2

Columbia University Press
New York Guildford, Surrey

Printed in the United States of America

For Meg

ACKNOWLEDGMENTS

IN WRITING this book I have enjoyed the help and support of several different institutions and foundations, as well as the assistance and advice of many students, friends, and colleagues. I began the book at Columbia University, where I taught for twelve years, and completed it at the Massachusetts Institute of Technology, to which I moved in 1978. At Columbia I first tried out many of the ideas presented in this book in undergraduate and graduate courses in the history of science. I am grateful to several research assistants who helped me with the project, particularly Michael Brainerd, Michael Perez, George Liber, and Mark Pinson. At MIT I joined the new program in Science, Technology, and Society, which has provided an ideal intellectual atmosphere for the consideration of the issues studied in this book. Donald L. M. Blackmer, director of the program at the time I was writing this book, has been a major source of assistance to me.

The most significant financial contribution to the study was made by the Rockefeller Foundation, which awarded me in 1976–1977 a Humanities Fellowship. That grant came at a time when the project was just beginning to develop into a major study and played a crucial role in its fulfillment. The last stages of the research and writing were completed with the support of a grant from the National Science Foundation (Grant No. SOC-7825584).

Along the way I have enjoyed the hospitality and intellectual encouragement of several other academic institutions, including the Department of the History of Science, the Program on Science and International Affairs, and the Russian Research Center, all at Harvard University. Among the many libraries where I have found elusive sources, I wish to mention, beside the MIT, Harvard, and Columbia libraries, the Library of Congress, the Cambridge University Library, the British Museum, Trinity College Library, and the Lenin Library in Moscow.

I have benefited from so many comments from colleagues and friends on talks I have given based on this manuscript that there is no way that I can possibly thank them all. Several people, however, deserve special praise, since they were good enough to read either the entire manuscript or major portions of it. They include, first of all, Patricia Albjerg Graham, a critic and scholar who has been immensely helpful at every stage. Peter Buck, my colleague at MIT, and Joseph G. Brennan, my former colleague at Columbia, also gave me many good suggestions. Linda Lubrano of American University was a sound and generous critic of an early version of the manuscript. Bert Hansen of the University of Toronto helped me by disagreeing with many portions of the manuscript and thereby giving me incentive to try to improve it. None of my good critics would agree with everything I have written, of course, nor did any of us expect such agreement to occur.

Portions of this book have appeared in other publications in earlier and different versions. Parts of chapters 2 and 3 will appear in a volume edited by Gerald Holton and Yehuda Elkana honoring the centennial of Einstein's birth, to be published by Princeton University Press. Chapter 8 was published in a different form in *The American Historical Review* (December 1977), and in *Morals, Science, and Society* (Hastings-on-Hudson: Hastings Center 1979), edited by H. Tristram Engelhardt, Jr. and Daniel Callahan. Chapter 10 appeared in another form in *Daedalus* (Spring 1978) and in *Limits of Scientific Inquiry* (New York: Norton, 1979), edited by Gerald Holton and Robert S. Morison. Several other small segments appeared in *The Hastings Center Report* (October 1977, and June 1979). Permission of these publishers is gratefully acknowledged.

Loren Graham
Grand Island, Lake Superior
November 1980

CONTENTS

BETWEEN SCIENCE AND VALUES

introduction

EXPANSIONISM AND RESTRICTIONISM

The Interaction of Science and Values

THE RELATIONSHIP of science and values is a subject that is frequently discussed, but rarely written about with care. And yet the topic is a very old one. It is an issue that has more often attracted the attention of philosophers than historians; philosophers have made the classical distinction between factual statements and normative statements and have emphasized the principle that "is" cannot be converted into "ought." Any person who confuses these categories may be accused of committing the "naturalistic fallacy." The fact-value distinction has been criticized recently by some philosophers and scientists, but it is still useful and important—the intellectual prohibition against confusion of "is" and "ought" still reigns. Without the introduction of an extraneous value statement somewhere along the line of reasoning, a factual statement cannot logically entail a normative one.

And yet in my opinion, this sort of logical analysis misses the main point about the historical relationship of science and values. When one turns from philosophical to historical analysis, one sees many interactions of scientific theories with value systems, in apparent defiance of philosophical principles. The prohibition against commission of the naturalistic fallacy has not, it seems, prevented scientific theories from impinging on sociopolitical and religious values. Of course, rigorous philosophical examinations of some of these moments of mutual influence might result in the conclusion that they were based on poor reasoning, and were therefore illegitimate in an intellectual sense. But just as illegitimate children need to be taken seriously by those who conceive them and by the society in which they live, so also must the interactions of the supposedly alien systems of science and values be taken seriously by those who wish to understand history. The first step, then, in a historical study of the relationship of science and values is to be willing to examine intellectually flawed arguments as well as intellectually sound ones in order to get a better understanding of the ways science and values have interacted in recent history. Our most rewarding efforts will not be to attempt to solve the fact-value dilemma on an abstract philosophical level, but to examine specific historical instances of interactions of science and values in order to gain insights into the various ways in which these interactions can occur and actually have occurred.

In choosing the episodes or examples in this book, I have asked, "What was the fundamental impact of science upon values during the twentieth century?" Although many possible replies are conceivable, it is my opinion that when historians attempt to answer this question they will see two outstanding transformations in thought in this century, both of which were provoked by traumatic disruptions by science of previous philosophical assumptions about the natural world and man's place in it. The first developed in the early decades of the century and originated primarily in physics. This transformation, which I call epistemological, arose out of relativity physics and quantum mechanics, where new developments caused a dramatic break with nineteenth-century assumptions about the materiality of the world, the significance of space and time as absolute frameworks within which natural events occurred, causality as a universal principle of scientific explanation, and determinism as a world view. Many writers attempted to show the significance of these events in science for the relationship of science to sociopolitical values.

The second transformation, which came primarily out of biology, developed gradually early in the century and then dramatically accelerated after the century's midpoint. This transformation, which I call ethical, revolved around the implications involved in the turning of science and technology upon man as an object. As a result of a series of developments in genetics, eugenics, physiology, psychology, animal behavior, and sociobiology, scientists gained capabilities for interpreting and sometimes altering human physiology and behavior that were so intimately connected with ethics that the traditional value systems appeared increasingly inadequate as guides. In response to this, there eventually arose institutions and curricula centering on such topics as biomedical ethics, and even the possibility of "forbidden knowledge" was raised.

In order to discuss the historical relationship of science and values, it is essential that we understand what we mean by the terms "science" and "values." Achieving such an understanding is a formidable problem, as is indicated by the fact that the questions "What is science?" and "What are values?" have served as themes for a library of scholarly volumes, and the disputes show no sign of abating. In a short space I have no hope of providing definitive answers that will be accepted by all critics, but only of providing working definitions that will serve the purposes of the limited analysis in this study.

Throughout this book I have taken science to be the systematic study of nature, including humans, as represented by the accepted academic disciplines. For reasons of convenience I have included the natural sciences (e.g., physics, chemistry, biology) and those areas of the social sciences which, in my opinion, are studying humans with methodologies and results highly similar to those of the natural sciences (e.g., experimental psychology, some aspects of anthropology). My omission of fields and areas that other people might consider proper subjects for my examination (history, sociology, psychoanalysis) is not an indication that I consider these subjects of less importance than the natural sciences (they are clearly of equal, and perhaps even of greater, importance); instead, these omitted fields are either sufficiently far from natural science to be reasonably excluded (history), or sufficiently far from my own special knowledge to be excluded on personal grounds (sociology, psychoanalysis). Their exclusion may make me vulnerable to the criticism of incompleteness in my study, but I do not think that it will invalidate the analysis as a whole.

The question, "What are values?" is even more controversial than "What is science?" The term "value" was until the nineteenth century almost entirely connected with economics or political economy, and meant "the worth of a thing." Beginning a little over a hundred years ago, however, the meaning of the term became much broader, and this new definition, or set of definitions, has become quite popular. The term is used in this book in the sense of one of these broader definitions, namely "what people think to be good." The term thus encompasses the broad range of normative judgments which all human beings make, and which can be further broken down into various types of sociopolitical values (e.g., moral, political, esthetic, religious).

Although my approach in this book to the relationship of science and values is historical, my interests here are somewhat different from those of the majority of historians of science. The main concern of historians of science has been, understandably and appropriately, the growth of scientific knowledge. Many observers of the work of recent historians of science have divided them into two groups—the "internalists" and the "externalists". The first pay the greatest attention to the seemingly autonomous growth of concepts and theories within a particular scientific discipline; the second place the greatest emphasis on the social context of that growth. If one accepts, for a moment at least, this overworked distinction between internal and external historians of science and then asks what the central concerns of these scholars are, it becomes clear that, for all their differences, they are united in one purpose: to understand the sources of science and the forces that mold science. This goal, important and legitimate as it is, has had the effect of screening out investigations of science-value interactions whenever the results of those interactions do not seem significant to science itself. The internalists have emphasized the logic of scientific thought and the significance of empirical findings as the major influences on the growth of scientific knowledge. The externalists, on the other hand, pay attention to the social, economic, and nonscientific intellectual influences on the development of science. Both groups do valuable work in terms of their stated goals, but neither has taken the science-value nexus as the primary theme for historical research. Their self-assigned subject is the growth of knowledge, not the relationship of science and social values.

Studies of the relationship of science and values might appear to be a part of "external history of science," but a look at some examples of what we usually consider external history of science shows that in the consideration of the linked words "science-values" the emphasis is upon the first term and not on the interface between them. Whether one examines existing studies of the relationship between Puritan religion and science in seventeenth-century England, or the influence of German *Naturphilosophie* on the origins of field theory, or the significance of Platonic thought to Galileo, or the bearing of Marxism on Soviet science, or the influence of Weimar culture on quantum mechanics, the center of attention in most cases is the impact of extrascientific influences on science itself, not on interaction between science and cultural values.

Recent debate on biomedical ethics rarely includes examinations of the relationship of science and values. In these studies it is true that the arrows of influence are usually described as pointing in the opposite direction from those in external history of science; instead of worrying how *science* is being affected by societal influences, the concern is now often about the influence upon *society* of new advances in medicine or biology. But if we examine our words carefully, we will see that most of the recent studies of biomedical ethics are not about science and values but about technology and values. The growing sophistication of new machines for the support of life in extreme situations or of new possibilities in areas such as genetic engineering and psychosurgery raise fundamental questions deserving great attention, but we should recognize that in most instances we are discussing the impact of medical technology much more than we are discussing science-value interactions.

Throughout the history of science a great many attempts have been made to draw conclusions about sociopolitical values on the basis of science, and these attempts have differed greatly in approach and in the logic of argument. Without attempting to classify exhaustively all of these efforts, I would like to point to two distinctly different classes of arguments about the relationship of science and values, whose supporters I will call the "expansionists" and the "restrictionists."

Expansionists cite evidence within the body of scientific theories and findings which can supposedly be used, either directly or indirectly, to support conclusions about sociopolitical values. The result of these efforts is to expand the boundaries of science in such a way that they include, at least by implication, value questions. A historically well-known type of interpretation in this category is the "argument from design" for the existence of God; the architecture of the physical universe, the structure of biological organisms, or the form of individual organs may be cited as evidence for the existence of some sort of Supreme Architect. This type of argument—"the revealing of the glory of God in the heavens and the earth"—was very popular in Europe and America before the nineteenth century. Numerous prominent examples could easily be given, from Isaac Newton's belief he was revealing the architecture of God, to the English churchman William Paley's similar convictions about evidence of a divine pattern in biological and physical nature. Such expansionist arguments are not, however, unique to supporters of religious belief, nor have they disappeared in our century. An atheist who counters religious belief by citing scientific evidence—as, say, Clémence Royer did in the anticlerical introduction to her 1866 French translation of Darwin's *Origin of Species*—is also an expansionist, since evidence found in the body of science is brought to bear on value questions. Expansionists are obviously of very different types; most twentieth-century Marxists are expansionists, but then so was the unorthodox Jesuit priest Teilhard de Chardin.

Among expansionist viewpoints several different subclasses of arguments exist. The linkage between science and values constructed by expansionist authors can be either direct or indirect. A direct linkage is one where science is said to relate to values in a way that is not merely by suggestion or implication, but in a logical, confirming, or denying fashion. To take a simple example, if a person is a biblical literalist who accepts the Genesis story as factually true, and sees this story as an integral part of his or her religious system, then the sciences of geology and biology, with their secular and historical accounts of the origins of the earth and of life on it, should speak to that person in a direct way. Gillispie's *Genesis and Geology* contains much discussion of controversies in the late eighteenth and early nineteenth centuries which resulted from this sort of direct expansionist argument.[1]

On a more sophisticated and contemporary level, psychological behaviorists who believe that values are environmentally formed and can be created and controlled at will, once science is refined, are not only working with an expansionist argument but are making direct linkages between science and values. An example of such a linkage is B. F. Skinner's statement:

> When we say that a value judgment is a matter not of fact but of how someone feels about a fact, we are simply distinguishing between a thing and its reinforcing effect. . . . Reinforcing effects of things are the province of behavioral science, which to the extent that it is concerned with operant reinforcement, is a science of values.[2]

There is, of course, a significant difference between an expansionist argument "from design" of the type previously discussed and an expansionist argument like B. F. Skinner's. The first attempts to find evidence favoring a religious value in scientific findings; the second attempts to create or mold social value in an individual by controlling his or her environment; both are expansionists, however, in the sense that science is made to impinge directly on social values.

E. O. Wilson in his book *Sociobiology* opened the door leading to direct expansionist linkages when he called for a "biologicization of ethics," but one remarkable aspect of that book was that, with only a few exceptions, Wilson did not walk through the door; he opened it, pointed through it, and occasionally put a toe across the threshold. The reason, however, that the fields of sociobiology and animal behavior have excited interest among the educated lay public is that members of that public correctly see these academic fields as efforts to extend natural science to human behavior, and as attempts to present explanations of sociopolitical values, including ethics.

Expansionists may rely on arguments belonging to a second subclass, that of indirect linkages. These expansionists do not try to bring a particular piece of scientific evidence or a particular scientific theory into immediate logical relationship with values, but instead work indirectly with the instruments of analogy, simile, or metaphor. Social Darwinists of the nineteenth century who made apologies for unregulated or unsoftened industrial capitalism by pointing to the analogy between the struggle for existence in the biological world and competition in the economic one were following the line of argument of the "indirect linkage" of science and values within the expansionist approach. So was the critic of capitalism Friedrich Engels when in his book *Anti-Dühring* he pointed to similar dialectical laws in chemistry and economics. And the astronomer James Jeans playfully pursued a similar indirect expansionist argument in his popular writings when he spoke of the "finger of God" that started the planets in their orbits.

The logical alternative to expansionism is restrictionism, an approach that confines science to a particular realm or a particular methodology and leaves values outside its boundaries. Although there are many types of values other than religious ones, restrictionism is best known in debates about religion; advocates of the restrictionist approach often say "science and religion cannot possibly conflict because they talk about entirely different things." An agnostic or an atheist may make the same point about science and his or her ethical values, believing that values can be drawn from a realm that is neither scientific nor religious.

A strict adherence to restrictionism would mean that the relationship of science to ethical, sociopolitical, and religious values is neutral. Science can be used to support neither human selfishness or human altruism, nor can it affirm or deny either religious belief or atheism. Science is simply neutral with respect to values. Period.

The restrictionist point of view was particularly strong among English-speaking intellectuals in the first half of this century. In an article entitled, "Can Science and Ethics Be Reconnected?" Stephen Toulmin observed that the normative neutrality of science was expressed in an exaggerated form by his professors and senior colleagues during the time of his education in England before World War II. Among the factors he described as strengthening this view were "the factual, unemotional, antiphilosophical, class-structured, and role-oriented attitudes of the English professional classes between the two world wars."[3] The main characteristic of this attitude toward science, he continued, was "to select as one's profound center of attention the purest, the most intellectual, the most autonomous, and the least ethically implicated extreme" on the spectra of science-value interactions.[4] And Toulmin further implied that the main error of this approach was not that it *never* is correct, but that it ignored almost everything that was happening on the other ends of the spectra, those topics in science, particularly in the biological and social sciences, where some concepts are irretrievably value-laden.

The restrictionism that was popular in England and America in the first half of this century contained a paradox. As we have noted, strictly speaking this restrictionism should not have supported any particular value system, for it was based on the assumption that science and values belong to entirely separate realms. But in order to understand the effects of restrictionism we need to turn from abstract analysis to chronological and social analysis. Seen from this historical point of view, the restrictionism of the thirties and forties had a considerable impact on values, for its actual function was to protect two systems of values: the professional values of scientists and the predominant nonscientific ethical and sociopolitical values of society as they existed at that time. For if science and values could not interact, then scientists were safe from incursions by critics who tried to submit scientific ideas and the scientific profession to social criticism; and ethicists and spokespeople for political or religious values were safe from attempts by scientists to show the relevance of science to their concerns. The historical result of this demarcation of boundaries was to support existing institutional expressions of the defenders of science and social values. Because I am a supporter of the scientific enterprise and also believe that society cannot exist without value systems, I believe that at least some of the effects of this demarcation were positive; the exaggerated restrictionism that embodied the demarcation was, however, a temporary historical product based on assumptions no longer tenable. Indeed, the negative effects of this compromise are now increasingly clear.

The interpretation of the relationship of science and values I am presenting in this book emerges from the illustration of how different authors and interpreters of science have used different variations of the expansionist and restrictionist approaches in talking about science and values. One may ask why the expansionist-restrictionist categories are analytically useful—i.e., what do we learn by feeding various examples through this particular grid? In my opinion, these categories illuminate what is one of the central unresolved issues of scientific civilization, namely the question of whether the study of nature and the study of ethics and other social values can continue to be held in separate boxes to the degree that they have during the last one hundred years. The effort to compartmentalize these realms is fairly new in history, and coincides (not accidentally, as I will discuss in the next section), with the rise of professionalized science. In the final analysis, this compartmentalization is not intellectually defensible, but an alternative approach that does not run the risk of doing more damage to society, or to science, than is done by the present compartmentalization is still not apparent. Yet the growing relevance of science to values makes the denial of their relationship ever more artificial. Many of the expansionists and restrictionists whom I discuss in the following chapters were aware of this dilemma, addressing it either implicitly or explicitly, and they wrestled with it in different fashions. The expansionist-restrictionist dichotomy is, then, a method of revealing the deep contradictions that have arisen in society as the result of this unresolved dilemma about the relationship of science and values.

It is obvious that I cannot deal with every aspect of the subject of science-value interactions in the twentieth century; the topic is far too vast for that. I have limited myself, therefore, to a few of the major controversies, and I have chosen episodes or examples in the recent history of science that illustrate a variety of arguments among both expansionists and restrictionists.

We will see, for example, that Albert Einstein was a restrictionist who criticized those of his colleagues who tried to show the relevance of modern physics to ethics and other nonscientific areas of human activity. His colleague Niels Bohr, on the other hand, was an expansionist who believed his concept of complementarity had relevance for ethics and human culture. Other examples in the book of expansionists include V. A. Fock, Teilhard de Chardin, and Konrad Lorenz; other restrictionists include Arthur Eddington and Henri Bergson.

The book is organized into four parts (after an introduction), the first two of which concern the physical sciences and the biological sciences. Each of these parts is broken down further into two sections, the first dealing with new developments in that area of science in this century and the ways in which several of the most eminent scientists in the field related their work to values. The second section of each of these parts deals with "knowledge and society," and explores how scientific conceptions in physics and biology became involved in questions of human values.

In discussing physics, I have put great emphasis on "authoritative popularizers," scientists who were both eminent specialists in their field and who also made serious efforts to interpret their work (and its bearing on social values) to the educated public. I have placed less emphasis on such individuals in the biological sciences, since I think that the authoritative popularizers who exist in biology (and I have discussed several of them, such as Jacques Monod and Francis Crick) did not play quite the equivalent role in their time as physicists like Arthur Eddington and Werner Heisenberg did in theirs. Relativity physics and quantum mechanics seemed so baffling to educated readers of the generation from 1920 to 1950 that almost every person with some pretension to intellectual curiosity turned to the popular writings of one of the giants in the field for enlightenment. Although these readers did not often learn much physics, they frequently picked up the views of the authors on the meaning of the new physics for social values. Biology, for all the wonderment it engendered, never provoked quite the same curiosity and frustration among lay readers as did physics and consequently never produced quite the same galaxy of widely read authoritative popularizers. But it became entangled more deeply in social controversy than physics, as I show in the discussions of eugenics, sociobiology, and biomedical ethics.

Parts 3 and 4 of the book deal with public reactions to the dilemma of the relationship of science and values, and attempts by scholars interested in the philosophy of nature to define their connections. In the concluding chapter I briefly survey the analysis of the earlier parts of the book, summarize the types of linkages that have been made between science and values, and venture my own conclusions about the expansionist-restrictionist dichotomy that is the central focus of the book.

The general interpretation presented in this book can be illustrated in more specific terms by giving a preliminary glimpse of two of the following sections: the analysis of the views of Arthur Eddington and the discussion of the eugenics movements in Weimar Germany and Soviet Russia. The first example is taken from the debates over the "epistemological transformation" coming out of physics; the second example comes from the early stages of the "ethical transformation" emanating from biology. The examples also illustrate various uses of expansionist and restrictionist arguments in ways that affected social values.

The great British astrophysicist Arthur Stanley Eddington strongly supported the restrictionist approach and, simultaneously, found it a useful foundation on which to support existing social values. He was well aware of the naturalistic fallacy and—contrary to the opinions of several of his critics—he never tried to support religion with the findings of science. Although he was a devout Quaker, he wrote "I repudiate the idea of proving the distinctive beliefs of religion either from the data of physical science or by the methods of physical science."[5] Eddington realized that to give scientific arguments in favor of ethical or religious systems was simultaneously to provide the theoretical base for scientific arguments pointed in the opposite direction. Thus, he affirmed that "the religious person may well be content that I have not offered him a God revealed by the quantum theory, and therefore liable to be swept away in the next scientific revolution."[6] Eddington found restrictionism a source of great security, for it left his religious preferences undisturbed.

His motivation for relying on restrictionism emerges in the following quotation:

> If you want to fill a vessel with anything you must make it hollow. . . . Any of the young theoretical physicists of today will tell you that what he is dragging to light on the basis of all the phenomena that come within his province is a scheme of symbols connected by mathematical equations. . . . Now a skeleton scheme of symbols is hollow enough to hold anything. It can be—nay, it cries out to be—filled with something to transform it from skeleton into being, from shadow into actuality, from symbols into the interpretation of symbols.[7]

Eddington was trying to create a thirst in his readers for values derived from nonscientific realms, and he was accomplishing that purpose by maintaining that science was a system of symbols with no relevance to the major questions of human existence. Far too sophisticated and subtle a person to engage in proselytizing for his own religion of Quakerism, he nonetheless pointed out that "Quakerism in dispensing with creeds holds out a hand to the scientist." Eddington confined science to a small realm of man's concerns and he then invited his readers to fill the remaining space with values based on religion. We thus see that in Eddington's hands restrictionism was turned in on itself and became a justification for certain kinds of values.

If Eddington is an example of a person who made a specific use of restrictionism to serve his values, in the debates over eugenics in Weimar Germany and Soviet Russia in the 1920s we can find examples of the use of expansionism to support two different and contrasting sets of value goals. In the early twenties, before either Germany or Soviet Russia developed official ideological attitudes toward human genetics, controversies occurred over the value implications of differing genetic theories. In both countries there were eugenics movements; the fate of German eugenics is rather well known but relatively little is known of the Soviet eugenics movement. In both Germany and Russia the early eugenists came from rather diverse political backgrounds. Only later did eugenic theories acquire a clearly conservative political reputation. In Soviet Russia for a while it even appeared possible that eugenics might be compatible with Soviet ideology and the principles of socialism. The Soviet government was, after all, committed to the creation of "a new Soviet man," and eugenics might be one path to that goal.

An important question in the debates was, "Which of two contrasting theories of heredity—Lamarckian or Mendelian—is most desirable from the standpoint of social and political values?" The question was based on an expansionist assumption, the belief that the science of human heredity *does* impinge upon human values.

The course of the debate in the two countries was far more complex than has usually been understood, but eventually the governments of the two countries made opposite decisions: Germany and Russia embraced contrasting theories of heredity on the basis of different value systems and labored mightily in later years to show that there was a natural fit between the particular scientific theory and the particular set of values chosen or approved by each. For all the differences between the political ideologies of the two countries, however, they were united in their support of expansionism.

As one studies examples of restrictionism and expansionism, the question naturally arises, "Which approach is the more adequate interpretation of the relationship of science and values?" At first glance, the restrictionist approach appears to be superior; it is certainly safer. The most blatant vulgarities and abuses in the history of the relationship of science and values have occurred as a result of expansionist analyses—whether of the type of Social Darwinism of the nineteenth century, or of National Socialist eugenics of the twentieth. The temptation to run to the safety of restrictionism, therefore, and in that way to confine science and values to separate realms is very strong indeed.

The problem with restrictionism is that this position is no longer tenable. When science was largely confined to the study of the inanimate world, or to the biological world other than man, restrictionism was a defensible and convenient doctrine for people who wished to protect both science and existing social values. As science moves more and more toward the study of the behavior of man, including his value systems, restrictionism seems less and less persuasive. When we discuss the scientific study of human beings the question is no longer whether the theories or findings could conceivably affect our value beliefs, but whether the particular theories and findings being presented to us at any given time must be accepted as valid. While the history of the misuse of scientific theories for political or ideological goals should make us suspicious of some of these theories and pieces of evidence, we also must recognize that the apparent refuge of restrictionism is illusory.

What seems to be needed now is a fine-grained analysis of the relationship between science and values, one that will avoid the no longer acceptable polarities of "value-free science" and "science-free values." In the last chapter, I will turn to this problem, although I do not hope to solve it. My goal in the main part of this book will be simply to give illustrations of the historical tendencies in our changing interpretations of the relationship of science and social values and to provide a glimpse, on some essential points, of the possible future courses of those interpretations.

When one discusses the historical relationship of science to social values, the identification of the differences between various types of expansionists and restrictionists is helpful; nonetheless, this distinction by itself is still inadequate for a deep understanding of the relationship. In order to go further it is necessary to see that the term "science" is too global to provide a secure basis for meaningful analysis. The term must be taken apart in a definitional sense. "Science" needs to be seen as a spectrum of subjects which range from the relatively value-free fields of physics and mathematics to the inextricably value-laden social sciences. Furthermore, when we speak of "science," we often mean very different things; significant distinctions must be made between terms, theories, empirical knowledge, and practices, although all are sometimes understood under "science." In the concluding chapter of this book I intend to examine the terms "science" and "values" to illustrate some of their multiple meanings, and then refer to the examples already discussed in the book to illustrate some of the interactions of science and values in terms of the separate meanings. My goal will still be, however, a better understanding of historical instances of science-value interactions, and not an effort to solve the problem of science and values in purely philosophical terms.

Traditional Western Views of Science and Values

Before the advent of modern science Western thought made no sharp division between statements of fact (knowledge about the natural world) and statements of value (norms of human attitudes and behavior). The concept that man is a part of the cosmos and therefore governed in some sense by law—natural or divine—can be traced back to the Greeks, who believed that human morals were grounded in the harmony of nature.

Thomas Aquinas refined the concept of natural law by dividing the *ius naturale* of the Romans into "eternal law" (*lex aeterna*) and "natural law" (*lex naturalis*). In the formulation of Aquinas, natural law applied only to man, but it was still integrated in the eternal law of the cosmos, which came from God. When man acted correctly, he obeyed natural law and, hence, eternal law. Proper human behavior was, to Aquinas, based on laws that were natural, absolute, and divine. But since man has the power of will, it was possible for him to act against his own nature. The moral person was one who chose voluntarily to live in accordance with natural law.

Any difference that Aquinas saw between laws governing human behavior and laws governing nature was, then, only in man's position toward them, not in their source. Both forms of laws emanated from God, and it was therefore impossible for the two kinds of laws to conflict. A true description of natural phenomena and a true description of the proper beliefs and behavior of human beings were merely different expressions of divine law.

Medieval thinkers who operated within this framework sometimes treated issues that we would consider "ethical" or "aesthetic" in ways quite similar to the ways in which they examined questions we would call "scientific." As one modern philosopher studying the relationship of science and values wrote, "In the medieval period there was no need to urge that the methods of physics should be extended to the study of human behavior, because there was no basic disparity between the two."[8] One of the most interesting examples of a person operating within this framework was Nicole Oresme, a fourteenth-century French Scholastic who developed a remarkable graphical system which permitted representing the motions of physical bodies in ways that prefigured the concepts of acceleration and motion in Galileo and other later natural philosophers. Oresme's configurational method was a geometric adaptation of the so-called Merton Rule by which the results of uniform acceleration can be described in terms of the average velocity of the object. But Oresme applied his method not only to the motions of physical bodies—the aspect of his work most interesting to contemporary historians of science—but also to aesthetics and emotions. To him, beauty, friendship, and anger were also describable in terms of the same intensity of qualities that were applicable to moving objects. Speaking of the emotions of "love," "hate," and "imagination," Oresme remarked, "In this matter one ought to speak of motions of the soul in the same way as we spoke of the motions of the body."[9] As Marshall Clagett, a historian of science who has studied Oresme extensively, remarked, "Oresme holds that pain and joy are qualities of the soul which are extended in time and intensifiable by degrees and accordingly the configuration doctrine is applicable to both."[10]

Popular literature of the Middle Ages also conjoined issues of morality with those of factual knowledge. Medieval bestiaries and herbals were not only rudimentary descriptions of animals and plants, but also contained lessons for moral edification. A. C. Crombie, historian of medieval science, has written, "The phoenix was the symbol of the risen Christ. The ant-lion, born of the lion and the ant, had two natures and so was unable to eat either meat or seeds and perished miserably like every double-minded man who tried to follow both God and the devil."[11]

Perhaps the most explicit form of this mixing of values and natural philosophy was the "doctrine of signatures," according to which every plant, animal, or mineral displayed a mark which indicated the value of that substance to humans. Thus, the plants with yellow flowers were often considered valuable for the treating of jaundice, lichens shaped like a skull were seen as medicine for epilepsy, and the bloodstone was recommended for hemorrhage. Not only were these plants and minerals seen as being useful to humans, but it was believed that nature *indicated* these uses by the shape or coloration of the plants and minerals. Nature was supposedly constructed according to anthropocentric values.

This perception of symbols in nature was a part of the popular culture of the Middle Ages; while it was often a component of the Christian works of holy fathers, such as St. Ambrose, who used animals to illustrate Biblical moral teachings, the same doctrine occurred in astrological and alchemical writings, which were often viewed suspiciously by the Church. The objection of the Church to astrology and alchemy was not, however, to the identification of human values in nature but to the pagan arrogance of the adherents of these fields; the astrologers and alchemists often believed they could interpret nature independently of the Church, and they sometimes boasted they could perform creative and transforming acts which the Church reserved for God.

There were always a few Christian writers who discerned the danger of relying on religious inspiration alone for knowledge of nature. St. Augustine had recognized the issue early in the Christian era when he wrote:

> In points obscure and remote from our sight if we come to read anything in Holy Scripture that is, in keeping with the faith in which we are steeped, capable of several meanings, we must not, by obstinately rushing in, so commit ourselves to any one of them that, when perhaps the truth is more thoroughly investigated, it rightly falls to the ground and we with it.[12]

Augustine was not calling here for a division of the normative from the factual, but merely asking for caution in permitting an apparently cogent argument from faith to prevail over arguments from fact when much is still unknown.

In the writings of the twelfth-century Scholastic philosopher Adelard of Bath one can see the beginning of a turn away from faith as the primary access to knowledge:

> I do not detract from God. Everything that is, is from Him and because of Him. But [nature] is not confused and without system and so far as human knowledge has progress it should be given a hearing. Only when it fails utterly should there be a recourse to God.[13]

As mentioned previously, Aquinas left room for this argument when he spoke of the laws governing the cosmos and the laws governing men as both being derived from God. Man could study the laws of the cosmos and know that the knowledge gained thereby could not contradict God's plan.

The incorporation of Ptolemaic astronomy and Aristotelian physics into the core of the late medieval Church's moral teachings was not, therefore, a logical necessity, since man's knowledge of God's law was necessarily fallible, as Augustine had emphasized. One could stress the unity of moral and physical law and still leave room for changes in man's understanding of the universe. But history rarely proceeds by logic, and by the late Middle Ages the Church had become wedded to the Ptolemaic and Aristotelian versions of natural explanations, which were the prevalent ones throughout Europe. Ptolemaic astronomy not only provided a good calculating method for calendars and astronomical events, but its depiction of the changing and corruptible earth and the immutable heavens was a convenient vehicle for lessons of morality. Aristotelian physics as interpreted by the Scholastics intermixed the factual and the normative, for it imputed purpose to nature and described the "proper place" of physical objects. Together, Ptolemy's universe and Aristotle's physics provided a coherent scheme of nature that also gave meaning to theological teachings about man.

Copernicus was a faithful Christian who realized that his alternative description of the universe would appear absurd and heretical to many of his readers. In the first sentence of the prefatory letter to Pope Paul III which he prefixed to *De Revolutionibus* he predicted that "certain people" will cry out that he should be "hooted off the stage with such an opinion." Copernicus prepared a careful defense against this eventuality that was not only technical but also normative. He accepted thoroughly the view that knowledge of the universe should also be morally enlightening and should inform us of God's work. After affirming that a great distance separates the stars from the planets, Copernicus observed that such immeasurable distances were an illustration of the "divine work of the Great and Noble Creator." And his oft-quoted description of the central sun shows that despite Copernicus' differences with descriptions of the universe by the medieval Scholastics, he was in agreement with the medieval tradition of the anthropomorphic interpretation of nature:

> In the middle of all is the seat of the Sun. For who in this most beautiful of temples would put this lamp in any other or better place than the one from which it can illuminate everything at the same time? Aptly indeed is he named by some the lantern of the universe, by others the mind, by others the ruler. Trismegistus called him the visible God, Sophocles' Electra, the watcher over all things. Thus indeed the Sun as if seated on a royal throne governs his household of Stars as they circle round him. Earth also is by no means cheated of the Moon's attendance, but as Aristotle says in his book *On Animals* the Moon has the closest affinity with the Earth. Meanwhile the Earth conceives from the Sun, and is made pregnant with annual offspring.[14]

The rejection by the Catholic Church of the Copernican system as propagated by Galileo came sixty years after the publication of *De Revolutionibus*. Copernicus' hope that Rome would see the logical possibility of accepting the theory without contradicting the Church's most sacred teaching was not realized; sensitive to the charges of corruption and doctrinal laxity arising in the Reformation, the Church was engaged in an effort to purify itself. The Church reasserted a narrow interpretation of the relationship of science and values in which the Ptolemaic system retained its special moral significance.

The subsequent embarrassment and damage to the Church resulting from the episode of Galileo were so great that the leaders of the Catholic Church and of other religious faiths became cautious about declaring scientific theories false merely because these theories had been interpreted by some commentators in ways that seemed offensive to religious authorities. Links between physical and moral laws were still often assumed to exist, but more recognition was given to the fact that these links were subject to a variety of conflicting interpretations. If one identified connections between science and religion, it was clearly much safer to affirm that science supported religion (e.g., arguments from design) than to assert a contradiction between religion and a given scientific finding or theory. The embarrassment of being wrong about a particular aspect of science was more damaging than the mere awkwardness of shifting one's argument from design to another aspect of nature if one's previous emphasis no longer seemed scientifically persuasive. Prudent religious leaders turned to science for sustenance, and they avoided condemning it if possible.

An intellectual preparation for a separation of the jurisdictions of science and religion was made by Descartes in the seventeenth century, even though the actual separation occurred much later. Descartes described "mind" and "body" as two almost completely separate entities, united only tenuously in the action of the pineal gland. Within the system of thought developed by Descartes it was possible to explain anatomy and physiology with the same principles used for the explanation of terrestrial and celestial mechanics. Religious and moral values, however, were secure from such examination. This division of authority opened up the possibility, only slowly realized during the next two centuries, of placing religion and science (natural philosophy) on separate noncompeting planes: science would look after the natural world and religion would concern itself with moral values.

This division of realms was more a potentiality than a reality. Throughout the last part of the seventeenth, through the eighteenth, and into the early nineteenth centuries many investigators of nature continued to write—even in their professional works—about the philosophical or religious meaning which they found in their research. Robert Boyle wrote that what he saw when he looked through telescopes and microscopes often forced him to exclaim, "How manifold are thy works, O Lord? In wisdom has thou made them all!"[15] Isaac Newton considered his system of the universe an illustration of the grandeur of God's works, and many other scientists expressed similar views. A prominent group of English churchmen seized upon the Newtonian world system and transformed it into an ideology that, in the words of a recent historian, "had enormous social and political implications not only in their own time but throughout the eighteenth century."[16] The clergymen believed that the harmony and order of the Newtonian universe was a valuable model for an English society torn recently by revolution and social chaos. They described Newton's universe as one ordered by God, and they underscored the relationship between the "world politic" and the "world natural."[17]

This attribution of values to science was increasingly an individual decision, however, rather than one enforced in detail by political or religious institutions. In most instances it would be more accurate to speak of individuals finding religious values in nature, rather than religious authorities commenting on the validity of science. As Charles Gillispie described England in the seventeenth and eighteenth centuries, "The mutual tolerance of scientist and theologian did not reflect indifference to one another's calling. Neither had ever regarded the other's truths as irrelevancies. On the contrary, ever since the time of Bacon, scientists had been congratulating themselves upon unfolding a divinely ordained system of nature."[18]

In England in the eighteenth century it was common for writers to take established religious and moral principles as givens, and to seek support for these principles in the architecture of the heavens and earth revealed by science. Another type of linkage was attempted on the Continent, and particularly in France of the Enlightenment, by thinkers who hoped that science might be used to reform or even revolutionize the moral order. Several mathematicians, from the Bernoullis to Condorcet, tried to use probability theory to rationalize political, legal, and economic decision making. In many ways, they were the founders of social science. The most ambitious effort was made by Condorcet, who was described by a biographer as a person who "set out to demonstrate . . . that the moral and political sciences are susceptible of mathematical treatment."[19] One of Condorcet's best-known examples of "social mathematics" was his effort to answer on the basis of statistical analysis the question, "Under what conditions will the probability that the majority decision of an assembly or tribunal is true be high enough to justify the obligation of the rest of society to accept that decision?"[20] The answer he produced, divested of its mathematical trappings, was that more enlightened assemblies make truer decisions than less enlightened ones. Unremarkable as Condorcet's conclusion seems, it had serious implications for him. Democratic government was, therefore, not wise, he thought, if the citizenry was unenlightened (uneducated). This viewpoint did not fit well with the opinions of the more radical adherents of the French Revolution, a political event which Condorcet supported but which eventually overwhelmed him.

Underlying Condorcet's analysis here was the assumption of a link between what is "true" and what is "just." Many people would maintain that the essential issue of democratic government is justice, not truth. But Condorcet was convinced that eventually a connection will be established between what is and what should be. He affirmed that a "small number of facts [is] necessary to establish the first foundations of ethics" and that these "facts are as general and as constant as those of the physical order."[21]

In the nineteenth century serious conflicts arose in Europe and America over the impact upon moral and religious principles of the new sciences of historical geology and evolutionary biology. An essential issue was the question of the role of divine Providence in nature, and it was an issue over which scientists themselves were in disagreement. In geology, the question took the form, "Was the process by which the earth developed self-sufficient, or was divine intervention an occasional, or even a constant, element?" By the time of the debate over evolutionary biology several decades later, the issue of divine intervention had lost much of its earlier force, but the question of reconciling a naturalistic origin of man with the concept of his unique place in a divine order was still capable of arousing great popular interest.

Transformation of the Debates on Science and Values

The representatives of religion were the losers in these debates, and we now justifiably dismiss the positions they took. In the last decades of the nineteenth century, however, a different sort of issue arose, one which in retrospect is much more interesting than the much better known conflict between religion and science. A few spokesmen in the nineteenth century saw that the value questions upon which biological evolution touched were not all religious, or, at least, that an agnostic or an atheist could find interest in the debates on purely ethical grounds. The significance of that perception for later issues arising in the twentieth century is so great that we need to pause for a moment to examine its emergence. What was happening can be summarized as follows: Before the late nineteenth century, discussions of the relationship of science to values almost always became debates about religion. In the late nineteenth century there appeared the first significant discussions of the modern period of what would become a far more important question, one that is entirely secular: Can science explain the values that influence human behavior, treating those values as phenomena for scientific examination not essentially different from the other objects of scientific investigation? E. O. Wilson would much later comment, "Scientists are . . . not accustomed to declaring any phenomenon off limits."[22] To such scientists, human values are just another set of natural phenomena awaiting examination and explanation. And if science can potentially explain values, can it eventually tell us how to modify them? If it does someday tell us how to change our values, are not the goals of modification also values, and therefore susceptible in turn to scientific research and possible modification? The relationship between science and values—once one admits that values are susceptible to scientific explanation (whether on genetic or environmental grounds makes no difference in principle)—becomes one of infinite regression.

These issues began to become visible in the late nineteenth century among some of the adherents of a current of thought often described as evolutionary naturalism. Members of this school of thought looked upon the universe as a developing cosmic process. Since all aspects of reality were included in the process, ethical values were also often seen as evolutionary products. Some of the essential features of evolutionary naturalism may be found in the writings of Herbert Spencer, but it was only after the publication of Darwin's *Origin of Species* that the school of thought took on a scientific tone filled with promise of approaching insights. The movement was strongest in England, where Darwin, W. K. Clifford, L. T. Hobhouse, and Francis Galton were among its supporters, but it had prominent adherents on the Continent as well, in Edward Westermarck, Ernst Haeckel, and J. F. Guyau. As the full significance of Darwin's theory sank in, it became clear that his method might be applied to ethical behavior as well as to physical morphology. Darwin himself said as much in the *Descent of Man* when he maintained that the emotions of men, including love, "may be found in an incipient, or even sometimes in a well-developed condition, in the lower animals."[23]

It is not surprising that in the writings of many of the evolutionary naturalists an important motivation seemed to be to justify ethical systems already accepted by the authors as preferable, rather than to question all existing ethical systems, including their own, on the basis of scientific research. In many instances the authors' ethical preferences were connected with their social and economic status. Herbert Spencer was justly described by his biographer J. D. Y. Peel as a person who "did not start off from a phenomenon to be explained, but from ethical and metaphysical positions to be established."[24] Many of the evolutionary naturalists were trying to demonstrate not only that values are products of natural processes but that one particular set of values—namely, the liberal values so closely connected with nineteenth-century industrialism—is bound to triumph, since it is the necessary result of inevitable evolutionary forces.

One sees, then, that discussions of the relationship of science and values always have a socially and historically contingent quality. One may, therefore, be tempted to conclude that the discussions are all epiphenomena of much deeper economic and political causes, and that science, qua science, can have no independent role in examining values. Such a conclusion would be rash. One should not surrender so easily the basic assumption of science, namely that a real physical and biological world exists; that although humans will never possess absolute truth of that world, they can gain ever more accurate information about it and ever more comprehensive explanations of it.[25] Science, built by humans, includes knowledge about humans themselves, and there is no reason in principle why that scientific knowledge cannot include knowledge of the sources of human values and the influences that change them. Since values are more controversial than anything else humans study, it is inevitable that theories attempting to explain them will, themselves, be value-laden and will reflect economic and political ideologies. Ideology serves as a powerful brake or accelerator on value interpretations of science. That fact, however, is not adequate reason for rejecting the search for knowledge of human values but is, instead, reason for including ideology within the spectrum of evidence that one examines when considering naturalistic explanations of values.

One of the most interesting moments in the debate over evolutionary naturalism came in 1893, when T. H. Huxley delivered the Romanes Lecture entitled "Evolution and Ethics," and in the following year when he wrote an additional essay on the subject as an introduction to the printed lecture. At this time Huxley was sixty-eight years old, and his credentials as a scientist and as a Darwinist could not have been more authentic; he was a former president of the Royal Society, and everyone knew him as the brilliant polemicist who had, thirty years earlier, bested Bishop Wilberforce in the famous debate over evolution at a meeting of the British Association for the Advancement of Science. He had given a decade of his life to the battle for the acceptance of

Darwinism, winning for himself the sobriquet of "Darwin's Bulldog" in the process. Huxley was never happier than when belittling the clergy and their views of the world. He once observed, "Extinguished theologians lie about the cradle of every science as the strangled snakes beside that of Hercules."[26] As for his own religious views, it was Huxley who invented the term "agnostic," and when asked "Does God exist?" his consistent reply was "I do not know."

If Huxley felt perplexed over the question of the relationship of science and values, it was not, then, because some inner religious principle or urge resisted the growing expansionist claims of science. Indeed, if that had been the case, his perplexity would probably have been less, for his religious beliefs would have shielded him from the starkness of the question of whether human values have entirely naturalistic origins which are, in principle, explicable by science. It was this question which Huxley faced in his lecture, and, after some waffling, he revealed a restrictionist tendency which had earlier been nearly invisible in his writings. The young Huxley of earlier decades had fought for the victory of science over theological dogma; the old Huxley, knowing that the earlier battle had long since been won, asked himself how much farther science would be able to push in explaining man. His answer was that science does not explain ethics. Indeed, the cosmic process and the ethical process are different phenomena that are "at war with each other." Science should submit to ethics, not ethics to science.

The metaphor Huxley employed to explain the place of civilization in evolution was that of a garden in which the gardener has eliminated the struggle for existence that still holds valid in the surrounding wilderness. The survival of the plants in the garden is not determined by struggle against each other but by the selective choice of the gardener. Similarly, the evolution of civilization has produced a system of ethics which makes men independent of the cosmic process. As Huxley observed, "Since law and morals are restraints upon the struggle for existence between men in society, the ethical process is in opposition to the principle of the cosmic process, and tends to the suppression of the qualities best fitted for success in that struggle."[27]

In posing an antithesis between nature and ethics, Huxley asserted a demarcation between science and values that many of his earlier opponents thought he had denied. He did not pursue the crucial question, "What are the sources of the ethical values that human beings take as their guides, and those which the gardener accepts in making his selective choices in his garden?" Surely those ethical values have sources, unless one thinks they fall from the sky; whether those sources are in biological evolution, as Herbert Spencer believed, or in the social environment, as Huxley thought, scientists will study them and in the process destroy once more the demarcation Huxley tried to establish. For in the coming century the sources of ethics would not be attributed only to biology, a simplified view which Huxley correctly criticized, but also to social conditioning, the purview of the emerging science of psychology.

Spencer expressed his disappointment at Huxley, remarking that his call for man to struggle against the process that created him was based on "the assumption that there exists something in us which is not a product of the cosmic process."[28] Huxley's wife Henrietta was probably closer to the truth of Huxley's ambiguity when she later said she saw a contradiction in the core of the lecture. Huxley rejoiced in the possibility that science explained man's origin; he recoiled from the possibility that it explained his behavior.

Huxley, of course, had good reason for his distaste for evolutionary ethics. He lived to see the outbreak of Social Darwinism, the belief that biology sanctioned ruthless competition and violence. He appealed for an ethics that came from outside of nature, since "the thief and the murderer follow nature as much as the philanthropist."[29] He illustrated the social significance of the separation of science from values, but his anguish showed that the expansion of science was making the separation more difficult. In the century that began a few years after the Romanes lecture, the difficulty would increase immeasurably.

The episode of Huxley's essay on evolution and ethics prefigures several later twentieth-century discussions, especially those over ethology and sociobiology. A different type of attack on Huxley's call for an "ethics from outside nature" would come from twentieth-century behavioral psychologists. Many decades passed between the controversies, however, and in the intervening period the attempt by scholars to separate science and values reached the apogee described by Stephen Toulmin, discussed earlier in this introduction. Paradoxically, in the early years of the twentieth century, exactly when research scientists were beginning to study the bases of human conduct in a variety of disciplines, the belief that science and values are separate realms became the explicit ethos of science in Western Europe and North America. This development will seem less paradoxical, however, when we see that it was precisely because science was beginning to touch ever more closely on values that scientists found it comfortable to speak of the value-free nature of their research. Many quarrels were avoided that way—or, to be more accurate, the day when the debates would have to be fully faced was postponed.

A powerful impetus to this emerging restrictionism (a term I prefer to "separationism," since a pure separation is never possible) of the late nineteenth century was the professionalization of science. In the nineteenth century amateurs and dilettantes in science gradually lost out to the salaried professionals, and in the process the tone of scientific writing changed. Before the nineteenth century, the journals of learned societies often included speculative articles in which normative and factual questions were intermingled. By the end of the century such writing was all but absent from the prestigious professional journals. Membership in the societies became increasingly restrictive, often requiring higher education and the concomitant introduction into the ethos of research. A mark of a serious professional scientist was a sober and factual tone.

Professionalization and specialization were important factors in the effort in the late nineteenth century to eliminate explicit normative statements from scientific discussions, but they were not the *reasons* for the effort, they were merely the *modes*. As a contemporary biologist might say, the effort to construct a body of scientific knowledge free from anthropocentric concerns had an "adaptive value." Restrictionism protected science. Scientists learned that mixing issues of science with those of social values invariably led to controversies, as both the Galilean and Darwinian episodes so clearly illustrated; they also knew that science thrives best when it stays clear of social criticism. It was in the interests of scientists, therefore, to assert that science and values belong to different realms. But this emphasis should not be seen as merely a self-serving maneuver of the scientists. If one keeps in mind the analogy of adaptation in biology, it is not necessary to assume that all scientists who avoided value questions did so in order to protect themselves, even though such protection was one result of their avoidance of these issues. All that was necessary for the trend to become stronger was for the specialists in a given area to hold the opinion that the discussion of values in one's scientific work was "unprofessional."

But restrictionism served interests much broader than those of scientists. Restrictionism also protected society by making its values imperturbable by science. Defenders of social values—whether they were religious or secular, political or ethical—also benefited from this clear demarcation of realms, for they were thereby protected from incursions by scientists into value realms. If you insist that science and values do not mix, then the antecedent values of society are protected.

The severing of the explicit links between the study of nature and the study of values in the late nineteenth century can be illustrated by briefly comparing the ethos surrounding the founding of Harvard College in the seventeenth century and that of the Johns Hopkins Univerity in the nineteenth. Each institution, in its own way, represented a paradigm of its type: Harvard was the first institution of higher education in the New World, and Johns Hopkins was one of the first universities in the United States pursuing advanced research.

Historian of Harvard Samuel Eliot Morison described how the early college was based on the principle of the unification of knowledge and of Christian morality:

> An important, and, as many will think an essential ingredient of this early liberal education was the religious spirit of Harvard. Our founders brought over this medieval Christian tradition undiluted. . . . Harvard students were reminded in their college laws, and by their preceptors, that the object of their literary and scientific studies was the greater knowledge of God; and that the acquisition of knowledge for its own

sake, without "laying *Christ* in the bottome, as the only foundation," was futile and sinful. "In Christam Gloriam" appeared on the college seal of 1650; and "Veritas" on the original design of the college arms undoubtedly meant (as it did to Dante) the divine truth.[30]

Compare this assertion of the normative goals of knowledge with the discussion of "fundamental principles" of education and research given in 1876 by President Daniel Coit Gilman of Johns Hopkins in his first annual report:

The Trustees have come to a substantial agreement as to the principles which must govern their earliest proceedings. . . . Here it is only necessary to place on record their desire that the University now taking shape should forever be free from the influences of ecclesiasticism or partisanship, as those terms are used in narrow and controversial senses; that all departments of learning—mathematical, scientific, literary, historical, philosophical—should be promoted, as far as the funds at command will permit.[31]

Harvard also came around to the explicit call for the relentless pursuit of knowledge for its own sake so well represented by Hopkins, but not without a struggle. In the late 1840s President Edward Everett rejected the effort of his predecessor Josiah Quincy to enthrone the motto "Veritas" without mention of its Christian basis; he restored the seal of 1693, with its motto of "Christo et Ecclesiae" and he castigated Quincy's Veritas seal as "fantastical and anti-Christian." Not until 1886 was the present seal of Harvard adopted, with its secular call for "Veritas" emblazoned on three open books. By this time, the new specialized graduate education was well established.[32]

The growth of professionalization and the establishment of the research universities were accompanied by an accumulating belief that scientific knowledge is value-free. More and more frequently academic researchers assumed that "as scientific problems were solved the ethical issues would take care of themselves."[33] American colleges which in earlier decades had required courses in "moral philosophy" of all undergraduates either abandoned such courses or converted them into electives. And the philosophers themselves became much less interested in the relationship between ethics and science. The influence of G. E. Moore's 1903 *Principia Ethica* was widely felt in England and America. Moore revived Hume's distinction between "is" and "ought" in a new form by labeling any effort to derive ethical conclusions from the study of nature as "the naturalistic fallacy." Many people failed to notice that Moore, like Hume, used his philosophy to combat particular values, rather than all value systems. Hume was interested in criticizing morality based on religion, not in opposing the view that rational morality might be connected with human nature;[34] Moore was concerned to defend the autonomy of ethics against naturalism while shoring up the "good" as an "indefinable, immediately intuited reality."[35]

The move toward restrictionism in the late nineteenth and early twentieth centuries came not only because of growing specialization, the development of professional graduate education, and changes in philosophical interests; it also came because the relevance of science to values seemed to be changing in a way that made such protection desirable. In the eighteenth and early nineteenth centuries science could be rather easily used as an apologia or justification of the values most widely accepted in society. "Arguments from design," whether based on biology or astronomy, were essentially the employment of science for the buttressing of orthodox value positions. When this kind of argument was readily available and fairly persuasive, it was in the interests of natural philosophers or scientists to advance such views in an explicit fashion. However, when science began to undermine existing beliefs (e.g., Darwinian evolution or behaviorist psychology versus a priori moral systems) the motivation for being explicit about links between science and values disappeared. More and more, research institutions employed only working scientists who avoided value questions; it was much safer that way. There were always a few rebels among the scientists who wished to cause trouble by raising accursed questions, but their number greatly diminished after the end of the classical debates over Darwinism.

Nonetheless, even after restrictionism became the reigning ethos in science in Europe and America, implicit links between science and values continued to pile up, like interest in some secret bank account, as science continued to develop. One day the dimensions of the reserve would demand discussion, and restrictionism would be placed under heavy strain. Twentieth-century science moved heavily into the fields of behavioral psychology, human genetics, biomedicine, and ethology; the impossibility of keeping the links between science and values outside the concerns of scientists and their institutions became increasingly apparent. We must now reckon with the account that was accumulated, and we must define our position on its future growth.

The emergence in the nineteenth century of the principle of unlimited secular knowledge was a great victory for academic freedom, one of the most significant events in the development of civilization. But we should recognize that this intellectual victory concealed several important unresolved issues that are now growing in potency. Even if we try to make a distinction between science and technology (increasingly difficult to do), is it really possible to separate science from social and ethical values? If all the universe, including humans, is the object of study, will not an understanding of that universe also give us an explanation of human value systems and in the process refute the still reigning late nineteenth-century restrictionism that gave us the freedom to start the search? We should remember that restrictionism was based on the assumption that a division between scientific knowledge and ethical principles could easily be made, and that people concerned with ethics and other values need not fear the results of science. But if we should ever be able to explain scientifically why people possess the values that they do, then the demarcation would break down. The fact that logically an "ought" could still not be derived from an "is" would seem unimportant if it became clear that any given "ought" possessed by real human beings had a naturalistic origin and development explicable by science.

My goal in this book is not to prescribe a particular relationship between science and values, or even to attempt to answer the question of whether, on a philosophical level, knowledge of the sciences is ultimately "value-free." My goal is, instead, to analyze some of the interesting kinds of relationships between science and values that scientists in the twentieth century have identified. In the first part of that century the most common discussions of the relationship of science to values centered not on biology, but on physics. It is with physics, then, that I shall begin, but I shall discuss in subsequent sections of the book a variety of topics in both the physical and biological sciences.

PART ONE

THE PHYSICAL SCIENCES

PHYSICAL KNOWLEDGE AND VALUES

chapter one

THE NEW PHYSICS

DURING THE first thirty years of the twentieth century the mechanistic picture of the universe based on nineteenth-century science was thoroughly demolished by revolutions in the two areas of physics which came to be known as relativity physics and quantum mechanics. The first revolution, associated largely with the work of Albert Einstein, overturned the concepts of absolute space and time which formed the framework within which the laws governing the behavior of matter were described in Newtonian physics. By disproving the existence of temporal simultaneity, demonstrating the variability of the lengths and masses of bodies moving at high velocity, establishing the equivalence of mass and energy, and tying together space and time in a four-dimensional manifold of varying curvature, Einstein created a world picture that differed radically from that of Newton in its theoretical principles. This world picture, furthermore, defied trustworthy popular interpretations, and in their absence a great variety of dubious interpretations vied for attention. Some of the best known and respected of these personal interpretations will be considered in subsequent chapters of this book.

The very term "relativity" seemed to many nonscientists to deny the existence of any absolutes—even truth itself—and the equivalence of matter and energy was interpreted by numerous authors as the end to the materialistic world view that underlay nineteenth-century science. Such conclusions, if accepted, had immense implications for sociopolitical values since many existing ideologies had been grafted onto the Newtonian framework and were seen as inextricably linked to it.

The upheaval in quantum mechanics was even more radical.

The work, which extended from 1900 to the late twenties, and which is connected most prominently with the names of Max Planck, Niels Bohr, and Werner Heisenberg, called into question two of the most sacred methodological principles of modern science: causality and objectivity. By showing that rigorous scientific causality (often defined as the existence of one predetermined outcome of any given situation) did not hold in the microworld of quantum mechanics, and by illustrating that the observing physicist cannot be abstracted from the microbodies he is observing, physicists seemed to be undermining not only the basic principles of nineteenth-century physics, but what had been widely regarded as the assumptions of all of science since its birth. Some commentators believed that the advent of quantum mechanics also meant the end of determinism as a sociopolitical world view.[1]

The two scientists whose names were most closely connected in the minds of educated lay people with relativity physics and quantum mechanics were Albert Einstein and Niels Bohr. Since both men lived long lives during which they wrote on many subjects, the educated nonscientist who was seeking an authoritative interpretation of the new physics and an answer to the question of whether the new physics impinged on sociopolitical issues often turned to their writings. The following chapter centers on the attitudes of these men on these questions, as recorded in their published works.

Einstein and Relativity Physics

Albert Einstein (1879–1955) was one of the greatest scientists ever to have lived. Born in Ulm, Germany, into a modest family committed to education and culture, Einstein was from an early age alert to intellectual issues, but he had difficulty adjusting to formal educational institutions, which in his opinion concentrated their attentions on inessential matters. As a university student Einstein found his greatest pleasure in reading the fundamental works of great scientists and philosophers, from Boltzmann and Maxwell to Hume and Mach. Such reading was more important to him than the prescribed curriculum and as a result he established only a good, not distinguished, academic record. The story of how

Einstein was unable to find an academic position after completion of his studies, only to surprise the scientific world as a patent clerk writing articles on physics, is by now a part of almost every educated person's knowledge of the history of science. Actually, Einstein the patent clerk came to the attention of only a very small circle of theoretical physicists, but what he did in those years eventually gave him world fame.

In order to address the question of Einstein's views on the relationship of science to social values, it will be useful first to look briefly at his opinions on the philosophy of nature. These opinions influenced all his thought and writings.

Throughout his life, beginning in his twenty-second year in 1901, Einstein was a prolific author. The most complete bibliography of his works yet published contains 607 items, and includes works on a great variety of topics, scientific and nonscientific, written under strikingly different circumstances.[2] A person seeking contradictions in Einstein's thought by selectively quoting from these publications would be quite successful, and if he made use of the vast body of unpublished materials now being catalogued and prepared for publication, the room for finding contrasts would be even greater.[3]

The most helpful approach in understanding the basic features of Einstein's philosophical views is, however, first a discussion of one or two constant elements in his thought, and, second, a consideration of the most important long-term and permanent changes in his epistemological conceptions. In this way we hope to pay the greater attention to the fundamental aspects of his thought and less attention to the episodic and ephemeral.

The feature of Einstein's approach to the natural world which played the largest role in his thoughts and which was a constant from the days of his initial professional efforts was the search for unity and order in nature. In one of the very first letters contained in the Einstein archives, Einstein wrote to his close friend Marcel Grossman, "It is a wonderful feeling to recognize the unity of a complex of appearances which, to direct sense experience, seem to be separate things."[4]

At approximately the time that Einstein wrote this letter he submitted his first publication, an unsuccessful work that expressed his search for unity to a degree that can only be called audacious, particularly when one recalls that at that time Einstein was a completely unknown recent graduate of an engineering school with only an undergraduate degree. (He would receive a doctor's degree in physics from the University of Zurich four years later.) This first publication was no less than an attempt to seek a connection between two types of natural forces that seemed quite dissimilar: intermolecular and gravitational forces.[5] In later years Einstein spoke disparagingly of this work, but the striving for a vision of simplicity and coherence in the world around him which motivated the article was to stay with him forever.[6]

It is probably not too much to say that the same motivation that led Einstein to seek a unity between molecular and gravitational forces as a young man influenced him in the latter part of his career when he vainly sought a unified field theory which would unite electromagnetic and gravitational forces. Again the world would judge the effort as unsuccessful. Einstein was seeking once more that "wonderful feeling" of recognizing unity of which he wrote so enthusiastically to his friend Grossman in 1901.

This search for order was the most important and permanent of Einstein's scientific and philosophical drives, but he was also deeply interested in what kind of a physical meaning should be ascribed to that order once it was found. And here we find rather clear evidence of a gradual transition in Einstein's opinions, a series of shifts that the historian of physics Gerald Holton has aptly described as "the philosophical pilgrimage of Albert Einstein, a pilgrimage from a philosophy of science in which sensationism and empiricism were at the center, to one in which the basis was a rational realism."[7]

In his later years, Einstein repeatedly observed that the two philosophers who exerted the greatest influences upon him as a young man, before his major scientific achievements, were Ernst Mach and David Hume. Both of them supplied him with deeply skeptical, analytical philosophies which gave him powerful weapons for attacking the assumptions of the mechanistic physics of his day. These were corrosive viewpoints, and they helped Einstein to question concepts earlier physicists usually accepted as self-evident, such as absolute space and time. They provided a "philosophy of the minimum," as it were, a cutting intellectual edge with which Einstein could clear away the metaphysical underbrush of nineteenth-century physics. Yet, as we will see, Einstein would eventually find this frugal methodology unsatisfying, and would move beyond it. Once the destructive work of examining assumptions, eradicating the unnecessary concepts, and reestablishing physics on a new basis had been accomplished, Einstein found a philosophy of the minimum inadequate for a recognition of the place of hypotheses in science. He gradually moved away from positivism toward ontological assertions of the existence of a physical reality entirely independent of man. Physical reality and causality, concepts which Mach and Hume considered unfounded speculations, became important components of Einstein's philosophy of science. All these changes occurred gradually, and were more in the nature of changing emphases than they were absolute shifts. Indeed, at any point in Einstein's professional life, one can find elements of both positivism and realism in his thought, and a tension between them, but the chronological transition was definitely toward the realist end of the spectrum.

In addition to emphasizing the objective reality of matter, the older Einstein also placed more and more emphasis on the role of free, creative thought in the formulation of hypotheses. On this point as well he moved into greater disagreement with his earlier inspirers, Hume and Mach, who had been quite critical of anything approaching speculation or metaphysics. The mature Einstein believed that without some metaphysical speculation great advances in science would be difficult and fruitful philosophy impossible. Writing in 1944, Einstein observed:

> By his clear critique Hume did not only advance philosophy in a decisive way but also—through no fault of his—created a danger for philosophy in that, following his critique, a fateful "fear of metaphysics" arose which has come to be a malady of contemporary empiricist philosophizing.[8]

EINSTEIN'S ATTITUDE TOWARD APPLICATION OF SCIENTIFIC CONCEPTS TO NONSCIENTIFIC AREAS

It is well known that in addition to his works on physics, Einstein wrote prolifically on many other topics, from politics, to religion, to Zionism, to education. The comprehensive bibliography of his works mentioned earlier lists two hundred seventy-four publications under "Scientific Writings" and three hundred seven works under "General Writings."[9] A whole generation of the educated lay public became accustomed to reading Einstein's views on almost every conceivable topic. Underlying this attention from the public was undoubtedly the lay person's fascination with genius, especially when it occurs in such a warm and picturesque person as Einstein. This great scientist was so original and talented that his opinions were always interesting, whatever the subject; his interviews frequently contained innocent barbs so cleverly and good heartedly aimed at his questioners that everyone, including the interviewers, was entertained. And his general intelligence and deep humanity often led him to valuable conclusions in areas that were very far from his special knowledge.

It is important to notice that Einstein explicitly rejected the view that his science related directly to ethics and politics. He stated outright, "Science can only ascertain what *is*, but not what should be, and outside its domain value judgments of all kind remain necessary."[10] With very few exceptions, he kept his science separate from the other areas of his interests. He did not believe that the content of natural science could bear on subjects such as the social sciences or human behavior and he criticized those who did. Asked in an interview in 1930, "Then you would not hold with the behaviorists, or even the eugenicists, who would guide human conduct by the light of scientific teaching?" Einstein replied, "I think I have made that clear enough."[11] He even chided his fellow physicists, including ones he deeply respected, such as Bohr and Jeans, when they wrote in popular articles and books that the development of relativity physics and quantum mechanics had an impact upon topics such as religion, freedom of will, and the social sciences. Einstein once pointedly observed that when Bohr begins to write on topics such as complementarity "he thinks of himself as a prophet."[12]

Einstein was willing to write about issues involving social and political values even though he denied their relevance to his main labors, first of all, because he believed that scientists, like all other citizens, have a deep responsibility to society. All intelligent people, he thought, should try to exert the weight of their opinions to achieve solutions to social and political problems. He disagreed sharply with the viewpoint so strong in the German academic world he knew so well that professors should leave politics to the political authorities.

As Einstein's fame increased a second reason for his espousal of social and political causes began to emerge more clearly. He realized that, almost despite himself, he was now a public figure who could automatically command a readership whenever he chose. He regarded the accident of the acquisition of this authority as a heavy responsibility which he could not shirk, since it gave him the power to help rectify evils and relieve suffering. He knew that this influence came to few people. Therefore, he spoke out in favor of those causes he considered to be just and necessary. Thus he became a spokesman for Zionism, even though his earlier attitude toward Judaism had not prepared him for this role, and despite the fact that he had reservations about some aspects of the Zionist movement.[13]

Evidence for Einstein's rejection of direct links between scientific and nonscientific thought runs throughout his writings. This restrictionist opinion was undoubtedly strengthened during the first squabbles in Germany over relativity theory, when Einstein's critics connected physical relativity to ethical relativism, and considered Einstein a dangerous radical. It was known that Einstein was a Jew, a socialist, and a pacifist. Furthermore, popular opinion sometimes lumped all of the radical events of the early twentieth century together in one unholy "destruction of absolute standards:" the growth of socialism, the decline of Christianity, the outburst of the Russian Revolution (in which the role of Jewish revolutionaries was rumored to be strong), the development of modern art and modern music, the rise of Freud's psychoanalytic method, and the decline of nineteenth-century moral standards. Einstein correctly regarded such sweeping criticism as sloppy and dangerous thinking at its worst, and he carefully distinguished the new developments in science from all other radical currents. In his rejoinders to his critics Einstein emphasized that when he used the term "relativist" he meant a "physical relativist" and certainly not a "philosophical relativist," that is, one who attacks all absolute principles.[14] Thus, Einstein from early in his scientific career resisted efforts by others to find social implications in recent developments in physical theory.

Einstein's denial of such extrascientific links for his work was far more than a maneuver to protect relativity theory from attack, although it also served that purpose. It was based on a strict separation of statements of fact and value. In 1932 he commented, "I believe that the present fashion of applying the axioms of physical science to human life is not only a mistake but has something reprehensible in it."[15] He was particularly upset by the opinion of some writers, including physicists, that science could shed light on questions of human will and value. As he wrote:

> For the scientist, there is only "being," but no wishing, no valuing, no good, no evil—in short, no goal. As long as we remain within the realm of science proper, we can never encounter a sentence of the type: "Thou shalt not lie."[16]

In each of the above cases in which Einstein denied a connection between science and values, it is important to notice that Einstein was speaking of the *content* of science. On a more subtle level, Einstein did see some affinities between science and human concerns, and he often spoke of these affinities as ones of *method* and *source*. In the area of method, Einstein thought that the rigor and deductive clarity of science could be helpful to ethics as a model of procedure once the goals and value content of the ethical system had already been acquired on a nonscientific basis. As he wrote:

> Scientific statements of facts and relations . . . cannot produce ethical derivatives. However, ethical derivatives can be made rational and coherent by logical thinking and empirical knowledge. If we can agree on some fundamental ethical propositions, then other ethical propositions can be derived from them, provided that the original premises are stated with sufficient precision. Such ethical premises play a similar role in ethics to that played by axioms in mathematics.[17]

Thus, Einstein thought it was possible to take a question such as "Why should we not lie?" and by a process of logical analysis trace the question back to a basic ethical derivative, "Human life should be preserved," since integrity in the community is necessary for social cooperation and the preservation of life. But science alone could not supply the crucial initial assumptions that cooperation and survival are "goods."

Einstein, therefore, believed that the scientific method of thought was an important tool for discussions of human values even though he maintained that the content of scientific knowledge was not useful for that purpose. Einstein had such a high opinion of the methodological strength of rational, scientific thought that he believed that scientists ought to be able to go further toward a solution of social problems ridden with prejudice and superstition than could those not acquainted with the scientific method. In his opinion scientific ways of thinking helped one overcome national prejudice, as well as religious superstition.[18] Science, he observed, is a powerful influence "in drawing men a little away from senseless nationalism."[19]

It came as a particularly bitter disappointment to Einstein that during and immediately after World War I scientists fell prey to nationalistic passions.[20] He observed that scientists had become "representatives of the strongest national traditions and have lost their sense of community." How striking, he observed, that, contrary to all his expectations, the proposals for the League of Nations did not come from the scientists but from the politicians.[21] Still, he clung to his simple but attractively optimistic hope that the scientific method and the world scientific community would play unique roles in solving global issues. Scientific analysis was to him a model for social analysis, even though science itself could not make a direct contribution to social values.

Yet Einstein saw a link between science and nonscientific subjects in another area, that of the source of human motivation in all creative endeavors. In 1918 he commented:

> I believe with Schopenhauer that one of the strongest motives that lead men to art and science is escape from everyday life with its painful crudity and hopeless dreariness, from the fetters of one's own ever shifting desires. A finely tempered nature longs to escape from personal life into the world of objective perception and thought. . . . Man tries to make for himself in the fashion that suits him best a simplified and intelligible picture of the world; he then tries to some extent to substitute this cosmos of his for the world of experience, and thus to overcome it. This is what the painter, the poet, the speculative philosopher, and the natural scientist do, each in his own way.[22]

It is possible that the pessimistic references to the crudity and dreariness of everyday personal life bore some relationship to

Einstein's own affairs, since at the time he wrote his marriage was rapidly breaking up, but the effort to arrive at a superpersonal and objective view of the universe went to the very core of his beliefs. Indeed, one biographer of Einstein considered entitling his book "Flight from the Merely Personal."[23]

In later years Einstein ever more frequently discussed the relationship of science and religion in similar terms of human motivation. Advocates of religion took satisfaction in the fact that Einstein, in a manner common among restrictionists, saw no conflict between the two, and, indeed, spoke of his own "cosmic religion." These religious spokesmen usually conveniently ignored the fact that the writings in which Einstein discussed religion contained sharp criticisms of all established religions. When Einstein spoke of God, it was a God in Spinoza's sense of a belief in the existence of an orderly harmony in the universe. Einstein was opposed to all confessional religions, and refused to grant religion a unique authority even in the realm of ethics, where he believed accumulated experience was far more important. He did believe, however, that there exists an affinity between science and religion on the deepest motivational level: "I am of the opinion that all the finer speculations in the realm of science spring from a deep religious feeling, and that without such feeling they would not be fruitful."[24] And, at another moment he continued in the same vein: "While it is true that scientific results are entirely independent from religious or moral consideration, those individuals to whom we owe the greatest achievements of science were all of them imbued with the truly religious conviction that this universe of ours is something perfect and susceptible to the rational striving for knowledge."[25] Without this religious conviction, Einstein observed, these scientists would not have been capable of the devotion necessary for their achievements.

While these statements may seem a concession to expansionism at first glance, they are not expansionist in a true sense, since Einstein was not granting a genuine purchase of science on religion, or vice versa. The relationship which he saw between the two areas was one of motivation, not one of overlap of substance. Science could not, according to Einstein, explain a person's ethical values, and certainly not tell him or her what to do. Indeed, if we take Einstein at his word, we must judge him to be a restrictionist, a person who carefully separates the world of science from the worlds of ethics and values. His affirmation of support of the classical separation of facts and values could not have been clearer. And yet one senses in Einstein a tension between his intellectual and logical conclusions on this question and his intuitive and emotional viewpoints, a tension somewhat similar to the one he experienced on the question of physical causality. Einstein knew that, strictly speaking, Hume was correct when he said that causality in nature could never be proved; the most that could be observed was a sequence of events that some people might choose to call causality. An actual causal link was a hypothesis, and one unsusceptible to verification. Despite Einstein's intellectual agreement with this point, as he grew older his intuitive allegiance to causality outweighed his rational appreciation of Hume's point and he eventually so strongly supported the causal hypothesis that he battled his friend Niels Bohr on the issue in some of the greatest debates in the history of science.

Einstein was similarly in agreement with Hume's distinction between relations of ideas and matters of fact, and he believed that empirical evidence from science could not be directly linked to questions of value. Yet here also, Einstein's own life is a witness to a relationship between his scientific views and his social ones. His hopes for order, justice, and rational explanation in the social dimension were too similar to his striving for order and simplicity in the world of physics to be coincidental. Furthermore, Einstein's commitment to a causal explanation of physical nature seems highly consonant with his belief in socialism, an explanatory scheme that unifies human history in terms of underlying economic causal relationships.[26] If "God does not play dice" with subatomic particles, as Einstein affirmed, was He more likely to do so with humans? And when Einstein refused to give religionists a monopoly even over ethics, saying that "accumulated experience" was more important as a guide to behavior, he was pointing to the importance of empirical evidence (in principle, amenable to scientific analysis) as an influence on value systems. Deep down, then, there *were* some links between Einstein's scientific and nonscientific views, despite his warnings that the areas should be kept separate. But we should remind ourselves that Einstein never made these relationships explicit—and, in fact, was highly skeptical of such linkages—and, therefore, we too should recognize them as speculative.

Einstein was a revolutionary in physics, but his attitude toward the relationship of science and values was representative of the restrictionism that was a common viewpoint among the academic professions in the first half of the twentieth century. Einstein was more aware than most of his colleagues of the positive functions that this restrictionism fulfilled: it protected society from abuses such as those that came from the eugenists and the Social Darwinists, while it also insulated science from the attacks of critics who believed that the new physics undermined traditional standards of ethics and politics. Einstein, furthermore, made a contribution toward understanding the relationship of science and values by his differentiation of the substance of science from its form and its motivational origins. But Einstein's belief that science does not study or explain values would become more difficult to defend a few decades later when science turned more and more successfully to the study of man, his behavior, the genetic bases of some of that behavior, and the environmental and biological origins of some of man's values.

Bohr and Quantum Mechanics

BOHR'S VIEWS ON CAUSALITY AND COMPLEMENTARITY

Niels Bohr (1885–1962) was a person with a deep interest in epistemology and natural philosophy, and to a very large degree the same issues that were important to him as a young man were still present in his mind in later years: questions of the nature of life and of matter, of mechanism and teleology, of causality and of the possibility of a unified view of nature. As one reads through Bohr's over one hundred publications seeking to identify the philosophical assumptions contained in his work and paying particular attention to those writings in which he gave broad interpretations to scientific achievements, one is struck again and again by the consistency of the man's thought.[27]

Nowhere in his writings do we find evidence of swerves in his philosophical assumptions. Indeed, Bohr appears to us now as a person who by intellectual background was from an early date almost uniquely prepared for clarifying and absorbing the revolutionary philosophical implications of quantum physics. In fact, in later years an anecdote was told among Bohr's Copenhagen associates to the effect that Bohr had already "said everything" twelve years before quantum mechanics arose.[28] The meaning behind this complimentary jest was that Bohr's philosophical disposition prepared him, even as a young man, for the passing of the mechanistic world picture, and that he saw the implications of the new physics more rapidly than did most of his colleagues.

We also know that the links between Bohr's philosophy and his physics were mutual and complex, with each being conditioned by the other throughout his life in a symbiotic relationship too intricate ever to be entirely unraveled. We know, too, that Bohr's emphases and interests shifted somewhat during his career, as the life of physics and the life of politics changed during his generation. All these shifts, however, appear quite small when they are situated within the framework of his strikingly constant set of philosophical interests and assumptions.

Niels Bohr grew up in a liberal academic family in Copenhagen, where his father, Christian Bohr, was a professor of physiology at the university. The father, too, was interested in the philosophical issues of science, and we have good evidence indicating that the arguments over the nature of living organisms that took place around the turn of the century in the Bohr home were formative influences on the young Niels. Not only did Niels in later years return to these same issues again and again, but in a lecture in Copenhagen almost fifty years after his father's death he quoted his father's publications on the nature of life and remarked that they "express the attitude in the circle in which I grew up and to whose discussions I listened in my youth."[29] The point at issue was the adequacy of the mechanical conception of nature, a question that was at the center of many of the concerns of the son as well.

Christian Bohr as a postdoctoral student in the early 1880s had worked in the Leipzig laboratory of Carl Ludwig, who was a member of the famous school of Berlin medical materialists. The basic assumptions of this school were that all living beings could be explained in terms of physical and chemical principles, and that a complete understanding of the behavior of a whole organism can be attained by an analysis of its individual parts. Within this reductionist viewpoint, no place was granted to teleology or purposiveness in living beings, and the physical and chemical conceptions that were believed to be the keys to an exhaustive explanation of life were, at least in principle, thought to be picturable in terms of physical models acting in absolute space and time. The ultimate goal of the physiologist was to describe all the functions of the human body and brain in terms of these physical models.

Already by the time of Christian Bohr's most active years at the end of the nineteenth century a reaction against the Berlin medical materialist school had begun among European scientists, not so much because the assumptions of the school had been proven incorrect, but because most of the gains that could be made within this framework, in terms of the science of the time, had already been attained, and interests were beginning to move in other directions. Science often proceeds along a path of alternating intellectual fashions, with each viewpoint being pursued until the returns inherent in that approach begin to diminish, at which point another attitude begins to prevail. At these moments social and extrascientific intellectual concerns often play roles too, and by the time of Niels' childhood progressive and liberal thinkers like his parents and their friends no longer considered it obligatory to espouse radical materialist views of the type that had been so much in fashion in such circles a generation earlier.

As did his teachers, Christian Bohr remained convinced of the primary value of physics and chemistry for an explanation of physiology and he successfully applied these methods in his own research on respiration, but he was much more willing than the leaders of the Berlin medical materialists to use words such as "purposiveness" in explaining the behavior of organisms. To attribute purposiveness to organisms, said Christian Bohr, is "quite natural as a heuristic principle" and can prove "not only useful, but indispensable."[30] Christian Bohr was trying to reconcile here what an earlier generation had often considered irreconcilable: belief in reductionist, mechanical explanations of life on the one hand and the attribution to life of overall guiding purposes, on the other. Christian's son Niels would later frequently refer to this duality as an example of his principle of "complementarity," the origin of which was in his work on atomic physics.

The young Niels Bohr was also influenced by several other older scholars in Copenhagen, including the philosopher Harald Høffding and the physicist C. Christiansen. Both of them were involved in the discussions in the Bohr house about the nature of life and the validity of mechanism. Høffding was also interested in the philosophy of religion, the thought of Kierkegaard, and the writings of William James.

One student of Bohr's thought has suggested that Høffding introduced Bohr to the work of James and that James was subsequently an important influence on Bohr's development.[31] Since Bohr did not refer to James in his publications it is difficult to trace the influence, but from several of Bohr's associates we know that Bohr certainly knew and valued James' writings. In conversations and in his publications Bohr made reference to numerous philosophers and writers, including those of ancient Greece, China, and India.[32] Against this diverse background we must conclude that the James thesis—while one of the most interesting interpretations of Bohr recently advanced—is only one of several hypotheses concerning the still largely unknown intellectual morphogenesis of Niels Bohr.

In the development of Niels Bohr's views on science as expressed in his writings, several phases can be discerned. In the first phase, lasting until about 1925, Bohr confined himself to the problems of atomic physics proper, and did not discuss, at least in print, the philosophical implications of his field for the concepts of causality, determinism, and epistemology. In the second phase of his writings, lasting from about 1925 to 1929, Bohr developed his views on a broader philosophical level, but still confined himself almost entirely to physics, i.e., he discussed the philosophical implications for physics in general of the new developments in quantum mechanics but did not extend these implications to fields outside physics itself. In the third phase, beginning about 1930, he became an explicit expansionist, and wrote rather extensively on what he saw as the implications of the new physical views for areas of study far afield from physics itself, including biology, psychology, and even anthropology.

From a very early point in his scientific studies Bohr was well aware that quantum physics implied a radical break with the conceptions of classical physics even though the exact nature of that break was still unclear. His younger colleague and biographer Leon Rosenfeld tells us that even as early as 1911, in his doctoral dissertation on the electron theory of metals, Bohr possessed "the firm conviction of the necessity of a radical departure from classical electrodynamics for the description of atomic phenomena."[33] Bohr believed that Planck's discovery at the beginning of the century of a quantum of action was the starting point of this breakdown of classical physics, and he maintained earlier than his older colleague Rutherford himself that Rutherford's work on the nuclear structure of the atom had far-reaching implications.

In a less rigorous scientist this willingness to forsake traditional principles might have resulted in hasty speculation, but Bohr combined his radical expectations with the belief that the new physics should be carefully joined to the old, with quantum physics to be a rational generalization of classical physics, agreeing with it to the maximum extent possible. It is true that on one occasion in a famous article with two coauthors Bohr advocated what amounted to the abandonment of the sacred conservation laws of matter and energy, a conclusion that was soon contradicted by other work, and that Bohr quickly acknowledged in print to be "unpromising" (*aussichtslos*).[34] In general, Bohr remained faithful to his "correspondence argument," which he expressed as "the endeavour of utilizing to the outmost extent the concepts of the classical theories of mechanics and electrodynamics, in spite of the contrast between these theories and the quantum of action."[35]

What does emerge from Bohr's attitude toward traditional physics is that, rigorous and careful as he was, he was less disturbed than many of his colleagues—Rutherford, Planck, Einstein, de Broglie, and Schrödinger are examples—by the possibility of the abandonment of principles that had been considered tantamount to the foundations of science itself.

In the first phase of his work, the period until 1925 mentioned above, Bohr's publications were concerned with detailed studies of the structure of atoms and molecules, the spectral lines of helium and hydrogen, and the relationship of atomic structure to the periodic table of elements. Not a single one of Bohr's publications before 1927 contains sections explicitly discussing the broad topics of "causality" or "determinism," the subjects of much later debate. Of course, Bohr realized that these questions were not far removed from his research, and in an article submitted in 1922 he remarked that the application of the quantum postulate to heat radiation processes leads to the conclusion that probability rather than causality operates on that level. "We do not inquire about a 'cause' for the appearance of radiation processes," he wrote, "but we simply assume that they are governed by laws of probability."[36] And in a 1925 article he wrote, "In the general problem of quantum theory, one is faced not with a modification of the mechanical and electrodynamical theories describable in terms of the usual physical concepts, but with an essential failure of pictures in space and time on which the description of natural phenomena has hitherto been based."[37]

The event that precipitated Bohr's elaboration of the first steps of a broader interpretation was probably the Solvay Conference of October, 1927, in Brussels, the first one of these famous meetings of the elite of the world's physicists which Bohr attended. For that meeting Bohr was asked to prepare a paper on the epistemological problems confronting physicists in quantum physics. Shortly before the Solvay Conference Bohr gave a lecture at a Volta celebration in Como where for the first time he wrote a long semiphilosophical section entitled "Quantum Postulate and Causality." Referring to Heisenberg's formulation of the indeterminacy principle earlier that same summer, Bohr announced that in quantum mechanics "there can be no question of causality in the ordinary sense of the word," since the new physics portrayed the behavior of an atomic particle in a way that showed a "complete rupture in the causal description of its dynamical behaviour."[38]

Drawing a parallel between quantum mechanics and special relativity, Bohr wrote: "Just as the relativity theory has taught us that the convenience of distinguishing sharply between space and time rests solely on the smallness of the velocities ordinarily met with compared to the velocity of light, we learn from the quantum theory that the appropriateness of our usual causal space-time description depends entirely upon the small value of the quantum of action as compared to the actions involved in ordinary sense perceptions." And Bohr continued his appeal to the similarity of his work to Einstein's in the conclusion of his lecture: "We find ourselves here on the very path taken by Einstein of adapting our modes of perception borrowed from the sensations to the gradually deepening knowledge of the laws of Nature."[39]

When Bohr wrote these words he had not yet had the opportunity of trying out his new concepts on Einstein himself, although they first met in 1920, when quantum mechanics had been in an earlier, less definite stage of development. In describing his later, 1927 meeting with Einstein, Bohr observed that "at the Solvay meetings, Einstein had from their beginning been a most prominent figure, and several of us came to the [1927] conference with great anticipations to learn his reaction to the latest stage of the development which, to our view, went far in clarifying the problem which he had himself from the outset elicited so ingeniously."[40] To Bohr's dismay, and all of his references to the methodological similarity of relativity and quantum mechanics notwithstanding, Einstein differed radically with him on the philosophical meaning of quantum mechanics, although, of course, he greeted as enthusiastically as Bohr did the new advances in understanding in atomic physics and its mathematical descriptions, advances which Bohr used as a foundation for his broader interpretation.

Einstein's major objection at the Solvay conference was not to the current state of quantum theory, but to the view that it meant a permanent renunciation of causality in nature. He expressed the opinion that future developments in physics might make it possible to take into account the interaction between microbodies on the atomic level and measuring instruments in such a way that a causal, deterministic description would be possible, at least in principle. Einstein believed that causality was too fundamental a principle in science to be forsaken, while Bohr was already convinced that the only way to give the mathematical formalism of quantum theory a physical interpretation was by abandoning causality.[41]

Bohr thus differed with Einstein on questions of principle, but he acknowledged that "Einstein's concern and criticism provided a most valuable incentive for us all to reexamine the various aspects of the situation as regards the description of atomic phenomena."[42] In the period 1927 to 1929 Bohr developed his physical interpretation of quantum mechanics ever more clearly, presenting what quickly came to be known as "the Copenhagen Interpretation," a viewpoint that remains to the present day the predominant one among physicists. Several of the essays which Bohr wrote in this period and which served as foundation stones in the Copenhagen Interpretation were reprinted in 1931 in German and in 1934 in English under the title *Atomic Theory and the Description of Nature.*

Bohr believed that physicists' experience with atomic phenomena had brought about a "complete revision of the foundations underlying our description of natural phenomena."[43] Probabilistic quantum mechanics was not, he asserted, a temporary expedient, a means of dealing with microbodies which on a deeper level obeyed causal laws yet to be discovered, but was an expression of the fact that nature inherently was acausal on the atomic level. Indeed, the very concept of causality was to Bohr a conclusion drawn by man from a limited range of experience, the range familiar to him in the world of ordinary objects, which obeyed the laws of Newtonian physics. The law of causality which physicists and nonphysicists alike had come to take for granted was expressed in Newtonian physics by the belief that in any given physical situation, if the positions and velocities of all objects are known at any given moment of time, the future state of those objects at any specific time could be predicted to any desired degree of accuracy. What atomic physics shows, said Bohr, is that when science reaches the level of atomic phenomena, this expectation proves to be mistaken.

The concept that enabled Bohr to find a noncontradictory, coherent interpretation for atomic processes was "complementarity," the view that for material particles on the atomic level and for light itself a set of strikingly different conceptual pictures are necessary to account completely for the phenomena—pictures that would be quite contradictory if one tried to unify them simultaneously. Thus, photons can be described both as waves and as particles. The way to ease this apparent contradiction, said Bohr, is to recognize that the only reason it arises in the first place is because humans have no resource for the description of atomic processes than the use of classical concepts and ordinary language, even though these concepts and this language were developed for a level of experience far from that of the atomic phenomena for which they are being used, and therefore are not adapted to it. Instead of trying to abandon the classical concepts or to develop a new language, scientists should recognize the hopelessness of such tasks and describe atomic processes with their ordinary concepts and language, recognizing that the price they must pay for this approach is the use of methods of description which, taken together, would be contradictory.

Bohr emphasized that man's knowledge at any one point of time is thoroughly conditioned by the range of experience encountered until that time, and, therefore, "we must always be prepared . . . to expect alterations in the points of view best suited for the ordering of our experience. In this connection we must remember, above all, that as a matter of course, all new experience makes its appearance within the frame of our customary points of view and forms of perception." Atomic physicists were making this transition but "we have been forced step by step to forego a causal description of the behavior of individual atoms in space and time, and to reckon with a free choice on the part of nature between various possibilities to which only probability considerations can be applied."[44]

Fruitful as Bohr's approach was to understanding and dealing with atomic phenomena, there were both scientists and philosophers who were disturbed by the radicalness of his proposed break with causality and with his use of phrases such as "free choice on the part of nature," as in the above passage. Since the time of discussions of atomism and determinism by Democritus and Epicurus, physics had often been related in Western thought to questions of social values. Bohr's new approach opened up all of these old issues and some new ones as well.

BOHR'S ATTITUDE TOWARD APPLICATION OF SCIENTIFIC CONCEPTS TO VALUES

In the years after 1929 Bohr began to expand his interpretation of physical theory into areas outside of physics itself. Intellectually, this was an enormous step, one often discounted by his physicist colleagues, but evidently important to Bohr. If the attribution of physical terms to the mathematics of quantum theory was already an interpretation, the expansion of the physical interpretation to other fields, including the social sciences, was obviously an "interpretation of an interpretation" on a level several times higher, with infinitely greater risks. Bohr was much more willing to expand his interpretations in this way than was Einstein.

The first article in which Bohr identified what he saw as the philosophical impact of quantum mechanics for areas of thought outside physics was a jubilee pamphlet of June 1929, published in honor of the fiftieth anniversary of Planck's doctorate. In this article he referred to the problem of complementarity in atomic physics and observed that "the epistemological and psychological questions which arise here lie perhaps outside the range of physics proper."[45] The root difficulty, he maintained, was that of distinguishing between the perceiving subject and the perceived object. In classical physics, this problem could be ignored because of the negligibility of the effects of measurement, but the problem had already been familiar to man for centuries in another area, that of human psychology; in psychology such well-known characteristics as "emotion" and "volition" are not only "incapable of being represented by visualizable pictures," but also cannot be studied in a reductive, analytic way without a shift of awareness occurring that essentially alters the effort to measure and understand human psychology.

Bohr also extended his arguments to maintain that the developments in atomic physics affected arguments about the classical question of the freedom of will:

The opinion has often been expressed that a detailed investigation of the brain, which, although not practicable, is, nevertheless, thinkable would reveal a causal chain that formed a unique representation of the emotional mental experience. However, such an idealized experiment now appears in a new light, since we have learned, by the discovery of the quantum of action, that a detailed causal tracing of atomic processes is impossible.[46]

Thus, Bohr believed that "in the facts which are revealed to us by the quantum theory . . . we have acquired a means of elucidating general philosophical problems."[47] In subsequent publications he elaborated his views of these problems, touching intimately upon ethical and even religious questions in many different forms.

Again and again in his writings, Bohr returned to the question of the relationship between the subject and the object in man's attempt to gain knowledge. He saw here a deeply dialectical relationship, either side of which could be isolated only by losing an important contribution of the other side. Man is forced to take these distorted one-sided views sequentially—this is the nature of complementarity—but he should notice the greater undefinable truth that is contained in the unified relationship.

When a person has the sense of free will and says "I will" he has momentarily restricted himself to the subjective view, but that view is equally as valid—and equally as restricted—as the opposite objective view which attempts to explain human volition as a causal physiological process. Bohr clearly believed that behavioristic psychology represents a one-sided approach to human actions. As he observed, "The place left for the feeling of volition is afforded by the very circumstance that situations where we experience freedom of will are incompatible with psychological situations where causal analysis is attempted."[48] According to Bohr, "the word volition is indispensable in an exhaustive description of psychical phenomena."[49] This was a point of view that behaviorists sharply rejected. Yet Bohr was not condemning the causal analysis of the human psyche, merely emphasizing the limitation of that approach in accord with his view of complementarity: "The decisive point is here the recognition that a state of consciousness, in the description of which words like 'I will' find application, is complementary to a state where we are concerned with an analysis of motives for our actions."[50]

Bohr also believed that his concept of complementarity had

application to the study of different human cultures, a rather breath-taking jump from the level of atomic phenomena. Speaking in 1938 to the International Congress of Anthropological and Ethnological Sciences, he remarked, "The new lesson which has been impressed upon physicists regarding the caution with which all usual conventions must be applied as soon as we are not concerned with everyday experience may, indeed, be suited to remind us in a novel way of the dangers, well known to humanists, of judging from our own point of view cultures developed within other societies." Elaborating on the analogy between anthropology and physics, Bohr said that not only is cultural relativism "helpful in promoting a more objective attitude as to relationships between human cultures," ensuring that one does not assume a natural superiority of one type of society over another, but also that in the studies of cultures of primitive peoples, ethnologists were becoming aware—just as atomic physicists—of the "risk of corrupting" and the "problem of reaction" in the process of measurements. Bohr was thus skeptical of the enterprise of comparative anthropology, since it seemed to him to involve viewing one culture from the standpoint of another. As he observed, "We may truly say that different human cultures are complementary to each other."[51]

Taken literally, this viewpoint might make it seem impossible for true communication across cultures to occur. Two different cultures could be seen as different in kind, and as unbridgeable in description, as the wave and particle descriptions of a photon. Undoubtedly, Bohr did not wish to be held so strictly to his description of physical complementarity when he spoke of anthropological complementarity. He was probably led to a slight exaggeration of the strength of the analogy by the intellectual climate of the late thirties. Two factors in that climate should be mentioned which help to explain his position and which indicate that he may not have meant to be advancing pure cultural relativism. First of all, in 1938, when he spoke to the anthropologists, a view of a rigid hierarchy of cultures was being advanced in Nazi Germany—Bohr opposed this with a plea for tolerance based on the validity of contrasting complementary views. And, second, among non-German cultural anthropologists, particularly in the English-speaking world, the twenties and thirties were a time when the differences, not the similarities, between primitive cultures were being emphasized. The pioneering studies of those decades—in the South Pacific, Africa, and South America—revealed how strikingly differently family and social life could be organized in separate societies that were, nonetheless, quite viable. Against this background, the view of cultural relativism inherent in complementarity seemed quite valuable. Many members of a later generation of anthropologists who increasingly sought the common features of different cultures, emphasizing linguistic structure, biological determinants, and environmental constants, would find Bohr's emphasis on the literal message of complementarity too strong. But in his time his analogy played a valuable political role.

Some observers of Bohr's philosophy feared that the emphasis he placed upon the equal validity of apparently contradictory explanatory concepts was an undermining of the unity of knowledge itself, a fracturing of man's framework of explanation into a set of competing nonreconcilable schemes. Bohr's participation, therefore, in a 1954 symposium on "The Unity of Knowledge" attracted some attention.

In his contribution to the symposium, Bohr repeated his belief that atomic physics helped to show that terms like "volition" are essential for an explanation of human behavior, and he continued that "we may both practically and logically speak of freedom of will in a way which leaves the proper latitude for the use of words like responsibility and hope, which themselves are as little definable as other words indispensable to human communication." But he emphasized that he did not see anything mystical or irrational in nature on the atomic level, and that he was not calling for a "personification of nature."[52]

Bohr then turned to the question of "whether there is a poetical or spiritual or cultural truth distinct from scientific truth." The answer which he gave was consistent with his earlier emphases on the subject-object duality and the complementary nature of cognition. Science and religion, he said, take "essentially different starting points." Science "aims at the development of general methods for the ordering of common human experience," while religions start from "endeavors to further harmony of outlook and behavior within communities," each of which may have different histories and traditions. Therefore, any strict evaluation of religion from the standpoint of science is impossible: "The description of the position of the individual within the community to which he belongs presents typically complementary aspects related to the shifting separation between the appreciation of values and the background on which they are judged."[53]

Bohr did not give a literal answer to his question of whether there exists poetic or spiritual truth distinct from scientific truth, but, judging from what he did say, it seems apparent that his answer was that neither a spiritual nor a scientific truth exists alone, but that they are complementary one-sided views of a nonanalyzable unity that is truth.

On one other occasion Bohr addressed himself directly to the problem of religion, and that was in an article in honor of the seventieth birthday of the Old Testament scholar Johannes Pedersen. We know that from youth Bohr had been no friend of organized religion, but in his article he was not concerned with institutional aspects of religion, but with the intellectual question of the place of religion. He reasserted his belief that religion and science emphasize different sides of the human psyche, the emotive and the cognitive, sides which he saw as deeply complementary. One of the best examples of the practical importance of both the emotive and the cognitive was what he called the complementary relationship of justice and love:

> It must be recognized that, in any situation which calls for the strict application of justice, there is no room for display of love, and that, conversely, the ultimate exigencies of a feeling of love may conflict with all ideas of justice. This situation, which in many religions is mythically illustrated by the fight between the divine personifications of such concepts, is indeed one of the most striking analogies to the complementary relationship between physical phenomena described by different elementary concepts which were combined in the mechanical conception of nature, but whose strict applications in wider fields of physical experience exclude each other.[54]

In the above passage Bohr was again pointing to the heuristic value of the analogy between physical concepts and social concepts, just as he had done in anthropology. And here also it would be narrow-minded to hold Bohr to a strict and literal reading of the analogy. Bohr turned to physics for help in understanding social phenomena, but his conclusions were often suggestive rather than demonstrative.

The nature of life was the philosophical topic that concerned Bohr most of all. We have a particularly good opportunity to observe the evolution of his views on the topic, since we have two essays on the subject written thirty years apart, entitled "Light and Life" (1932) and "Light and Life Revisited" (1962), as well as several other essays on closely related issues.

Bohr believed that the new physics could be brought to bear on the problem of the nature of life in a direct way. On this topic he shifted from analogical arguments to identification of what he considered to be substantial links between quantum physics and biology. Therefore, the relevance of atomic physics for understanding life and social processes was, he thought, considerable. He pointed out, for example, that "in the carbon assimilation of plants . . . we are dealing with a phenomenon for the understanding of which the individuality of photo-chemical processes must undoubtedly be taken into consideration." The new physics had shown the irreducibly individualistic behavior of atomic particles. The acausal nature of quantum mechanics played an instrumental role in the study of life. Bohr believed that the recent developments in physics supported the view that the essential characteristics of living beings have "no counterpart in inorganic matter."[55]

Furthermore, since the sensitivity of the retina of the human eye is such that "the absorption of a few light quanta, or perhaps only a single quantum . . . is sufficient to produce a light impression," human perception is affected by the acausality of atomic physics.[56] Bohr was persuaded that the other organs of the human body were equally finely tuned, and, consequently, "the feature of individuality symbolised by the quantum of action, together with some amplifying mechanism, is of decisive importance." Here Bohr was making an expansionist argument for a direct, nonmetaphorical relevance of atomic physics to the problem of understanding life and free will and he was, as did his father Christian Bohr, arguing that a mechanistic conceptionalization was inadequate to physiology.

Bohr was also impressed by the fact that in order to study an organism in an exhaustive, reductive way it was necessary to kill it, which meant that the processes of life could not be studied, only its structure. He saw a resemblance here to the problem of the measuring instrument in atomic physics, and he took satisfaction in the fact that:

> In every experiment on living organisms, there must remain an uncertainty as regards the physical conditions to which they are subjected, and the idea suggests itself that the minimal freedom we must allow the organism in this respect is just large enough to permit it, so to say, to hide its ultimate secrets from us.[57]

Bohr loved this sort of paradox. He believed that it meant that it was impossible for life to be explained exhaustively on a physical or chemical basis, although he admitted that no "well-defined limit can be drawn for the applicability of physical ideas to the phenomena of life."[58] He was not arguing for the side of idealism or vitalism versus that of materialism or mechanism, but for a complementary approach to biology that avoided "the extremes of materialism and mysticism."[59]

Bohr remained faithful to his complementary approach to the problem of the nature of life throughout his life, but there was a time in the late forties and early fifties when he seemed considerably more enthusiastic about the vitalistic half of the complementarity than the mechanistic half. He definitely resented any intimation that life could be fully explained on a physical basis, and he used the word "vitalism" without apology in those years, describing it as one of the necessary approaches to biology. Writing in 1952, he said, "Although science will of course strive for ever more detailed knowledge of the physical mechanism underlying the functions of organisms, a description of life corresponding to the ideal of mechanicism will only constitute one line of approach. . . . In actual biological research, a vitalistic approach is equally indispensable, since the primary object must often be the studies of the reaction of the organism as a whole for the purpose of upholding life, a point of which we are not least reminded in medical research."[60]

At the time of his death in 1962, Bohr was working on a revision of his 1932 essay on "Light and Life," and the unfinished manuscript was published after his death. By this time, events in molecular biology had thrown a whole new light on the problem of reductionism and mechanism in biology. The determination of the structure of DNA in 1953 had led to a powerful resurgence of the view that life could, indeed, be entirely explained in terms of physics. Francis Crick, physicist turned molecular biologist, and one of the leaders of the new breakthrough, would choose as the introductory slogan of one of his publications the statement, "Exact knowledge is the enemy of vitalism."[61] No room for Bohr's complementarity between vitalism and mechanism here.

In this last essay of his life, Bohr did not use the terms "vitalism" or "vitalistic," as he had in several earlier essays, and he made it clear that he did not see "any limitation in the application to biology of the well-established principles of atomic physics." Yet the basic assumptions of Bohr's older arguments were still present, and it is quite clear that he remained convinced that an exhaustive, reductionist explanation of life was in principle not possible. He spoke of the importance of paying attention to the overall life functions of organisms in which teleological terms are used. He observed, "Surely, as long as for practical or epistemological reasons one speaks of life, such teleological terms will be used in complementing the terminology of molecular biology." And he did not believe molecular biology was anywhere near the explanation of such complicated phenomena of life as "consciousness" or "sentiments." To speak of them, it was still necessary for human beings to speak of "mutually exclusive experiences" in a typically complementary manner.[62] Bohr had retreated a bit in his claims for the importance of a combination of the vitalistic and mechanistic views for an adequate explanation of life, but his philosophical preferences were as clear as ever.

Einstein and Bohr: A Comparison

If we compare the opinions of Einstein and Bohr on the applicability of scientific concepts to questions of human values, we must conclude that their opinions were—at least on the first level of analysis—strikingly different. Einstein maintained that the content of scientific theory could not inform us about human behavior or social values, and we classify him, therefore, as a restrictionist.

Although he wrote frequently on social and political questions, he never attempted to relate physical theory to them. Bohr, on the other hand, wrote many articles in which he took the concepts and discoveries of physics and turned them to questions of human psychology, sociology, and religion. Often he used analogical arguments, but occasionally he tried to apply atomic physics directly to the ancient question of freedom of will, saying that the indeterminacy of atomic particles was the actual cause of what appeared to be arbitrary human volition. Bohr, then, was an expansionist.

Both of these great scientists grappled with the questions of the relevance of science for the larger questions of human existence, and of course they had reasons for the positions they took. Einstein's reasons were the more traditional; his form of dualism, of a sharp separation between science and values, was a very familiar one in modern thought. Bohr's approach was more radical, in the sense that it assumed that dualism was not, in the final analysis, valid; because of the nature of Bohr's philosophy, however, the effect of Bohr's projection of science into the realm of human behavior and values was very small indeed. If Bohr, on the other hand, had possessed the philosophy of nature of Einstein, Bohr's application of science to human questions would have had, at least potentially, very great effects indeed. One is tempted to say that Bohr could afford to be an expansionist because his view of science was such that it could impinge on value questions only in a subtle and nondetermining way; Einstein resisted expansionism, and a part of the reason for his resistance was surely that he saw that his philosophy of nature could have considerable effect, even mischief, if applied to human values. In order to understand this contrast more fully, let us look more closely at the philosophies of nature possessed by the two men.

Relativity theory is, despite its name, based on an absolute principle. The "relativity" in relativity theory has meaning only with reference to Newtonian physics. The old absolutes of "space" and "time" have been shown to be relative to frames of reference, and therefore have been discarded as unchanging standards. However, Einstein was just as interested in a unified world system based upon an absolute standard as Newton had been; his distinction was that he realized that only a different foundation for the system of physics could be secure. He found that new absolute in the concept of space-time, the conjoining of Newton's terms into a four-dimensional manifold. The theory of relativity could thus be just as easily called "The theory of absolute space-time," and a strong argument could be made that the new title would be more accurate. Paradoxically, the theory of relativity is not a purely relativistic theory. The framework upon which it is based provides a method for an exact and absolute description of any event anywhere in the universe.

Once we see this essential characteristic of relativity theory clearly, it becomes apparent that had Einstein been an expansionist, a person who believed that physics can be extended to the explanation of human behavior, the old fearful visions of a deterministic and exhaustive explanation of all natural phenomena—including human behavior, ethics, religions, etc.—would have returned. The nondogmatic Einstein was well aware of the damage that can be—and has been—created by such attempts, and he relied on a traditional dualistic conception of the universe which solved the dilemma created by his vision of nature. He was so committed to the ideals of order, unity, and coherence in nature that he was not willing to accept an alternative solution, the fractured picture of nature presented by Bohr, with its denial of causality and its apparently conflicting explanations in physics.

Bohr, as an expansionist, believed that physics can be helpful in understanding the most essential question of existence—human values. But what a latitudinarian philosophy of nature Bohr's was! He granted a universal scope to the regularities of physics, so that they could be extended to human behavior as well as to inanimate objects (here was his contrast to Einstein), but that scope included within it a concept of contradictory and complementary aspects of truth. Therefore, to Bohr physics did not iron out the paradoxes of life and thought in accordance with some rigid plan, it explained and informed those paradoxes. Physics made the riddles of human existence plausible and tolerable. It provided a grounding for freedom of will, rather than an attack upon it. Bohr thus believed that he had found a new way out of the dilemma between mechanism and vitalism constantly posed in the dining room discussions of his youth, and this sense of discovery must explain the zest with which Bohr celebrated the apparent tensions inherent in the concept of complementarity, a concept that always discomfited Einstein. While Einstein saw complementarity as a rather hopeless muddying of his picture of simplicity in (physical) nature, Bohr saw it not only as a helpful aid in understanding atomic phenomena, but as a method of successfully coping with the most complex issues in (physical *and* biological) nature. Bohr was, thus, not a dualist, but the monism he espoused contained a flexibility—perhaps a contradiction—so elastic that it accomplished much the same purpose as dualism.

Thus, Einstein and Bohr agreed that science does not impinge directly on religion, but they agreed *for different reasons*. Einstein believed that science and religion do not conflict because they belonged to different realms, and could be related to each other only on the indirect motivational level, not in a direct, substantive way. Bohr believed that science and religion were complementary aspects of the *same* reality, but that both were necessary for an understanding of that reality, just as wavelike and corpusclelike models are necessary for an understanding of atomic particles.

Just as Einstein worried about the implications of Bohr's interpretation of quantum mechanics, so also did Bohr have misgivings about the possible broader implications of relativity theory; Bohr thought relativity theory was too similar to Newtonian physics in its aspiration toward a single description of all natural phenomena. Bohr on one occasion expressed openly his fear of the monolithic and absolutist character of relativity theory when used as a guide to thought. In 1938, when discussing the problem of the anthropological study of cultures and societies, Bohr warned that while at first glance relativity theory might seem to underline the subjective importance of the observer, it actually opened the door to an authoritarian and absolutist interpretation of human behavior:

> The unity of the relativistic world picture, in fact, implies the possibility for any one observer to predict within his own conceptual frame how any other observer will coordinate experience within the frame natural to him. The main obstacle to an unprejudiced attitude towards the relation between human cultures is, however, the deep-rooted differences of the traditional backgrounds on which the cultural harmony in different

societies is based and which exclude any simple comparison between such cultures.[63]

The restrictionist Einstein would, of course, have been totally unimpressed by this argument because he would never have dreamed of suggesting that relativity theory could be helpful for the study of human cultures. And for praising quantum mechanics as a more helpful approach for talking about human societies, Einstein would have poked fun at Bohr for once again "acting like a prophet."

We see, then, that the attitudes taken by Einstein and Bohr toward questions of human values can be related to their views of physics. When we bring these two realms together, we are not suggesting that these two scientists' views on value questions were determined by their interpretations of science. We are merely suggesting that their views on science and their views on society can be related to each other, and that the expansionism of Bohr and the restrictionism of Einstein fit well with the overall patterns of their thought.

PHYSICAL KNOWLEDGE AND SOCIETY

CHANGES AS fundamental as the ones occurring in early twentieth-century science had effects that went far beyond science itself. Leading natural scientists and philosophers of different nations and political backgrounds grappled with the new events in science, attempting not only to settle questions in their own minds about the philosophical significance of these scientific developments but also to provide the educated public with at least a preliminary appreciation of what was happening. A striking fact that emerges when one considers these efforts is that a science that was essentially identical in the leading nations received different popular interpretations in the different geographic areas that reflected the various social and political environments.

In the next four chapters I shall examine the viewpoints of five different interpreters of science who were highly popular in their particular political cultures. They are Arthur Eddington in England, Vladimir Fock in the Soviet Union, Werner Heisenberg in Germany, and Henri Bergson and Jacques Monod in France. The first three of these men were primarily interested in interpreting the new physics, while Bergson is usually considered to be an interpreter of biological evolution. However, the concepts that lay at the core of Bergson's analysis of nature came from his critique of physical science, and his extensive writings on both physics and biology included a book on relativity physics. Although he formed his views before the new physics emerged, he also responded to the events in science as they developed. Monod, coming long after Bergson, differed from him dramatically, and advanced ideas based on the development of the new field of molecular biology.

Monod, then, serves as an introduction to Part 2, The Biological Sciences.

All five of these men were concerned with the relationship of science to sociopolitical values and it is that concern that serves as the focus of my attention. Rather than attempting to analyze their philosophies of nature in a comprehensive way, I shall seek an answer to the question, "What did each of these writers see as the connection, if any, between science and social values?" We shall see that two of these men, Eddington and Bergson, can be classified as restrictionists, although each defended that position in his own distinct fashion. Both agreed that science, in principle, is valid in only a limited realm that does not include values. Fock was an expansionist and made an effort to use the new physics to defend his own values and those of his society. Monod also was an expansionist, although he described science's effect on values as primarily destructive rather than constructive. Heisenberg is the most difficult person to classify as an expansionist or restrictionist since his views and emphases on the issue of science and values varied during a long life in which the political arena changed radically several times. Nonetheless, we will see in his maturity a gradual growth toward a mild and somewhat reluctant restrictionism.

The first two of these men, Eddington and Fock, provide the most apposite comparisons among them since they were talking about almost exactly the same issues while drawing opposite conclusions. Fock was aware of Eddington's views and was, in a sense, responding to them. Against this background, I will link these two chapters together more closely than the others. Eddington and Fock popularized physical theory in virtually antithetical political cultures, England and the Soviet Union. At the end of the chapter on Fock, I shall make a direct comparison of the two men, showing that while they differed on most philosophical issues, they shared one or two basic philosophical commitments that played important roles in their respective interpretations of the relationship of science to values.

chapter two

EDDINGTON AND THE ENGLISH-SPEAKING WORLD

ARTHUR STANLEY EDDINGTON (1882–1944) was one of the most distinguished astronomers and astrophysicists of the twentieth century. He was also one of the most influential popularizers of science in the first half of that century, a man who interpreted the new physics for the educated Anglo-American public. His mathematical genius was widely recognized from the moment he won the position of "First Wrangler" in the mathematics tripos examination at Cambridge University; he was the first second-year student to do so in the history of the university. In later years his pioneering work on astrophysics included a wide range of problems: stellar motions and constitution, radiation pressure, interstellar matter, the mass-luminosity relationship, and relativistic cosmology. Between 1914 and 1946 Eddington published thirteen books (several of which are still in print) and approximately one hundred sixty articles.[1]

Eddington first came to the attention of the public in 1919, when he led an expedition to Principe, a small island off the west coast of Africa, in order to try to measure the deflection of light by a gravitational field during an eclipse of the sun; in that way he hoped to provide empirical evidence in favor of Einstein's general theory of relativity. The successful outcome of this expedition became known throughout the world.

Eddington's associates knew how interesting his expedition was to science, but they could hardly have sensed the depth of its significance to him as an individual. Among the papers of Eddington deposited after his death in Trinity College Library is a handwritten school essay on solar eclipses written when he was fifteen years old, a full twenty-one years before the Principe expedition. In that paper, the schoolboy wrote that whenever a solar eclipse occurs, "expeditions are sent out from different countries of Europe, frequently accompanied by some of the greatest astronomers in the world in order to make observations."[2] When Eddington led the expedition of 1919, he was quite literally fulfilling a boyhood dream. But much more than that, the expedition in which he was participating was not just another observation of a solar eclipse, but perhaps the most intriguing observation of an eclipse of the sun in the history of astronomy, since its purpose was to yield evidence in support of a revolution in the foundations of physics. No wonder that Eddington later referred to the 1919 observations as "the greatest moment of his life."[3]

Eddington's achievements in succeeding years were recognized in a long series of public honors, including knighthood, the Order of Merit, and the presidencies of the Royal Astronomical Society, the Physical Society, and the Mathematical Association. Einstein once commented that Eddington's technical treatise *Mathematical Theory of Relativity* was the finest presentation of the subject in any language. Harlow Shapley of Harvard wrote upon Eddington's death that "it is probably correct to say that during the past twenty-five years Eddington's influence on fellow astronomers has been greater than that of any other man."[4] If Eddington's more popular writings sometimes caused his scientific colleagues to wince, they could hardly blame the public for believing that Eddington should know what he was talking about. If there were ever an "authoritative popularizer" of science, Eddington was one.

Eddington wrote four or five books that enjoyed large circulation, especially *The Nature of the Physical World* and *Science and the Unseen World.* One of his goals in these books was popularization—to make the new physics comprehensible to a large audience. The British physicist J. J. Thomson commented in 1930 that Eddington "by his eloquence, clearness and literary power persuaded multitudes of people in this country and America that they understand what relativity means."[5] Eddington was a master of metaphor and literary craftsmanship, and he often included poetry and citations from the classics of English literature, which he knew well. Beyond popularization, however, Eddington had another goal—the exploration of the significance of the new science for social values, and, particularly, for religion. The emphasis upon religious mysticism was the aspect of Eddington's writings which distinguished them most clearly from the works of the host of other popularizers of relativity physics. And it was this emphasis that attracted the attention of the public.

Eddington believed that science offered a very incomplete picture of reality, and he thought that as a scientist studying the newest developments in physics he was in a better position than many to point out just how limited the scientists' view was. One of his metaphors illustrated his opinion clearly: "The voice that comes to us over the telephone wire is not the whole of what is at the end of the wire. The scientific linkage is like the telephone wire; it can transmit what it is constructed to transmit and no more."[6]

Eddington expressed the view that when he wrote nontechnical books he was illustrating the greater whole at the other end of the telephone wire, and he considered himself to be writing for both scientists and nonscientists, not merely the latter. Far from providing a dilution of science that was designed only for nonscientists, as the word popularization would indicate, Eddington believed, correctly or not, that he was enriching science with an element that was crucially needed; he thought that the lack of that element was made particularly clear by the latest developments in physics:

> I think we may say that, although the physicist has carried his work to greater perfection than formerly, he now puts it in a form which does not hide its incompleteness. Implicitly, if not explicitly, he advertises for someone to complete it. And we who are interested in the non-material aspects of experience are not butting in; we are answering his advertisement.[7]

Thus, Eddington saw himself in a double role: as a physicist he helped issue the advertisement that science was incomplete and demanded a new form of supplement; as a writer of nontechnical books he tried to answer his own advertisement.

The Worlds of Arthur Stanley Eddington

Eddington's interpretation of nature was based on the supposition of the existence of three important and different worlds with which he thought every educated person in the twentieth century must deal: 1) the world of everyday experience, 2) the world of physical science, or "symbolic knowledge," and 3) the spiritual world, or "intimate knowledge." All three of these worlds were important actualities to Eddington, and he believed that modern science pointed to the necessity of distinguishing among them. One of the most common types of errors in thought, he maintained, was recognizing only one of these worlds and assuming that only that domain existed. All *three* existed simultaneously.

The first world, that of everyday experience, was the one most familiar to the ordinary person—the world of substantial objects, of colors, sounds, and scents. People wedded to this domain of thought believed that the correct explanation of any phenomenon was based on telling what things are "really like," i.e., giving an explanation that stated what entities exist, how they interact, and what their essential natures are, using descriptive terms that are familiar to all humans from their personal knowledge of nature. A physical scientist who attempted to extend this mode of thought to his own work would insist on the use of models and pictorial representations in order to understand phenomena on all levels of reality, from atoms to the universe.

At one time, Eddington observed, there was not much difference between the familiar world of everyday experience and the scientific world. It was possible to hope that scientific terms, including their mathematical frameworks, could be assigned referents that could be understood in terms of the familiar world. The symbol m in equations represented an actual entity, mass, and it was the obligation of the physicist to describe the characteristics of mass in everyday language.

With the development of relativity theory, such hopes disappeared irretrievably. Eddington believed that the implications of this development were significant in all other areas of thought, including philosophy and religion.

Relativity theory had demonstrated, Eddington continued, that the secondary world of physical science is built upon a much smaller part of our total experience (the first world) than earlier had been thought. Many of our experiences are nonmetrical, but science is based only upon the metrical: "The cleavage between the scientific and the extra-scientific domain of experience is, I believe, not a cleavage between the concrete and the transcendental, but between metrical and nonmetrical."[8] In fact, physical science so restricts the scope of human experience that human beings have perceptions that are superfluous from a scientific standpoint. For example, "There is no colour in the physical world. . . . Everything that we [scientists] assert can be verified by a colour-blind person."[9]

Eddington continued that we can speak of a scientist as a person who has "eliminated all superfluous senses."[10] What senses would be left to such a person? Only enough for him to read a pointer on a scientific instrument: "When we stripped our ideal observer of most of his sense organs we left him part of an eye that he might observe coincidences,"[11] since physical measurements resolve themselves into pointer readings. All observations will be reduced to coincidences between a pointer and a graduated scale.

Eddington knew that his description of the ideal scientist as a person stripped of all sense organs but part of one eye would seem a gross exaggeration to many of his readers, and surely it was, but Eddington attempted to ground his views in the general scientific community:

> I should like to make it clear that the limitation of the scope of physics to pointer readings and the like is not a philosophical craze of my own but is essentially the current scientific doctrine. It is the outcome of a tendency discernible far back in the last century but only formulated comprehensively with the advent of the relativity theory.

Of course, Eddington knew that physical science goes far beyond its initial starting point of pointer readings. The fundamental influence in this process of the formulation of scientific laws and generalizations is, however, the human mind itself, which must be sharply differentiated from the physical world. Man builds his world out of *relations* and *relata*, but "ultimately it is the mind that decides what is lumber—which part of our building will shadow the things of common experience, and which has no such counterpart." The result is the world of science, or symbolic knowledge, which gives a "shadow performance of the drama enacted in the world of experience."[12]

The selective influence of the mind is the essential forming force in the world of science, since it is " . . . the innate hunger for permanence in our minds which . . . directs the world-building." Thus, the mind has by its power fitted the processes of Nature into a system of laws "largely of its own choosing" and during the process the "mind may be regarded as regaining from Nature that which the mind has put into Nature."[13] To Eddington physical science was a series of human creations based on epistemological principles, not a series of discoveries based on facts. He maintained that " . . . all the laws of nature that are classified as fundamental can be foreseen wholly from epistemological considerations. They correspond to *a priori* knowledge, and are therefore *wholly subjective*."[14] As an untracked moor has no paths until a person creates one, the physical world has no configuration until a scientist gives it one. "Without the mind," he observed, "there is but formless chaos."[15] And, indeed, he went so far as to observe that the "stuff of the world is mind-stuff," although he insisted that he did not use the words "mind" or "stuff" in the normal ways. It is not surprising that of all philosophers, he believed his thought to be closest to that of Kant.

By portraying the scientist as a person who concentrates on a very small part of the total spectrum of human experience, Eddington deemphasized the role of science in modern life, but he in no way felt deprived by this restriction. He gained sustenance by pointing out that the world of physics coexists with the world of the spirit, the world of intimate knowledge whose existence is as real and probably more important even than the world of science. Thus, by defining science in such restrictive terms, maintaining that it relies on only the most fundamental metrical properties of nature at the base of relativity physics, Eddington was able to point to the vast areas of experience that could not be described in these terms. The dimensions of his proposed extra-scientific realm were thus widened.

Eddington's third world was a spiritual realm that lay beyond the world of physical knowledge; it was an "unseen world," a sphere of "intimate knowledge." This form of knowledge "will not submit to codification and analysis; or, rather, when we attempt to analyze it the intimacy is lost and it is replaced by symbolism."[16]

What justification did Eddington have for believing that this third world actually exists? The same as for believing that the world of science exists:

> Whatever justification at the source we accept to vindicate the reality of the external world, it can scarcely fail to admit on the same footing much that is outside physical science. . . . We recognize a spiritual world alongside the physical world. Experience—that is to say, the self *cum* environment—comprises more than can be embraced in the physical world, restricted as it is to a complex of metrical symbols. . . . Those who in the search for truth start from consciousness as a set of self-knowledge with interests and responsibilities not confined to the material plane, are just as much facing the hard facts of experience as those who start from consciousness as a device for reading the indications of spectroscopes and micrometers.[17]

And Eddington believed that there were ways of learning about this third world, although these modes of cognition were entirely separate from the procedures of science. These other channels were the "eye of the body or the eye of the soul," and when we use them we should dismiss the feeling that "we are doing something irrational and disobeying the leading of truth which as scientists we are pledged to serve."[18]

In order to learn more about this third world, we should rely on our sense of "mystical contact with Nature," and we should recognize that for a modern man these modes are "no longer blind alleys but open out into a spiritual world—a world partly of illusions, no doubt, but in which he lives no less than in the world, also of illusion, revealed by the senses." Eddington called for the building of an intimate world taken from the human personality that would be equal in standing to the scientific world taken from the symbols of the mathematician. "We all know," he wrote, "that there are regions of the human spirit untrammeled by the world of physics. In the mystic sense of the creation around us, in the expression of art, in a yearning towards God, the soul grows upward and finds the fulfillment of something implanted in its nature."[19] We should assert the validity of these yearnings, for without them the problem of experience ends in a veil of mathematical symbols. With their help, on the other hand, we lift "the veil in places; and what we discern through these openings is of mental and spiritual nature."[20]

Eddington's Quakerism and Its Relationship to His Science

The character of Eddington's own religion, that of the Society of Friends, was important in forming his attitude toward science. Each of his three worlds—the world of everyday experience, the world of science, and the unseen world of intimate knowledge—was an experiential reality to him, with its own separate mode of cognition. Eddington's approach to these worlds was related to his Quaker background.

Eddington's roots in Quakerism could hardly have been more deep or authentic. His mother, Sarah Ann Shout, was a descendant of John Camm and John Audland, missionaries of Quakerism who participated in the very beginning of the "Quaker Explosion" in the seventeenth century; they had worked directly with George Fox, the man whom William Penn called "the first instrument" of Quaker religion. Eddington's father, Arthur Eddington, was also a member of a strong Quaker family, and was, at the time of the birth of his son, headmaster of Stramongate School of Kendal, an institution controlled by Quakers since the late seventeenth century. Arthur Stanley Eddington remained a loyal Quaker throughout his life, and frequently attended Friends' meetings and summer schools.

Quakers traditionally based their religion on experience, not on dogma or on speculation. George Fox emphasized in the very first years that the Quaker Way was empirical. He believed that he could come to know Christ on the basis of his inner consciousness, not through reasoning or analysis. Referring to his religious perception, Fox wrote, "This I knew experimentally."[21] The word "experimentally" here would probably be rendered today as "experientially." Eddington echoed Fox's sentiments when he wrote that those who believe in the unseen world of intimate knowledge "are just as much facing the hard facts of experience" as those who build their knowledge on the indications of spectroscopes. Eddington was often moved to a mystical experience by an object of natural beauty, and he loved arduous hikes through scenic parts of England and Wales. But his mysticism went far beyond reverence for nature—it had truly religious inspiration. His loyalty to the Quaker way of knowing was indicated by his observation that "in the mystical feeling the truth is apprehended from within and is, as it should be, a part of ourselves."[22]

Defenders of Quakerism have often maintained that their religion is particularly compatible with science, since it possesses no creed and therefore, in a logical sense, no possibility exists for a direct conflict with scientific findings. An example of this view was expressed by a Quaker professor who claimed, a few years ago, "Quakerism begins with an experience rather than with a dogma and therefore can appeal to scientists because it is itself scientific-minded. It is no accident that so often the most impressive departments in the Quaker colleges are the departments of science."[23]

Eddington believed that the nonverbal character of Quakerism was one of its most attractive features, and he found it totally compatible with his devotion to science. In Quaker meetings he was drawn most strongly to the silent worship periods; after attending a Quaker summer school in July and August, 1908, he complained in his journal, "There was no doubt that the school splendidly fulfilled its serious purposes. The devotional meetings were good but there was no silence, all far too many little speeches."[24]

Eddington's need for a mystical supplement to his scientific activity sometimes caused him to write so critically of science that the quotations, taken out of context, could mislead one into believing that he had little love for his profession. This conclusion would be incorrect, but even more incorrect would be the view that his religious sentiments were mere embellishments on his life as a scientist. He once wrote:

> The only instrument that I, a physicist, can manipulate is a bludgeon which, it is true, crushes illusion; but at the same time crushes everything of non-material significance and even reduces the material world to a state of uncreatedness. For I am convinced that if in physics we pursued to the bitter end our attempt to reach purely objective reality, we should simply undo the work of creation and present the world as we might conceive it to have been before the Spirit moved upon the face of the waters. The spiritual element in our experience is the creative element, and if we remove it as we have tried to do in physics, we must ultimately reach the nothingness which was the Beginning.[25]

Eddington and Restrictionism

Most writers about the relationship of science to values can be classified as either expansionists or restrictionists. Expansionists cite evidence within the body of science to support their values; restrictionists believe that such an effort is improper. Looking at Arthur Eddington, it is obvious that he was much closer to a restrictionist approach than to an expansionist one. Indeed, he exerted a truly remarkable effort to restrict science to the narrowest of conceivable realms, that of pointer readings, supporting this approach by showing how much the advent of relativity theory depended on the individual operations of the processes of physical measurement. Again and again he emphasized that now that science "omits so much that is obviously essential, there is no suggestion that it is the whole truth about experience."[26] He observed that if the restriction of science to pointer readings "has taken away all reality," then "I am not sorry that you should have a foretaste of the difficulty in store for those who hold that exact science is all-sufficient for the description of the universe and that there is nothing in our experience which cannot be brought within its scope."[27]

Although Eddington was a deeply religious man, he opposed all arguments that fall under what I have called the expansionist approach. He made no effort to justify religion on the basis of science itself. He believed that in the past such efforts had led to a series of harmful collisions between science and religion, events which he hoped could be avoided in the future. He rejected outright the attempt to ground the beliefs of religion on evidence from science.[28] He remarked, for example, "I have not suggested that either religion or free will can be deduced from modern physics."[29] Eddington, it is clear, was an articulate spokesman for the restrictionist approach to the relationship of science and values.

Yet there was something very strange about Eddington's restrictionism. Restrictionism was supposed, by definition, to avoid bringing science to bear upon values, yet Eddington somehow seemed to pull it off. Restrictionism in its pure form says that nothing that happens in science can legitimately be brought to bear on the realm of values. If Eddington had steadfastly supported that principle he could hardly have written several of his books, since their purpose was to show how recent events in physics (relativity theory and quantum mechanics) affected the relationship of science and values.

The eagerness with which Eddington pushed science back into a corner of the totality of human experience is an indicator that he was actually clearing the way for values—and not just any values, but particular ones which he preferred—to occupy the space which had been vacated. If a strict adherence to restrictionism would have led to his support of a relationship of neutrality between science and values, the stance of Eddington can only be described as the most "benevolent neutrality" imaginable.

Throughout his writings on the topic of science and religion, Eddington wrestled with a major problem: on the one hand he wished to affirm that science had nothing to do with religion, to work within the restrictionist approach. On the other hand he wanted to show that the new physics harmonized with religion. The difficulty of overcoming the problem is evident in his writings, although in his lectures he felt a bit freer to support his inner religious beliefs.

Several of Eddington's most popular books began as lectures. The Gifford Lectures of 1927 became *The Nature of the Physical World*; the Swarthmore Lecture of 1929 was published as *Science and the Unseen World*; the Messenger Lectures at Cornell in 1934 appeared later as *New Pathways in Science*; and the Tarner Lectures of 1938 were published as *The Philosophy of Physical Science*. Usually Eddington thoroughly revised his lectures (which were written out) before publication. He acknowledged the difficulty of changing the lectures into books. In the preface to *The Nature of the Physical World* (originally the Gifford Lectures) he observed, "In the oral lectures it did not seem a grave indiscretion to speak freely of the various suggestions I had to offer. But whether they should be recorded permanently and given a more finished appearance has been difficult to decide."[30]

I have not been able to find all the original lectures, but I have found enough original and manuscript material to indicate that in them, the relationship of science to religion was one of the problems that demanded revision. In the original handwritten manuscript of *The Nature of the Physical World*, located in Trinity College Library, one can compare portions that were later struck out with the published version. For example, at one point in the manuscript Eddington wrote:

I do not claim spiritual power only for the religious mystic; but the closing of the eyes to the overwhelming supremacy of that power over material and even intellectual power is the gross error of the widespread materialism of this and other ages—a materialism far more deadly than the philosophic doctrine of that name which I have been combatting in earlier parts of these lectures.[31]

Eddington struck a line through this sentence, and it did not appear in the published version, although the terminology indicates it may have been included in the Gifford Lectures themselves. These sentences are far more militant and assertive than the cautious Eddington normally permitted himself to be in print. The sentence was undoubtedly, in his mind, more strongly supportive of his form of religious mysticism than he wished to be in published form.

In yet another section of the same manuscript the following lines were deleted before publication:

The positive evidences for mystical religion, being in the part of our experience outside physical science, are unaffected by the recent changes in physical theory, which have formed the main theme of this book; they remain as adequate or inadequate as they used to be, and we have nothing essentially new to say about them.[32]

It is my opinion that the reason that Eddington deleted this section was the opposite of the reason for the deletion of the previous one; while the previous quotation asserted too much in favor of his religious viewpoint, the latter did not assert enough. The last quotation is a defense of the pure restrictionist approach, and it did not give Eddington enough room to say what he wished to say, namely, that the tenability of religious beliefs *was* affected by the recent changes in physical theory.

How was he to point to the effects of the new physics on religious values if he was not to abandon restrictionism entirely? He found a very intelligent answer: the new physics did not directly support religion, free will, and idealism (three of his preferred beliefs in the value realm), but they removed obstacles to them, they cleared the way for them. Eddington illustrated his answer by facing directly the vexed problem of the relationship of science and religion as postulated by restrictionism:

> This brings us to the view often pontifically asserted that there cannot be any quarrel between science and religion because they belong to altogether different realms of thought. The implication is that discussions such as we have been pursuing are altogether beside the mark. But the statement is one which requires a careful examination even if in the end we give a guarded assent to it. To avoid a quarrel they must confine themselves to their proper sides of the boundary and that involves a definite understanding of the boundary.[33]

Perhaps no scientist ever insisted as strongly as Eddington that science dealt with so little of what is really important to humans. He restricted science to a small portion of human concerns and he then beckoned everyone to fill the remaining space with value systems. He realized that the space could be occupied by all sorts of value systems, including that of the atheist or the religious dogmatist, but he asked his readers to consider the virtues of his own.

Limitations of Eddington's Approach to Values

We have seen that Eddington devoted a large part of his writing on the relationship of science and values to an effort to press science into the smallest of possible domains and in that way enlarge the realm where values could operate on an independent and legitimate basis. The main source of inspiration and methodology for his effort was the development of relativity physics, an event whose revolutionary significance Eddington fully realized. Einstein had insisted that each physical quantity be defined as the result of certain operations of measurement, and Eddington took this precept as his guide for the understanding of the nature of science. Science, henceforth, should be understood as the realm of the metrical.

As science continues to develop we see ever more clearly that Eddington overachieved in his effort to limit the boundaries of science. He strove to make the realm of values invulnerable to the incursions of science by binding science to metrical operations.

Yet by restricting his definition of science to physics, the actual result of his writing was ultimately to make values even more vulnerable to incursions coming from scientific fields that did not base themselves on metrical operations. This characteristic of his understanding of science can best be understood by seeing that Eddington's definition of science left many present academic disciplines outside its boundaries, and, indeed, excluded much of biology and psychology.

Eddington was rather careless in his writings about the distinction between "science" and "physical science," and the more that one reads his works the more the suspicion grows that to him only physics was truly scientific. Sometimes when he spoke of the metrical basis of scientific research he carefully said "physical science," while at other moments he spoke merely of the "scientific" and "extrascientific" domains of experience, distinguished by whether the basic operations within them were metrical. In the latter part of his life, he realized that he had not been careful enough in his definition of science; for the title of the last book published during his lifetime (there was one posthumous book), he carefully chose "The Philosophy of *Physical* Science," saying to his friend and critic Herbert Dingle that he had done so as an "attempt to keep him quiet."[34] The change, however, did nothing to strengthen his case for an understanding of the relationship between science and values, for he had based that case upon a definition of the "scientific world" as one derived from metrical operations. Indeed, the change illustrated the weakness of his argument, for it pointed to areas of science impinging on values that could not be held back from their incursions by the system of categories erected by Eddington.

Some of the areas of science that have in the last few decades (largely since Eddington's death) treaded most closely on human values cannot very well be described as metrical. And yet they are clearly within the bounds of science. To take one example: the science of ethology, now recognized by the award of Nobel Prizes to three of its practitioners, attempts to explain animal behavior, including human behavior. Its links to value questions are transparently clear. What would Konrad Lorenz, who rarely made measurements, once boasted that he had never drawn a graph in his life, and found mathematics largely incomprehensible, say to Eddington's assertion that the division between the scientific and extrascientific realms is the same as the cleavage between the metrical and the nonmetrical?

A biographer of Einstein wrote that the first half of the twentieth century would likely be called by future generations "the age of Einstein," and that relativity theory would probably be seen as having as pervasive an effect on arts, philosophy, and politics as Newton's work did in his day on the same fields.[35] Regardless of the accuracy of this description, Eddington's defense of values on a basis inspired by relativity theory is certainly one example of an effect of relativity theory outside physics. Even more interesting, however, is the rapidity with which that defense has aged. By insisting on a version of the restrictionist approach to the science-values relationship in which so little actual content was placed within the boundaries of "science," Eddington produced a scheme which was exceedingly vulnerable to the attack of a new form of expansionism in which science burst out of Eddington's boundaries and made new assertions about human values.

The Significance of Eddington's Interpretation of Science

It is difficult today for us to appreciate fully the widespread popular impact of Eddington's writings on science. Starting from the time of the 1919 expedition and continuing on through the twenties, thirties, and early forties, he was a man very much in the public eye. In that pretelevision age lay people hungered for words from recognized scientists about the significance of the revolutionary events in physics. This eagerness extended from the general reading public to the most prestigious university levels. One witness described a lecture which Eddington gave at Trinity College, Cambridge, on December 2, 1919 (after the Principe expedition), in the following way:

> Fifteen minutes before the lecture began there was a queue half-way across the Great Court of men anxious to obtain admittance, and during the lecture the hall was entirely filled with dons and students listening breathlessly to hear an intelligible account, if one could be given, of the new theory. The keen interest was due, no doubt, largely to curiosity

> stimulated by the newpaper accounts of the subject, but also partly to the feeling, to which at least some hope of satisfaction can be given, that a further great unifying principle is needed in natural philosophy. Whatever the reason, however, the size and appreciation of the audience were no less extraordinary than the subject of the lecture and the brilliance of its exposition.[36]

Eddington lectured over the radio in the United States and England, and his more popular writings were published in large numbers, many editions, and in over a dozen languages. His Swarthmore lecture of 1929, *Science and the Unseen World*, was more widely read, quoted, and reproduced than any other lecture in the series since its initiation in 1908. When the lecture was given it was cabled across the Atlantic word by word.[37] At least one generation of American and British citizens were more familiar with Eddington's writings about physics than with any other material on the subject.

Eddington was not unaware of the popularity of his works and, although a modest man, he kept careful records of the sales and receipts of his publications. Raised in relative poverty (his father died shortly after his birth), he left upon his death in 1944 a fortune of £47,237, a sum surely built up from sales of his publications.[38]

The popularity of Eddington's writings on science was in part a result of the state of physics itself, where a basic transformation was in full progress, and certainly in part attributable to his great skill as a writer. He succeeded where many others failed in writing comprehensible accounts of the new physics. In explaining his popularity, however, I believe that it is necessary to penetrate beyond the state of physics of the time and the talents of Eddington as an author to a consideration of the particular interpretation of the science-value relationship that he gave, as well as the public needs and desires that interpretation satisfied. When Eddington died in 1944, *Time* magazine called him "one of mankind's most reassuring cosmic thinkers."[39] Let us consider for a moment what about Eddington's view was "reassuring" and what differences there may have been between what the public often took Eddington to be saying and what he was, in fact, saying.

Relativity theory was seen by many lay people as tending toward the denial of absolute standards and values of all sorts. Coming at a time when the global political and economic environment was exceedingly unstable, and at a moment when the successful Russian Revolution had thrown into question the entire range of assumptions and values upon which European order was based, the new physics seemed to be one more step in the demolition of reliable standards. The fact that Einstein was a political radical and a Jew did not help allay the misgivings of those who felt that the old order—political and scientific—was under attack from similar quarters.

Against this background, the message that Eddington carried was an exceedingly comforting one. An article in *Current Biography* observed of Eddington that "to the public he became important first for reconciling the average man to relativity and later for reconciling religion with science."[40] His insistence that science (particularly the *new* science) dealt with only a small part of reality increased the confidence of those people who understood nothing of that area of thought, but could see with Eddington's help that there was still a very large realm where they were as well prepared as anyone to handle the challenges that constantly appeared. True, there was a momentous change in physics, but life could go on elsewhere as before. Indeed, the new physics seemed to show more clearly than the old physics that traditional values were valid: the threats of determinism and materialism that had often been identified with the mechanistic Laplacian world view were now in retreat. When Eddington in 1932 gave a lecture entitled "The Decline of Determinism," the Smithsonian Institute asked and gained permission to reprint it, and gave it wide publicity on both sides of the Atlantic.[41] The American establishment thus threw its support behind Eddington's defense of philosophical idealism.

Eddington's interpretation was sophisticated enough to avoid the danger of directly linking science to religion, but simple enough to be understood as defending traditional values, including religion. An English-speaking public, which had learned from the debates over Darwin in the nineteenth century that bringing the content of science into direct contact with religion was a risky endeavor that might damage both, was ready to appreciate an interpretation of science that cleared the way for religion without making any direct assertions about it.

Eddington's emergence as a popular figure and as a defender of values contained its ironies. Quakerism has never been widespread in England and America, and is often seen as the little-understood idiosyncracy of a tiny minority. Furthermore, Eddington himself—conscientious objector in World War I, bachelor and social isolate, described by his contemporaries as often almost incoherent in conversation—was hardly a prime candidate for public adulation. He was a man who preferred being alone, or on a long hike with his intimate friend of forty years, C. J. A. Trimble, to being in crowds or at public ceremonies. But a similar adulation and distortion had come to Einstein himself, who could not let the words "God" or "The Old One" escape from his lips without reporters seizing upon them as evidence for his religiosity, ignoring his opposition to all existing creeds and much of what is usually considered to be religion. Religious faith was much more important to Eddington than to Einstein, and Eddington's beliefs fitted within an existing organized group, the Society of Friends. Furthermore, Eddington actively promoted his interpretation of religion, considering it an important part of his activities. Nonetheless, his religion sharply contrasted with the faiths of well over 90 percent of those who found comfort in his writings. He must have been surprised by his own popularity.

In the twenties and thirties of this century there was an unusual conjunction between the conception of science and values held by Arthur Stanley Eddington and the needs of his world audience. That conjunction of elements has by now passed, but without seeing how complementary and appropriate it was at the time, one cannot understand the role played by his writings.

Although Eddington's fame with nonscientists rested largely upon that temporary conjunction, there is no evidence that he molded his opinions to fit public demand. His basic views about science, values, and religion had been formed long before the advent of relativity theory (he was thirty-three years old in 1915). He saw relativity theory as merely strengthening his interpretation of the relationship between science and values, not in changing it.

In a sense the world came to him—it was ready to hear this odd restrictionist interpretation, rather than some innovative explanation. Herbert Dingle once observed of Eddington, "He did not know what he was doing, but what he did was supremely great."[42]

It might be more accurate to say that the *world* did not know what he was doing. His listeners believed he was offering them an interpretation of the true meaning of relativity, and they found that meaning satisfying for reasons that were rarely made explicit. What Eddington was giving the world was not the meaning of relativity, but a defense of a deeply held view of the relationship of science and values. Eddington reassured the educated public. Science (to him, physics) need not take away our religious beliefs because it deals with only one of three worlds—the world definable in terms of clocks and yardsticks; the abstract world.

chapter three

FOCK AND THE SOVIET UNION

EDDINGTON WAS an outstanding physicist who interpreted the new developments in his field in a way that had particular relevance to the prevalent sociopolitical values in the society in which he lived. If we turn to the Soviet Union and look there for a physicist who played a similar role, we find no single person who stands out with quite the clarity of Eddington (or James Jeans, who would also have been a relevant example) in England. A number of Soviet candidates could be mentioned: A. A. Friedman (Friedmann), O. Iu. Shmidt, S. I. Vavilov, V. A. Fok (Fock).[1] All were physicists or mathematicians who defended relativity theory against crude attacks. A. A. Friedmann was a relativist of the first rank whose 1922 paper in *Zeitschrift für Physik* won Einstein's belated admiration and laid the foundation for theories of the expanding universe.[2] His early death in 1925, however, precluded the possibility of his becoming a major influence in the shaping of subsequent popular interpretations of relativity theory. O. Iu. Shmidt, mathematician, polar explorer, and vocal Marxist, never undertook an extensive interpretation of relativity theory, except to affirm that it was in thorough accordance with dialectical materialism, thus dismissing those militant Marxists who said otherwise.[3] S. I. Vavilov, future president of the Academy of Sciences, defended relativity scientifically and criticized it philosophically; his work on this topic was only one of his many activities, however, and never won a pervasive influence.[4] Many other prominent Soviet physicists—Igor Tamm, Lev Landau, Peter Kapitsa—avoided philosophical discussions of physics whenever possible, although their support for relativity physics was never in doubt.

The person who emerges from this consideration as the most likely counterpart to Eddington is Vladimir A. Fock, a scholar known to all Soviet citizens interested in the philosophy of physics. Although sixteen years younger than Eddington, he performed the same task of interpreting the theory of relativity in a fashion that, because of the values he attributed to the theory, was particularly appealing to his culture. Furthermore, Fock, like Eddington, meets high standards of scientific quality. His works in quantum mechanics, the theory of electromagnetic diffraction, and general relativity are well known. His name is carried in the title of the Klein-Fock relativistic wave equation for a particle with no spin in an electromagnetic field. His most outstanding characteristic was his mathematical talent; the usually reserved Paul Ehrenfest once remarked that there was no mathematical problem with which Fock could not deal; he added that "*Fock kann einen Stiefel ausrechnen*" ("Fock could integrate even a boot").[5]

Fock, like Eddington, won acclaim from nonscientists in his society because his interpretation of relativity physics was carefully fitted to the values held by many members of that society. Fock's articles were featured in philosophical as well as physics journals, and he wrote on relativity in the main Soviet newspaper *Pravda*.[6] He was the author of the article "The Theory of Relativity" for the second edition of the *Large Soviet Encyclopedia*. He won a Stalin Prize, a Lenin Prize, and was given the order "Hero of Socialist Labor." In the late forties and fifties Fock was the Soviet Union's leading authoritative interpreter of relativity physics to the wider Soviet public; at certain moments he had to fend off attacks from the Stalinist philosophers who thought his interpretation was too partial to Einstein and not sufficiently Marxist, and at other times he faced the opposition of many of his physicist colleagues who considered him too critical of Einstein and too dogmatic in his own Marxism.

For different reasons, Eddington also faced criticism from both philosophers and physicists in his country: from philosophers who thought he was writing bad philosophy, and from physicists who disliked the religious mysticism that permeated his writings. Two additional similarities between Fock and Eddington deserve mention. First, they wrote well-known books with almost identical titles—Eddington's *Space, Time and Gravitation*; Fock's *The Theory of Space, Time and Gravitation.* Second, the philosophic interpretations of relativity theory advanced by both men no longer enjoy their earlier popularity; indeed, Eddington's effort to show how relativity theory complemented religious mysticism appears idiosyncratic to many English-language readers today, and there are signs that something roughly similar has happened in the Soviet Union to Fock's Marxist interpretation of relativity. Yet both men, at specific moments of history, were the best-known interpreters of relativity theory in their respective societies, and these contrasting confluences of science and sociopolitical values deserve the attention of intellectual historians.

Fock's Marxism and Its Relationship to His Science

Any discussion of V. A. Fock's interpretation of Einstein's theory of relativity must begin with the unambiguous recognition that Fock accepted the mathematical formulation of both the special theory of relativity and the general theory. Furthermore, in a series of publications spread over a forty-year period he defended and praised Einstein's achievements. In 1939 he wrote a popular article on Einstein's birthday calling him "the great physicist of modern times," and in the last years of Stalin's life, when ideological pressures in science were particularly strong in the Soviet Union, he castigated what he called "ignorant criticism" ("*nevezhestvennaia kritika*") of relativity theory and quantum mechanics.[7] "To question the validity of relativity theory," wrote Fock in response to Stalinist philosophers, "was on an intellectual level with questioning the roundness of the earth."[8]

Nonetheless, Fock was an authentic "internal Marxist," and he believed that it was his obligation to produce an interpretation of relativity physics that would counter those of Eddington, Jeans, and other physicists in the West who tried to depict relativity as the cancellation of materialist philosophy. "Many foreign physicists in their roles as authors and especially popularizers," he wrote, "love to provide philosophical commentaries of an idealistic spirit along with the physical theories which they expound."[9] His goal, he maintained, was to create "a correct materialistic understanding of the theory of relativity."[10] He believed that the way to do this was to show more directly than others had done that relativity theory was not merely a mathematical innovation, but a description of physical and material reality. By emphasizing that Einstein had indissolubly linked matter and energy and had created an objective description of the universe uniquely dependent on the distribution of this matter-energy, Fock hoped to rid relativity theory of the idealistic aura that many of his fellow Soviet Marxists lamented and many Western writers like Eddington celebrated.

Fock pursued this goal within the framework of an explicit Marxism. In its attempt to explain human behavior on the basis of scientific principles Marxism has traditionally been an expansionist world view. Historically, those scientific principles were rooted in nineteenth-century materialism. Fock knew that this form of materialism was now out of date; he further knew that if the expansionist aspirations of Marxism were to be rescued, Marxists would have to update their conception of what materialism is, taking into account the revolutionary new physics. In 1938 Fock wrote, "I am convinced that materialist philosphers will be able, sooner or later, to bring the newly discovered properties of matter into agreement with the basic formulations of materialism. But this has so far not been achieved." Fock continued that it was necessary "first of all, to strike from the hands of the idealists their basic weapon," the belief that modern physical theories such as quantum mechanics and relativity "contradict materialism." The path to this achievement, he asserted, was a study of Lenin's writings, and particularly his view that the valuable achievements of "bourgeois physicists" can be distinguished from their philosophical viewpoints.[11] Fock expressed the hope that Soviet philosophers would defend modern physics from the standpoint of Marxism, but he knew that by 1938 most of the Marxist philosophers in the Soviet Union who were capable of such a defense had been replaced by Stalinist dogmatists who did not understand science.[12] The defenders of relativity among the philosophers were being eliminated in the factional political struggles raging in Soviet academic circles. Therefore, said Fock, "the physicists must take the initiative themselves."[13] And Fock accepted his own challenge.

In order to achieve his goals Fock was willing to go so far as to suggest changes in Einstein's terminology and to supplement Einstein's theories with his own elaborations, while preserving the mathematical core of relativity. The aspect of Einstein's system to which Fock objected most strenuously was the term "general relativity," since it seemed to Fock to imply the absence of any objective reference standards in nature, a conclusion he insisted was not derivable from the essential part of Einstein's theories. Indeed, Fock asserted that the relationships of material bodies described so brilliantly by Einstein were "just as objective as all other properties of bodies."[14] But while Fock rejected the term "general relativity," he accepted the term "special relativity" as having concrete meaning with reference to Newtonian physics. Fock agreed completely that Einstein showed that the old absolutes of "space" and "time" were relative to frames of reference and therefore no longer could serve as reliable and unchanging standards. However, Fock noted that the new theory which Einstein presented was also based on absolute standards, just as Newton's had been. Echoing the view of Max Planck,[15] Fock remarked that "The theory of relativity, after showing that a whole series of concepts earlier considered absolute were actually relative, at the same time introduced new absolute concepts. The majority of critics of the theory of relativity forget this."[16] Einstein found new absolutes in the constant velocity of light and in the concept of space-time. The one thing that the theory of general relativity was not, maintained Fock, was a purely relativistic theory. In order to illustrate his opinion on this issue, Fock used the term "theory of gravitation," where most of the rest of the world used "theory of general relativity."[17]

Fock believed, furthermore, that relativity theory should be seen as an explanation of nature that reflects the properties of physical reality.[18] Referring to Eddington's beliefs that all that counts in physics is measurement, Fock countered that what was actually primary in relativity physics was not measurement but objective relationships of material bodies and objective properties of space and time. Fock continued that the positivistic emphasis on the results of measurement gave many "Western physical idealists" the opportunity to subordinate the concept of scientific theory as a reflection of reality to the concept of theory as an "auxiliary construction" for the coördination of sensations.[19] Thus, Eddington could say that the physical world has no configuration until a scientist gives it one, just as an untracked moor has no paths until a person creates one. To Fock, Eddington's point of view denied the independently existing material reality upon which Marxism relied in its effort to describe the world, both physical and social. As Lenin had remarked, the most important principle of Marxist epistemology was that "of a world *existing and developing independently of the mind*."[20] Fock responded to Eddington's interpretation of the new physics with the view that the word "relativity" in the "theory of relativity" has nothing to do with the form of philosophical relativism that says "our knowledge is so conditional and relative that it has a purely subjective character."[21] On the contrary, Fock stated, "the theory of relativity is a correct reflection of the essential properties of space and time."[22] And, finally, the "theory of relativity is a brilliant confirmation of dialectical materialism."[23]

If Fock had not gone beyond this purely philosophical rephrasing of the theory of relativity, he would not be remembered by Soviet physicists as a critic of Einstein (although he would still be remembered by Stalinist philosophers as a defender of Einstein). But Fock did go beyond a materialistic verbal framework for relativity theory. He thought that an insistence on paying attention to the physical content of Einstein's interpretation of inertial and gravitational mass would yield an insight different from that usually seen by scientists and philosophers. Fock believed that the equivalence of inertial and gravitational mass as defended by Einstein was mathematically valid, but physically true only in a local sense.[24]

The important conclusions from these considerations and the ones that reveal most graphically Fock's unorthodox position concern the question of preferred or privileged systems of coordinates.[25] His position here was connected with his desire to show that relativity theory was not entirely relativistic, that even it contained reference standards determined, ultimately, by matter. Both as a scientist and as a Marxist Fock believed that such reference standards were necessary if one were to be able to retain a concept of objective truth. Without such standards everything became relative, and there were no secure pegs anywhere upon which one could hang a worldview, scientific or political.

Fock devoted much of his research to the task of proving that in space uniform at infinity there is a preferred system of coordinates. As his choice, Fock defended harmonic coordinates, which, in accord with his commitment to the existence of physical reality, he believed reflected "certain intrinsic properties of space-time."[26] Yet it should be noted that Fock's reliance on harmonic coordinates was one of the most controversial aspects of his approach; many physicists who found his criticism of the term "general relativity" interesting remained dubious of the preferential status of harmonic coordinates.[27]

Fock's stout defense of the preferred status of harmonic coordinates is responsible more than anything else for the fact that he became known among Soviet physicists as a critic of Einstein. His insistence on harmonic coordinates is now widely regarded as an unsuccessful appendage to relativity theory emanating from his eccentric criticism of Einstein's general relativity. An insight into Fock's position among other Soviet physicists in the mid-fifties is given in Leopold Infeld's memoirs. Infeld described a 1955 discussion of general relativity in Moscow's Lebedev Institute:

> Most interesting for me was the discussion between Fock and the Soviet physicists, in which I also took part. . . . First Fock spoke and formulated his objections to Einstein's theory by defining his own conception of harmonic coordinate systems. Then I spoke in favor of Einstein's theory and was followed by Landau, Tamm, and Ginzburg—all well-known physicists who preferred Einstein's theory to Fock's. They said that his additional equations for defining the coordinate system are unnecessary and provide nothing of value. Fock still stuck to his interpretation, as he does today, but I learned during my visit to Moscow that he was fairly isolated at the time.[28]

Isolated as Fock may have been among Soviet physicists in the 1950s because of his attachment to harmonic coordinates, he was celebrated by the Soviet official press as the best example of a Marxist interpreter of relativity theory. Despite the controversial aspects of his interpretation of physics, Fock was lauded in the

Soviet Union for having defended against critics like Eddington the expansionist principle at the base of Marxism; that principle was the belief that the construction of a scientific social world view must begin with recognition of an objective physical world. Perhaps the high point in his influence with the general educated public in the Soviet Union came in 1955, when his interpretation of relativity was featured in the *Large Soviet Encyclopedia*. Two years before he had won his second Order of Lenin.

After the relative relaxation of the political atmosphere in the Soviet Union in the late fifties and sixties, Fock's influence rapidly declined. The reasons were the decrease in the importance of ideology in the Soviet Union and the continuing unpopularity of Fock's harmonic coordinates among Soviet physicists. Perhaps the most graphic illustration of that shift can be seen by comparing the articles on relativity theory in the second and third editions of the *Large Soviet Encyclopedia*, published, respectively, in 1955 and 1974.[29] The first article, written by Fock, starts and ends with quotations from Lenin and presents the theory within the materialistic framework already described here. The article does not have a section on the "General Theory of Relativity" at all, but instead on the "Theory of Gravitation." Nonetheless, it contains an accurate description of relativity, and treats the theory as fully established.

The 1974 article, written by I. Iu. Kobzarev shortly before Fock's death, contains no mention of Marx or Lenin. It drops Fock's insistence on the term "theory of gravitation," and accepts the international terminology of "special" and "general" to describe the theory. All imprint of Fock's interpretation seems to have disappeared except for two details: the article still places an unusual emphasis on viewing relativity theory as an expression of the real properties of space and time (no subjectivism here), and Fock is one of the authors cited at the end.

Fock from the Standpoint of Personality and Political Culture

As with Eddington's interpretation of the relationship of physics to values, we may view Fock's interpretation both as a personal creation and as a manifestation of political culture. On the personal side, we have fewer firsthand descriptions of Fock than we do of Eddington, but Fock was well known to physicists in many countries. Infeld described him in the following way:

> He is an outstanding Soviet theoretical physicist, on a very high level, who came out for a change in Einstein's theory while retaining its mathematical skeleton, and he believed that he had much improved the theory. . . . Who is this Fock? Black hair, fairly fat, around sixty, pleasant, somewhat like Einstein personally. He had an attractive way about him, laughed gaily, was deaf, and wore a hearing aid which—so the spiteful said—he turned off when the argument of his opponent did not please him. He was dogmatic and very much believed in his own convictions. I may say that, during this time, I became friendly with Fock. Our relations were better than correct. I argued with him, tried to convince him that he was wrong, but of course without result.[30]

From all reports, Fock was brilliant, dogmatic, brave, witty, and ambitious. He did not shrink from polemics, and he got involved in a priority dispute with Infeld. He confidently maintained that he had persuaded Niels Bohr to change his opinion on the philosophic question of realism.[31] He was proud of the West European origins of his family, and liked to tell people that he was descended from Dutch shipbuilders imported to Russia by Peter the Great. Like his ancestors, he also brought knowledge from the West to Russia. They had brought nautical crafts, and Vladimir Fock brought the new physics from Göttingen and from Paris, where he studied for a year in 1927–1928.

While there is no reason to doubt Fock's basic commitment to Marxism, it is clear that he used Marxism as a vehicle to promote his personal scientific viewpoints. We find in this complex physicist a shrewd awareness of both physical and political reality. Before we accuse him of opportunism, however, we should recognize that Fock was a brave leader in the effort to prevent Soviet physics from being subjected to the kind of perversion that occurred in Soviet genetics with the victory of Lysenkoism.

Fock first began discussing Marxism and physics in the 1930s after it had become evident that those philosophers who wished to defend relativity physics from the standpoint of Marxism (e.g., S. Iu. Semkovskii, A. A. Bogdanov, A. Goltsman, B. G. Gessen [Hessen]) were disappearing. Some of the best were eliminated in the purges that wracked Soviet society. A new breed of Stalinist philosopher was coming to power, including people who were often hostile to modern science.[32] A shift in Soviet political culture was occurring. Fock was one of several Soviet physicists who tried to develop a form of self-defense to make up for the loss of these "middle men," the sophisticated Marxists who appreciated modern physics. Some physicists emphasized the practical value of modern physics as a means of defense, an approach that became very powerful after the development of atomic weapons. Fock's method was different: he became a philosopher of science himself, developing and promoting his own materialistic interpretation of the theory.

Fock resolutely rejected the attacks on relativity theory by ignorant Stalinist philosophers and third-rate physicists such as A. A. Maksimov and R. Ia. Shteinman.[33] Maksimov wrote in the Soviet Union's leading philosophical journal that Fock's defense of "reactionary philosophical ideas" in the "so-called theory of relativity" was holding back Soviet science by chaining it to idealistic views inherited from physicists of the capitalist countries. Fock replied in print that Maksimov's call for a rejection of relativity theory was an example of confusion and error, and was based on an antiscientific spirit. And he asked Maksimov why he was so ready to take the word of Western physicists such as Eddington and Pascual Jordan that accepting relativity meant rejecting materialism. The theory is brilliantly confirmed by experiments, Fock continued, and points to a new understanding of materialism.

Even some of the Soviet physicists who were critical of Fock's harmonic coordinates recognized that Fock's defense of relativity theory was a defense of all Soviet physics at a moment when most other physicists refused to participate in the public debate. In later years some of them recognized this debt. After Fock's death in 1974 the prominent Soviet physicists M. G. Veselov, Peter Kapitsa, and M. A. Leontovich wrote of the situation in Soviet science in the late forties and early fifties:

> More than once there flared up obviously false scientific tendencies connected with opportunistic maneuvers. At this moment Fock displayed great decisiveness and undeviating devotion to principle; he bravely and openly stood up for the purity of scientific ideas.[34]

By not only defending relativity physics, but by developing an interpretation of it in terms of Marxism, Fock merged his personal preferences with those of his political culture.[35] In the late forties and early fifties Fock won the approval of the leaders of the regime, and he was rewarded with the right to travel widely in Europe and America. Soviet intellectuals were told in university classrooms and philosophical journals that V. A. Fock had cleansed relativity theory of the idealistic trappings placed upon it by bourgeois scientists, and had developed a version of the theory that accorded thoroughly with Marxism. He had done this, they were told, because of his careful reading of Marx, Engels, Lenin, and Stalin. And Fock agreed by praising Marxism in his scientific works.[36] Fock received benefits from the government and he repaid it with political praise and cooperation.

Comparison of Fock's and Eddington's Interpretations of Relativity

Einstein's theory of relativity was such a radical break with common-sense notions possessed by lay people of all societies, whatever the prevalent philosophies, religions, or ideologies, that it caused great public concern. In England and America in the 1920s, underneath the fascination with Einstein and the titillation with popular lectures on relativity, lurked an anxiety about how this new theory could be fitted with conventional beliefs, particularly religious ones. Did the theory of relativity mean that all secure intellectual and religious foundations were destroyed, all absolute standards invalidated? Citizens in those countries witnessed the comic adventures of newspapers offering prizes for the best explanation of relativity theory in a few hundred words, and of telegrams being sent to Einstein asking, "Do you believe in God?" and demanding an immediate reply. Against the background of such concerns, Arthur Stanley Eddington's writings on relativity theory accomplished a dual purpose, for they not only made the theory sound vaguely comprehensible, but they allayed these anxieties.

But the theory of relativity was impartial in its ability to raise the fears of orthodox thinkers. Many Marxists, too, worried about relativity. Its linking of matter and energy seemed to some people to mean the disappearance of matter and materialism; its relativism seemed to undermine the concept of steady movement toward truth in which Marxists believed; and its utilization by popular writers like Eddington in order to support religious mysticism alarmed atheists and agnostics. Against this background Vladimir Fock's interpretation of relativity theory helped make the theory intellectually acceptable to a Marxist audience.

Eddington's and Fock's interpretations of relativity were not true opposites—were not mirror images, so to speak. Nor would it be correct to say that "just as Fock reacted to his public, so did Eddington." While there are analogical qualities between Eddington and Fock, there are also some very large differences. First of all, Fock was living in an oppressive society where the relationship between science and ideology was dramatically different from that in England. No one would have been able to publish Eddington's defenses of religion in the Soviet Union, while some of Fock's writings on Marxism *were* published in the West. Furthermore, while Eddington was reacting to relativity theory, Fock was reacting *both* to relativity theory and to its idealistic interpretations like Eddington's. Thus, Fock's interpretation was, in a sense, a reflection of both cultures, while Eddington's was much more closely rooted in his own. It is true that Eddington criticized Marxist materialism in several of his lectures, but his concern with an adversary was much less direct than Fock's. Fock's interpretation did not make sense without the supposition of opposing idealistic interpretations. Eddington's views could stand independently on the foundations of traditional idealism, modern operationalism, and the Quaker faith.

Nonetheless, both men were concerned with the same intellectual issues, and these issues can be summarized as two dualities: relativism vs. absolutism and idealism vs. materialism.[37] The first duality concerns knowledge and certainty, i.e., epistemology. The second concerns the actual content of reality, i.e., ontology.

It is obvious that the two men radically differed on the question of idealism vs. materialism. Fock proclaimed himself a materialist and Eddington defended idealism. On the other duality, however, the men were much closer. They both displayed a yearning for the absolute. Again, Fock was the more confident. He affirmed that relativity theory supported the proposition that all that exists in nature is matter-energy. His preference for harmonic coordinates was based on a desire to rescue an absolute standard. But Eddington also reached for an absolute, what he called the "yearning toward God." His belief that all the laws of nature have an a priori mental character was another assertion of absolute idealism. Fock and Eddington, then, had one characteristic in common: an unwillingness to live in a totally relativistic world.

Fock's and Eddington's opinions of the proper way to show the relevance of physics to social values were, however, decidedly different. Fock followed the traditional expansionist aspiration inherent in Marxism, based on the beliefs that all of nature—physical and social, factual and normative—can be brought under the sway of a single scientific approach; this approach includes the principles that mind cannot exist without matter, that the external world is an objective reality ruled by laws, and that truth is an approachable goal in a material world. Eddington chose a restrictionist approach that had its own traditions: the beliefs that there are multiple ways of knowing, of which the scientific way is only one, that alongside the material world of the scientist exists a no less real world of the mind, and that since science tells us nothing about values, we must turn elsewhere in order to gain guidance.

In many ways the two men illustrate the dilemma of the relationship of science to values that runs throughout this book. As is the case with many expansionist analyses, Fock's examination of science and values was based on a more solid intellectual framework than that of Eddington. The major assumptions made by Fock (the existence of one objective, material world of which man is a part; the potency of science as the path to knowledge about this world) are close to the assumptions of science itself. The main assumptions made by Eddington (the existence of three real worlds; the inherent limitation of science as a cognitive mode) are more personal and arbitrary. Yet one is not sure that Fock's superior assumptions led him to conclusions about social values that were more promising for human fulfillment than were Eddington's. Expansionists, confident of the power of science and often oblivious to the imperfections of their own particular conclusions about the social implications of science, have frequently been more dogmatic than restrictionists, and we see the same tendency in both Fock and in the society in which he lived. Can a way be found to recognize the validity of the initial expansionist assumptions while preserving the safeguards against abuses in the name of science which restrictionism seems to provide? This question lies at the root of the discussions over science and values.

chapter four

HEISENBERG AND GERMANY

WERNER HEISENBERG (1901–1976) was one of the great figures of twentieth-century physics, a person who was a major actor in forming the revolutionary events in quantum mechanics in the nineteen-twenties and who, as scientist and popular author, continued his research and writing for fifty years beyond that time. During those years he published over two hundred articles and a dozen books.[1] His influence, pervasive within physics, also extended far beyond it to philosophy and public affairs. He eventually became one of the few "authoritative popularizers" of the new physics, and the particular extrascientific interpretations which he gave to science exercised enormous influence throughout the literate world. To the educated lay person of Europe or America in the generation from 1930 to 1960, Heisenberg's name was alongside those of Jeans, Eddington, Bohr, and Einstein as one of the few physicists whose popular writings were required reading. When we discuss Heisenberg's views on the relationship of science and values, then, we are not talking only about the author of some of the most brilliant pages in the history of physics; we are also discussing an influential philosopher and spokesman whose world view was related to physics but certainly not totally determined by it.

Heisenberg was born into an academic family in Würzburg, Germany, in 1901, and he died in Munich in 1976. A student of Arnold Sommerfeld in Munich, of Max Born in Göttingen, and of Niels Bohr in Copenhagen, Heisenberg very quickly became the colleague of his teachers, and at the age of twenty-four earned permanent fame by fundamentally shaping quantum mechanics in his article, "About the Quantum-Theoretical Reinterpretation of Kinetic and Mechanical Relationships." Two years later he published his "indeterminacy principle," which stated that certain pairs of variables in quantum mechanics, such as position and momentum, could not be known simultaneously to arbitrary degrees of exactness. He received the Nobel Prize in physics at the age of thirty-one, and he was professor of physics at Leipzig University from 1927 to 1941. During World War II he worked on the development of a nuclear reactor, a task which, he later wrote, he purposely slowed in order to prevent Nazi Germany from developing nuclear weapons. Opinions differ still on the accuracy of this description of his wartime activities, but Heisenberg's disagreement with many of the National Socialist policies is quite clear. After the war Heisenberg organized and directed the Max Planck Institute for Physics, first located in Göttingen, later in Munich.

The book that established Heisenberg as an authoritative popularizer and interpreter of scientific ideas was his *Wandlungen in den Grundlagen der Naturwissenschaft* (Transformations in the Foundations of Natural Science; published in English as *Philosophic Problems of Nuclear Science*), his only book for lay readers published before World War II. This book, originally published in 1935, is still in print, has now gone through ten editions, and has, over that time, tripled in length.[2] Heisenberg added chapters over a period of several decades, and he wrote under strikingly different social circumstances. We will see the influence of those circumstances on this best-known of his popular writings. This one book, in all its various editions and translations, was the single most influential carrier of Heisenberg's interpretation of physics to the educated world, and we shall look at it carefully. First, however, it is necessary to examine briefly his publications before this book appeared, since a number of its features are understandable only against that background.[3]

Heisenberg as an Authoritative Popularizer

Heisenberg in 1925 made a historic breakthrough in the development of quantum mechanics. Until that time Bohr and the other leading developers of quantum mechanics had attempted to retain classical descriptions of the motions of quantum systems, i.e., they still clung to Newtonian kinematics and dynamics even though they had modified the results of classical physics in certain ways in order to make them fit with experimental data.

Heisenberg took a much more radical approach. He believed that all attempts to describe quantum events in classical terms were mistaken. He preferred to approach quantum mechanics as he believed Einstein had approached the problem of simultaneity,[4] by eliminating all nonempirical components, all metaphysical commitments, and concentrating instead on "observable values" gained from experiment. As he wrote in the historic 1925 article:

> In this situation it appears advisable to give up entirely all hope for an observation of those values which have up to now been unobservable (such as position and velocity of an electron) and . . . to attempt to build a quantum-theoretical mechanics that is analogous to classical mechanics, and in which are to be found only relations between observable values.[5]

Heisenberg's approach here was highly reminiscent of the emerging positivism or logical empiricism of the early nineteen-twenties, and the historian of physics Max Jammer has maintained that Wittgenstein particularly influenced Heisenberg.[6] Indeed, Jammer stated that Wittgenstein's famous 1921 statement, "Whereof one cannot speak, thereof one must be silent" had found "an unexpected application in Heisenberg's new approach to the study of atomic phenomena."[7]

Whether or not Wittgenstein personally was an important influence on Heisenberg, it is certainly clear that philosophically Heisenberg's pathbreaking article of 1925 was permeated with the positivistic spirit of concentrating only on what is empirically verifiable, of reducing philosophical commitments to the absolute minimum. Only in that way could Heisenberg find his way through the apparent contradictions of experimental data to a new and consistent quantum theory. As with Einstein before him, a "philosophy of the minimum" was an invaluable aid at the moment of crisis; and, as did Einstein, Heisenberg would abandon this spare and forbidding approach for a richer view once the crisis had been overcome. In later years the motto "Whereof one cannot speak, thereof one must be silent" failed to satisfy Heisenberg's growing desire to interpret the new physics and to find meaning in it extending beyond physics itself.

The following year, 1926, Heisenberg wrote his first article to appear in a nonphysics or a nonmathematics journal. The place of publication was *Die Naturwissenschaften* (The Natural Sciences), the most prestigious general science publication in Germany, a journal widely circulated among scientists but not among the general public.[8] In this article, which had earlier been given as a lecture to a general meeting of German scientists, Heisenberg presented in more accessible terms the underlying approach for his recent breakthrough toward matrix mechanics, an early form of quantum mechanics.

Heisenberg remained true to the positivistic spirit of his classic article of the previous year. Questions of causality or physical reality, which agitated many philosophers of science and popular writers, were not important to him, since he saw no way of giving answers to such questions without going far beyond the available evidence. They were metaphysical issues, not issues of science.

To Heisenberg in 1926 the most important aspect of quantum mechanics was the fact that it required the abandonment of visualizibility (*Anschaulichkeit*) in the microworld. Trying to visualize microentities (electrons, photons) inevitably brought with it the attribution of everyday characteristics to these entities, and once we did that, we began to think of an electron behaving in space and time as a ball might in the everyday world. Heisenberg called for the surrendering of this attempt, and for concentrating attention on experimental evidence and its mathematical ordering, without making further demands.[9]

With the instrumental goal in mind of finding a way to create a new physics, it is no surprise that Heisenberg was at this moment not interested in such questions as the philosophical meaning of quantum mechanics. Indeed, the notorious word "causality" was not mentioned in the whole article, nor was it discussed explicitly in any other terms. Heisenberg *did* speak of physical reality, but only of the "restrictions" (*Einschränkungen*) imposed on the concept by the fact that no longer could one speak of the "place" or "velocity" of an electron. The question of whether "physical reality" was fundamentally tied to such Newtonian concepts or could be separated from them was not important to Heisenberg at this stage, and he therefore gave the impression, without quite saying it, that as the applicability of Newtonian concepts to the microworld diminishes, so also does the legitimacy of such concepts as "reality" and "causality," a big step, which, if actually taken, would go beyond the science of quantum mechanics. And yet, in fairness to the positivistic position Heisenberg was still espousing, there was no way he could explicitly deny the validity of these concepts, for such a denial would be as much a violation of the principle of not making metaphysical statements as was their earlier affirmation.[10]

And yet the strict adherence to the positivistic program was not possible, not even in physics itself. Bohr had emphasized that there was no escape for man from the language which he used, a language that included terms like "position" and "velocity," terms based on man's everyday experience in the macroworld. The solution to the dilemma of quantum mechanics, said Bohr, was not to hope to cleanse quantum mechanics of such terms, even though we know they contain metaphysical elements, but to restrict their applicability. Heisenberg's uncertainty relationship, which illustrated the impossibility of an arbitrarily exact definition of microentities in classical terms, was such a restriction. But within the boundaries of this restriction, the old terms must be retained. Heisenberg supported Bohr in this approach, which was rapidly becoming known as "the Copenhagen Spirit" (*Kopenhagener Geist der Quantentheorie*) and, once this path became clear, the question of the applicability of other everyday language terms, such as "causality" and "physical reality," became more pressing.

In 1931, Heisenberg gave his first extended analysis of the question of causality, and his answer was cautious.[11] This article was his first publication intended primarily for nonscientists and appeared in a German philosophy journal. As such, it was rather different in tone from his previous publications. It marked a transition from his earlier refusal to write on philosophical questions to a later stage in which such questions were to be one of his major concerns as an author.

Heisenberg approached the ideologically loaded topic of causality with a note of disdain for the past strenuous debates of the issue, observing that they had not been scientific discussions, but a question of dogma (*Glaubensatz*). And he announced at the very beginning of his article that the question of causality cannot be answered either "yes" or "no". The crucial point, he observed, is that the development of atomic physics had forced a revision of our causality concept. The earlier formulation of causality no longer makes sense. The correct approach to interpreting quantum mechanics, Heisenberg continued, is to make a distinction between statements that are "contentless" and say nothing (statements that in principle cannot be tested) and those that can be submitted to experimental verification or refutation. The earlier formulation of causality ("If we know the past, we can predict the future") is now contentless, because we cannot determine the past to an arbitrary degree of exactness. The new physics has shown, contrary to earlier assumptions, that the effects of observation upon micro-bodies cannot, in principle, be avoided, nor calculated away.

Heisenberg criticized those people who attempted to maintain the validity of the old formulation of causality by considering it a necessary assumption, a synthetic judgment a priori in a Kantian sense. These people were saying, in effect, "Even though causality cannot be proved, should we not assume its existence, since such an assumption is so closely linked to the very concept of an objective world so necessary for science? If we do not see determinism in the microworld, should we not assume that we are still ignorant of those processes at that level which will eventually illustrate its presence?" To this viewpoint Heisenberg replied that because of the uncertainty relationship such an assumption of microcausality was incapable "in principle" of ever being tested, and therefore "says nothing." Heisenberg admitted, however, that there might be a new way of formulating a principle of causality that yielded "a degree of determinism" even on the microlevel, but whether one should make the effort or not was "a question of taste." Furthermore, Heisenberg indicated that this question of taste was not very interesting to him.[12]

Thus, it is certainly not true, as some people have maintained, that at this time Heisenberg unequivocally denied the validity of causality. But it *is* true that the question that so agitated a large part of the nonscientific public—whether the new science discredited causality and the world view of determinism, thus taking on potentially great significance for social and ethical questions—was dismissed by Heisenberg as an uninteresting minor issue.

We are now ready to consider Heisenberg's book of 1935, which established him as a leading authoritative popularizer of the new physics. In looking at this book we will need to notice how it changed over time, as the various editions emerged, and the particular context in which each version was written. Not all the German editions were translated into English, so in order to see all the changes between 1935 and 1973 it is necessary to go to the original German publications.

The first edition of *Wandlungen*, published in 1935, contained only two chapters, both of which had earlier appeared as journal articles, the first entitled "Transformations in the Foundations of the Exact Sciences in Recent Times," the second, "On the History of the Physical Interpretation of Nature."[13] We will first consider the second chapter, as it was written first.

Heisenberg's essay "On the History of the Physical Interpretation of Nature," delivered as a public lecture in 1932, was a brilliant effort to explain to a largely uncomprehending world why an understanding of quantum mechanics was difficult for people who thought in everyday terms.[14] Heisenberg tried to lessen the difficulty by showing that, radical though quantum mechanics may be, it is only the next logical development, in epistemological terms, in the long-term trend of physics since its earliest days. In this way he hoped to diminish the dismay of nonphysicists who unsuccessfully tried to picture quantum phenomena.

Heisenberg noted that every great step forward in physics has been accompanied by a sacrifice in the ability of people to have an understanding of the physical world in perceptual terms. By perceptual terms, Heisenberg meant those qualities of experience which we have when we observe and handle physical objects: color, smell, taste, shape, and movement. The Greek natural philosophers already sacrificed color, smell, and taste when they developed an atomistic interpretation of the universe. These qualities were explained by the Greek atomists as only the secondary results of certain geometrical and dynamic configurations of atoms. Although this viewpoint was a great step forward, we should recognize, said Heisenberg, that there was no reason, in principle, why qualities like color and taste should be considered secondary and shape and movement primary. If color can be "explained away" by reliance on geometry (a form of mathematics) so might shape be explained away by interpreting it in terms of a form of mathematics more abstract than geometry. And this, said Heisenberg, is exactly what quantum mechanics has done:

According to Democritus, atoms had lost the qualities like colour, taste, etc., they only occupied space, but geometrical assertions about atoms were admissible and required no further analysis. In modern physics, atoms lose this last property, they possess geometrical qualities in no higher degree than colour, taste, etc. The atom of modern physics can only be symbolized by a partial differential equation in an abstract multidimensional space. . . . *All* the qualities of the atom of modern physics are derived, it has no *immediate* and *direct* physical properties at all.[15]

In this article, Heisenberg was popularizing difficult physics in a particularly effective fashion, but he was also doing something more than that: he was defending the new physics from those who maintained that it was a totally radical break from the past, and possibly a mistaken one. By illustrating the continuity of even the apparently radical aspects of quantum mechanics with a trend of development including Galileo and Newton and stretching back to the Greeks, Heisenberg attempted to lessen the bewilderment and hostility of lay people.

One sees in Heisenberg's description of quantum mechanics as "only the next logical development . . . in the long-term trend of physics since its earliest days" an example of the socially and historically contingent quality of science-value interactions. Even when arguments about science and values are cast in the language of apparently cosmic questions—that is, even when the argument takes the form, "but we have always done it this way"—in fact, the issues being addressed are often just "issues of the day," and the appeal to tradition will prove to have as much strategic as substantive merit. Heisenberg provides a particularly poignant illustration of this fact, for clearly he wanted to establish the opposite claim, that the way to understand his science and its value implications was to connect them to the perennial themes of Western thought, not to anything so parochial as contemporary German culture and politics. And yet Heisenberg could not rise above the social battle going on around him, and, in fact, did not wish to. He acknowledged as much when he said that one of his goals in showing how much the new physics was a part of much older trends was to "protect modern science against the accusations of one-sidedness and arrogance."[16]

The Influence of National Socialism

And he had good reasons for considering such protection important, for those accusations were growing. Heisenberg wrote the first chapter of *Wandlungen* after the National Socialists had come to power, and he referred in the first sentences of this chapter to his desire to present quantum mechanics to a general audience in a way "that is free from the distortions arising from the strife of public debate."[17] What lay behind this reference? Who exactly was distorting quantum mechanics in a fashion that alarmed Heisenberg? Was it merely the general debate occurring throughout the educated world over the radical new physics, or did Heisenberg have a particular interpretation, or a particular interpreter, in mind?

The answers to these questions can be found farther along in the opening chapter of *Wandlungen*. In this entire twenty-one page essay, Heisenberg included only one textual footnote, and that was to "J. Stark, *Nationalsozialismus und Wissenschaft* [National Socialism and Science], München, 1932." The reference immediately followed these sentences in the text: "Progress in theory may significantly affect the direction of physical research, and thus eventually of technical development. In this connection, the relationship of experimental and theoretical physics must be touched upon. This relation has of late been rather distortedly presented to the German public."[18] It was Stark, noted Heisenberg, who was mainly responsible for this misrepresentation, and Heisenberg hoped to correct that distortion.

In order to understand this motivation underlying Heisenberg's presentation in the first chapter of this most popularly influential of his works, we must briefly consider the background and activities of Stark.[19] Johannes Stark (1874–1957) was an experimental physicist who had in 1919 received the Nobel Prize for his work six years earlier on the splitting of spectral lines in an electrical field (the Stark effect). In the earliest years of his scientific activity Stark had been a defender of both Einstein's theory of special relativity and of Planck's quantum hypothesis. However, he later became increasingly critical of both relativity theory and quantum mechanics, and by the nineteen twenties he began to oppose what he called "Jewish dogmatic physics." A partisan of Adolf Hitler, he rose to positions of great administrative influence after the National Socialists took over Germany in 1933. At the time Heisenberg's 1935 *Wandlungen* appeared, Stark was at the peak of his power; he was president of both the *Physikalisch-technische Reichsanstalt* (National Physical-Technical Institute) and the *Notgemeinschaft der Deutschen Wissenschaft* (Emergency Association of German Science; renamed the German Research Association).[20] In these positions Stark was working toward a complete reorganization of German physics. As the historian Alan Beyerchen has commented, Stark wished to become "the dictator of physics in Germany."[21] A strong partisan of experimental, rather than theoretical, physics, Stark's prime opponents were Heisenberg, Max von Laue, and Arnold Sommerfeld. In the summer of 1935 Sommerfeld's retirement made his Munich chair vacant, and Heisenberg was the leading candidate to fill it. Stark was determined to stop Heisenberg from receiving this honor. In a 1935 speech Stark called Heisenberg "the spirit of Einstein's spirit," knowing well the anti-Semitism in Germany that could be turned against Heisenberg.[22] At a moment of great political stress and danger, Heisenberg stoutly defended theoretical physics, carrying the debate with Stark even onto the pages of the main National Socialist newspaper, *Völkischer Beobachter*.[23] By 1937 Stark was calling Heisenberg a "white Jew," meaning that he considered Heisenberg to be a Jew in everything but race. In an infamous article which appeared in the SS journal *Das Schwarze Korps* Stark remarked that Heisenberg and others of his orientation should be made to "disappear" like the Jews.[24]

In the mid-thirties it appeared that Stark might be able to carry out his threats. Heisenberg, in fact, never received the Munich chair. Stark, however, lost his struggle to take over all German physics. He was a poor politician, and cast his lot with losing factions in the National Socialist hierarchy. When Heisenberg wrote the first edition of his famous book, however, Stark was still in the ascendancy. Let us examine briefly Stark's attack on theoretical physics, which Heisenberg cited in that book.[25]

In *National Socialism and Science* Stark attempted to distinguish two different styles of science, the "Jewish" and the "German," and he assigned Heisenberg to the Jewish style, despite his very German (and even nationalistic) background. Heisenberg, he said, had been captured by the Jewish spirit emanating from the "Göttingen mathematical Jews" such as Klein and Hilbert, and in the middle of which he saw Einstein and Sommerfeld. (That Sommerfeld, like Heisenberg, was not a Jew did not affect Stark's argument.)

The way in which Stark described the essential features of science, from what he called a National Socialist point of view, was a challenge to many of the things Heisenberg had been saying about science. First of all, Stark based himself on a strong assertion of the existence of physical reality, governed by causal relationships, and he put a much higher priority on experimental physics than on theoretical physics. Science, said Stark, is:

> . . . the knowledge of lawful connections of facts; the duty of natural science particularly is exploration and, as far as possible, by performing carefully planned experiments. The spirit of the German prevails on things outside of himself, as they actually are, without insertion of his own ideas and wishes.[26]

But while the good German scientist is interested in how nature "really is," the Jews and Jewish-minded scientists, said Stark, apply themselves merely to theory, not to the observation of facts and their portrayal "true to reality." What is important to the Jewish scientist, said Stark, is not reality, but his opinion about it and the formal representation he can impose upon it. The Jewish-minded scientists are taking over German physics, Stark railed. He considered the previous two decades not as one of the most brilliant chapters in the history of central European physics (as the rest of the world will always see it), but as a great decline. What was occurring, Stark maintained, was that "from Germany great dogmatic theories have been brought onto the world market, such as Einstein's relativity theory, Heisenberg's matrix theory, and Schrödinger's wave mechanics."[27]

Although this attack appears to us today to be hardly worth recalling, the form in which Heisenberg presented his influential essay "Wandlungen," which even today is still frequently read, cannot be understood unless we see that it was a reply to this criticism; in 1935, Stark's polemic may have been nonsense, but it was very dangerous nonsense about which Heisenberg was deeply concerned, and which he took as a significant political threat. At this time many people, not just National Socialists, were interpreting Heisenberg's uncertainty relationship as forbidding the possibility of researching things "as they really are," or "outside of oneself," as Stark called for here. And it is certainly true that Stark's assertion of a causally related external real world of bodies and processes separate from the human mind was, in terms of the young Heisenberg's philosophy of science, an unverifiable statement and therefore a "contentless statement."

Heisenberg met this challenge with a very carefully constructed argument, making concessions where he thought them necessary or relatively harmless, but always defending the great achievements of relativity physics and quantum mechanics. And, at the same time, he departed ever more clearly from his earlier positivistic views. Stark's challenge undoubtedly was one of the factors leading him in this direction, but, since he never returned to positivistic positions, the transition represented a permanent shift in his popular interpretation of physics. In the postwar years he continued and even deepened his criticism of positivism.

As one reads Heisenberg's 1935 "Wandlungen" chapter today, the first characteristic that strikes one is its intellectually conservative character, its effort to pacify lay people who were alarmed by rumors that something startlingly new had occurred in physics—something that invalidated traditional philosophic concepts. Earlier developments in man's physical world view, such as the Copernican revolution, said Heisenberg, were much more radical than what is happening in quantum mechanics. Indeed, in an astounding understatement, he stated, "It is wrong to speak today of a revolution in physics. . . . Only the conception of hitherto unexplored regions, formed prematurely from a knowledge of only certain parts of the world, has undergone a decisive transformation."[28]

This transformation was not wrought by "revolutionary ideas which, so to speak, were brought into the exact sciences from the outside," Heisenberg continued. (Here he displayed his awareness of the German conservative fear of the very term "revolution from abroad," whether political or scientific.) Futhermore, this transformation was not the result of speculative inventions (certainly not "Jewish speculative inventions"), but, on the contrary, the changes in science had been forced upon scientists by the results of experiments.[29] And in citing the experimenters who had played a role in bringing about the new physics, Heisenberg was careful to give attention to his National Socialist physicist-critics like Stark and Philipp Lenard, even though in an earlier (1929) historical article reviewing the development of quantum mechanics he had not mentioned them.[30]

Heisenberg made even more adjustments to the political scene. As we have seen, in earlier articles he had considered the question of the validity of the concept of causality an insignificant topic; he had further sharply rejected the concept of "visualization" on the microlevel; and he had avoided a commitment to objectivity or physical reality, considering such principles metaphysical. He now reversed himself and conceded the importance of each of these concepts. As Heisenberg wrote in his 1935 book, the use of perceptual or visual forms and the concept of causality are

> the premises of every objective scientific experience even in modern physics. For we can only communicate the course and result of a measurement by describing the necessary manual actions and instrument readings as objective, and as events taking place in the space and time of our *Anschauung* (perception). Neither could we infer the properties of the observed object from the results of measurement if the law of causality did not guarantee an unambiguous connection between the two.[31]

Heisenberg here was making significant concessions to individual ideological critics, and also he was trying to accommodate himself to hostile reactions of a public believing in common-sense reality (a reaction now potentially supported by political power), but he was not capitulating to these critics. After getting his philosophical defense in order he supported the new physics very strongly, praised Einstein and other scientists disliked by the political conservatives, and stood up firmly in favor of theoretical physics as opposed to experimental physics. On his politically most vulnerable points (earlier disregard of experimental physicists as pathbreakers; disdain for statements about how "the world really is") he yielded or compromised, but he voiced full force his defense of quantum mechanics and relativity physics against the charges that they were speculative innovations of the Jewish spirit, and he criticized Stark by name for his support of this interpretation.

We see, then, that Heisenberg was giving a brave and calculated defense of modern science in a troubled environment. We also see that the first chapter of the book that marked Heisenberg's debut as an authoritative interpreter of the new science bore the marks of its time and already differed somewhat from writing of the younger Heisenberg of the previous decade. And it would differ somewhat from the later Heisenberg as well. As an example, in 1935 Heisenberg denied for obvious political reasons that it was proper to speak of a "revolution" in physics; after World War II, when the political situation was much different, he subtitled one of his books about the philosophical implications of physics "The Revolution in Modern Science."[32]

In the immediately following editions of *Wandlungen*, published in 1936, 1941, 1943, and 1944, the urgency of the defense of relativity physics and quantum mechanics against the crudest National Socialist attacks diminished. Stark was forced out of his positions of power in the Third Reich in 1936 and subsequently went into retirement.[33] Philipp Lenard, the other prominent German physicist most strongly opposed to the new physical theories, was, by the beginning of the war period, an eighty-year-old man of diminishing influence, although Hitler continued to hold him in high regard. On July 21, 1938, Heisenberg received a letter from Heinrich Himmler exonerating him from the personal attack in the SS journal of being a "white Jew." Several appeals on Heisenberg's behalf had been made to Himmler. The most unusual came from Himmler's mother, who was a friend of Heisenberg's mother.[34] Other factors favoring Heisenberg gradually emerged. After the possible significance of atomic physics for warfare began to dawn on German political leaders, the positions of physicists like Heisenberg became more secure.

The influence of these changes can be seen in the subsequent editions of *Wandlungen*. Heisenberg dropped his criticism of Stark in the third edition (1941), while retaining his defense of theoretical physics. In his own work, he made accommodations with the National Socialist government and the political atmosphere of his time, although he found some of these adjustments distasteful.

The Influence of Classical Thought

Heisenberg continued, however, to pursue the main line of his interpretation of modern science developed in the first edition, and there is every reason to believe that this interpretation, albeit molded by his external environment, was comfortable to him. The main feature of that interpretation was to show how modern physics was connected to traditional aspects of knowledge, particularly the natural philosophy of the Greeks. Here he could draw upon his own family history, for his father August Heisenberg was a classicist and Byzantinist at the University of Munich, and Werner knew ancient Greek and the classical sources. The younger Heisenberg increasingly drew on these sources for his interpretations of nature in the expanding editions of the *Wandlungen*; in addition, he connected modern physics to traditional aspects of German culture, such as the reverence for Goethe and his differences with Newton, and even the German interest in questions of vitalism, spiritualism, and romanticism. It hardly needs to be noted that these succeeding chapters marked his departure even further from the earlier position of "Whereof one cannot speak, thereof one must be silent." By connecting modern physics with the internal threads of German culture Heisenberg continued his task of interpreting and defending science to lay audiences; at the same time he adjusted to the mood of his time. Yet these adjustments are not evidence of hypocrisy; Heisenberg was, like many scientists in other countries, addressing the issues of the day and relating his work to those issues.

In 1937 Heisenberg published an article in *Die Antike* (Antiquity), a journal of classical studies, entitled "Ideas of Classical Natural Philosophy in Modern Physics." This article would later become chapter 4 of the third edition of *Wandlungen* (1941). In understanding the background of this article it is helpful to look at its setting in the place of original publication, *Die Antike*, and the editors' foreword for that issue. This journal had been founded twelve years earlier by Werner Jaeger, with the goal of interpreting and preserving the intellectual and artistic values of classical culture. With the issue in which Heisenberg's article appeared a triumvirate of new editors more sympathetic to National Socialism had taken over, and they announced their intention to abandon the "indiscriminate dissemination of individual research projects" and concentrate instead on making classical research relevant to the present day, especially to the "great, historical transformation and renewal (*Erneuerung*)" going on in German society. And they observed that "whoever today speaks to Germans about antiquity cannot talk in empty space."[35]

Heisenberg's approach in this essay was to try to show that modern physics presents in a "purer form" two of the most important ideas in ancient Greek natural philosophy: the belief in indivisible units, or atoms, drawn especially from Leucippus and Democritus; and the belief in the "purposely directive power of mathematical structures" drawn from the Pythagoreans and from Plato. Yet both atomism and Platonism had been modified by modern physics, he continued, not in the sense of an undermining of the essential principles behind these viewpoints, but in a carrying of the innermost principles of these viewpoints to their furthest extensions, and thereby creating the erroneous impression among some people that the original doctrines themselves had been discredited. Nothing, Heisenberg believed, could be further from the truth. The relevance of ancient philosophy for modern physics was still very great indeed.

Many interpreters of ancient atomism, for example, find its essence in materialism, but Heisenberg carefully steered away from this pejorative term; the essence of classical atomism, he said, was not materialism but the belief that perceptible qualities can be "traced back to the behaviour of entities who themselves no longer possess these qualities." Thus, Heisenberg reassured his readers, "modern theory embodies the principal and basic idea of atomic theory in a purer form than did ancient theory."[36]

Heisenberg concluded this essay by expressing his regret that many nonscientists are not able to perceive in modern science this sequence of faithful executions of the program of ancient natural philosophy and accuse scientists instead of creating an abstract mathematization of nature which "illuminates only certain and, at that, not the most essential, aspects of nature."[37] Heisenberg was referring here to the romantic rebellion against mathematical science that was such a strong undercurrent in European thought and especially in German culture. Goethe's famous criticism of Newtonian science was one of the best known of these rebellions, and Heisenberg knew that in the current nationalistic moment Goethe's attacks on Newton again had appeal. He believed that it was important to handle that challenge in an understanding and subtle way that protected mathematical physics from public accusation. He turned to that task more fully in the next chapter of *Wandlungen,* first published in 1941 in a journal appropriately entitled "Spirit of the Time," and also appearing in the third edition of *Wandlungen* in the same year.

Newton, Goethe and Wartime Passions

This fifth chapter of *Wandlungen* was originally delivered in Budapest on April 28, 1941, before the Society for Intellectual Cooperation.[38] For his topic Heisenberg chose a comparison of the teachings of Goethe and Newton on the theory of colors.

It is difficult for us today to recall how controversial this topic then was. Germany and Britain were at war (in fact, Britain stood almost alone against Germany at that moment) and blazing accounts of bombing attacks by both countries filled the newspapers. The Battles of Britain and of the Atlantic were current events. Goethe and Newton were culture heroes of the opposing nations. In Germany, furthermore, a great debate among political factions had been underway for many years over the legacy of Goethe; different factions had selectively quoted from Goethe's works in order to try to show how the great man's views could be interpreted as supporting a particular ideological position. The results were, of course, intellectual distortions of the grossest sort, since the genius Goethe, dead for over a century, clearly belonged to a different age. But the distortions became concrete factors in the polemics of the day, and by 1941 the National Socialists had fashioned a particular interpretation of Goethe for their own benefit. Delving into the poet's political and social views, the National Socialists quoted his most malicious statements about Jews (and there was fertile ground here), ignoring his more positive statements on the same subject. The National Socialist ideologists also quoted Goethe to celebrate the virtues of the "natural" countryside over the "artificial" city, of common sensuous experience over abstractions, of aesthetics over mathematics, of the virtue of willful elitism over weak and liberal democracy.[39]

Thus, the rich and contradictory repertoire of Goethe's writings was dishonestly converted into an ideological message. In 1937 the leader of the Hitler Youth Movement, Baldur von Schirach, gave a speech entitled "Goethe Speaks to Us" ("Goethe an Uns") which was later published by the National Socialist Party in several hundred thousand copies for distribution to Hitler Youth. It contained quotations from Goethe grouped under headings such as "On Physical Training," "On Daily Life," "On Women and Morals," and "The Revelation of Nature."[40]

A confrontation of Goethe and Newton was an old subject in the history of science. Goethe stood for an intuitionist, poetic approach to nature, while Newton was popularly (and incorrectly) associated with cold abstraction. Exact scientists like Heisenberg were often uncomfortable when faced with an apotheosis of Goethe, for they knew that Goethe had attacked the Newtonian theory of colors as an intrusive dissection of an original phenomenon into component parts in a way that Goethe considered alien to Nature. Goethe had advanced an alternative theory in which natural daylight is the simplest and purest concept, and colors are produced by daylight as "boundary effects" where light and dark meet. To Goethe, Newton's most simple and pure concept—monochromatic light produced by experimental mechanisms—was outside of "living" nature, and therefore unsuitable for a starting point in building a science of nature.

Heisenberg's highly mathematical formulations of quantum mechanics and his belief that a different science reigned in the microworld than in the everyday world were compared by some of his critics to the formalistic world view of Newton. The sort of person who found Newton's theory of colors bewildering because it was far from common experience was also likely to find Heisenberg's uncertainty relationship, indeed, all of quantum mechanics, disturbing. Therefore, Heisenberg's discussion of Goethe and Newton was not an entirely academic exercise, but one that could conceivably have some real effect on his life.

When it was announced in Budapest in the spring of 1941 that the great Nobel Prize laureate Werner Heisenberg was going to compare Goethe and Newton in a public lecture, the occasion became a significant social event prominently advertised in the daily newspapers. Every place in the auditorium was filled, and prominently seated before the speaker were Counsellor Struwe from the German legation and other National Socialist functionaries and party members, both from Germany and from the German colony in Budapest. After the lecture a banquet in honor of Heisenberg was scheduled for the Hotel Pannonia.[41]

The lecture was vintage Heisenberg, a skillful celebration of Goethe and his views of nature and at the same time a defense of theoretical physics against obscurantist attacks based on the vulgarized Goethe. The basic framework of Heisenberg's argument was a restrictionist approach (a stronger and clearer one than he had previously assumed) in which he gave both Goethe and Newton realms of operation for their respective methods so long as they did not interfere with each other.

An early participant in the German youth movement himself, Heisenberg took second place to no one in his admiration of nature. He said that he valued Goethe not only for his great literary achievements, but also for his view of natural phenomena. Heisenberg mollified his critics by admitting that Goethe's theory of colors was an attempt "to save the immediate truth of the sense-impressions from the attacks of science."[42] And, continued Heisenberg:

> Today, this task is more urgent than ever. The whole world is being transformed by enormous extensions of our scientific knowledge and by the wealth of its technical applications, but like all wealth, this can be a blessing or a curse. Hence many warning voices have been raised during recent years counselling us to turn back. Already, they say, a great scattering of intellectual effort has resulted from our negation of the world of direct sense-impressions and the division of nature into different

sectors. Further withdrawal from "living" nature will, so to speak, drive us into a vacuum where life will no longer be possible.[43]

Heisenberg agreed with the view that modern science presents many dangers, including the one of intellectual and emotional alienation, but he sharply opposed the attempted resolution of this problem by rejecting science itself:

To effect a real change would undoubtedly entail a complete abandonment of the whole of modern technology and science. . . . Such a break is impossible. We have to reconcile ourselves to the fact that it is the destiny of our time to follow to the end of the road along which we have started.[44]

Yet Heisenberg still believed that a path existed for relieving some of the tensions between modern science and the aesthetic vision of the world. What was that path? Not one of discarding Newton's view and replacing it with Goethe's, but one of recognizing the legitimacy of both and combining them: "It is certain that an acceptance of modern physics cannot prevent the scientist from following Goethe's way of contemplating nature, too."[45] Indeed, by combining the scientist's "manipulation" of nature with Goethe's "contemplation" of it, one can hope to create a method of understanding suitable for both the natural sciences (*Naturwissenschaften*) and the arts (*Geisteswissenschaften*). That the creation of such a form of understanding was Heisenberg's hope is clear not only from the text of the lecture, but also in a direct statement of this goal which he appended to the Hungarian version of the article, but which was not included in the German and English translations.[46]

But while Heisenberg praised Goethe's experential approach to nature, he carefully stated that this does not mean that Goethe was right about colors and Newton wrong. On the contrary, stated Heisenberg in straightforward terms: "Victory in that particular conflict has been Newton's."[47] But that victory was restricted to science itself, the objective realm. What quantum physics has shown, continued Heisenberg, was how separate the subjective, sensual realm is from the objective, scientific one. Indeed, quantum mechanics has shown that physicists must give up everyday sense perceptions as valid concepts for science, since the last two retained by Newton—shape and movement—have now been abandoned.[48]

This very victory of science, said Heisenberg, means that since science no longer uses these sense perceptions, their interpretation is now handed over in an unfettered way to the masters of the subjective realm—the artists and writers. We now see more clearly, said Heisenberg, "that the range of science and technology as we know it, is finite." By adopting this restrictionist position, Heisenberg said that we can see that Goethe and Newton were talking about "two entirely different levels of reality," and "dividing reality into different aspects immediately resolves the contradictions between Goethe's and Newton's theories of colors."[49] Thus, Heisenberg converted one of the great disputes in the history of science into a nonbattle, and people on both sides should be content.

Heisenberg's restrictionist approach allowed him to reconcile two views of nature that were important to him and his culture—the scientific and the aesthetic. The approach was not a hypocritical one, since Heisenberg's respect for Goethe was deep and genuine; nonetheless, the political situation was such that Heisenberg's reverence for Goethe, in science as well as in literature, was probably more carefully and fully stated than would otherwise be the case. This supposition is supported by his postwar writings on the subject.[50] Finding a way by which both Goethe and Newton could be termed correct was the ideal solution to all problems, both scientific and political.

We saw that Arthur Eddington a few years earlier had followed a somewhat similar course when he had tried to restrict science in such a way that it left an untouched realm that could be filled with religion. The goals of the two scientists were quite different—Heisenberg, the defender of German cultural ideals, Eddington, the defender of his personal variety of religion—but the architectures of their defenses were similar. They both knew that if science could be restricted, if it covered only a part of reality, considerable freedom of choice would be available for shaping the body of beliefs that could be slipped into the remaining hole. And the particular goals which each interpreter of science served—preserving a place for religion in England in 1935 and a place for Goethe's vision of aesthetics in Germany in 1941—fit well with popular aspirations in those countries at the times in which they wrote.

Heisenberg the Reconciler

Heisenberg carried further his goal of reconciliation between allegedly contradictory approaches to reality in the sixth chapter of *Wandlungen*, in an essay entitled "On the Unity of the Scientific Outlook on Nature" (a paper presented in 1941 at the University of Leipzig). In this essay Heisenberg saw three distinct stages in the development of scientific thought. The first stage, according to Heisenberg lasting up to Newton, was that of a unified scientific world view, a view based on religion and devoted to the recognition of God's work in nature. Kepler personified this approach perhaps best of all when he said that his work was designed to give insight into God's "way of creation." This underlying religious impulse gave to the study of the cosmos a unity of purpose and design. Science and values were united.

This unified world picture was destroyed, said Heisenberg, by the full development of mathematical physics in the eighteenth and nineteenth centuries. Newtonian mechanics, he observed, "could be applied to individual problems posed by nature and hence it was no longer a question of understanding a single interconnected whole but of a detailed analysis of many small specific connections."[51] This specialization led to a fragmentation of the world picture in this stage, a loss of harmony and unity.

It is true that many scientists, especially physicists and mathematicians, saw in Newtonian physics a potentially unifying method for all science, and some believed that even such complex phenomena as life and consciousness would be explained by the Newtonian physical framework. But, continued Heisenberg, "it was soon seen that it [Newtonian physics] was incapable of creating a durable unity."[52] Too many phenomena, particularly in biology, could not be explained satisfactorily in that way, and soon a variety of non-Newtonian, perhaps even nonscientific, approaches developed in an effort to explain what Newtonian physics was incapable of explaining. The result was an "atomization" of man's world view.

The very arrogance of Newtonian physics, on the one hand, and its lack of success in explanation of all complex phenomena, on the other, led to the "regretted division of mental activity into the sphere of science and the realms of art and religion."[53] Here

Heisenberg's discomfort with pure restrictionism emerged. Recognizing the failure of Newtonian science, European thinkers often turned to romantic natural philosophy in an effort to fulfill the need for a unified understanding of the world, but this effort, too, failed to provide an integrated and successful world framework. In biology the fragmentation took the form of the division between mechanists and vitalists, with the former struggling to remain loyal to the Newtonian ideology, while the latter recognized its inadequacy, but failed to find an alternative that was scientifically developable.

With the rise of quantum mechanics in the twentieth century, Heisenberg continued, a third stage has appeared, and "the various branches of science are beginning to fuse into one great unity." This unity brings the mechanist and vitalist views together in a way that is simultaneously esthetically harmonious and scientifically fruitful. Heisenberg saw an open path to a new unified world picture not by the direct extrapolation of physics to the biological and mental realms, but by the recognition that even in physics there are different levels of reality in which different types of laws or regularities are valid. Thus we can hope, against the background of this much broader and complex approach to nature than that offered by mechanistic physics, to find a new philosophical unity, and we can once again join with Kepler in "the hope for a great interconnected whole which we can penetrate further and further."[54]

The most interesting aspect of Heisenberg's analysis here is not the question of the validity of his three stages of scientific thought. Certainly his picture of Newton as the apostle of nonreligious rationality is vastly overdrawn. What attracts our attention, however, is the role of healer of cultural and intellectual rifts that Heisenberg continued to try to play in his popular writings on science. Although he was one of the most revolutionary of scientists, in his popular works he did not advocate intellectual revolution, but intellectual and cultural continuity, and he sought in science those metaphors and approaches that would bring an integration of the conflicting traditions of European culture. He longed for a world picture that unified the strands of science, art, and religion, but he aimed for this union not by reconciling the strands in their old heterogeneous forms, but by finding pure principles within each of them that could be extracted and unified within the framework of a world view drawn largely from the new physics. A later generation would judge that even the new physics was not powerful enough for this unification, but Heisenberg's effort tells us a great deal about him and his time.

No essay in Heisenberg's *Wandlungen* shows more clearly the marks of its origin, the temporal and geographic characteristics of its birth, than the chapter entitled "Science as a Means of International Understanding." This essay was originally a speech delivered to students at Göttingen University on July 13, 1946, only a little over a year after Germany's defeat in World War II and only a few months after Heisenberg's release from internment by the Allies. Heisenberg faced an audience of young students who distrusted his generation and the values for which that generation stood. What could Heisenberg say to them that would be credible? He held out science to them as the one activity still worthy of commitment. "While we cannot avoid in political life a constant change of values, a struggle of one set of illusions and misleading ideas against another set of illusions and equally misleading ideas, there will always be a 'right or wrong' in science." He acknowledged that "we also make mistakes in science and it may take some time before these are found and corrected. But we can rest assured that there will be a final decision as to what is right and what is wrong. This decision will not depend on the belief, race or origin of scientists, but it will be taken by a higher power and will then apply to all men for all time."[55]

The objectivity of science and the contributions and cooperation of scientists of many races and nations is evidence that science is truly international, he commented, and refutes the "opposite thesis, which is still fresh in our ears, that science is national and the ideas of the various races are fundamentally different."[56] Heisenberg pointed to the scientists of many different nations who had contributed to the development of quantum mechanics—Danes, Englishmen, Americans, Germans, Russians, Japanese, Swedes, Dutch, Norwegians—and called for the use of science to strengthen understanding among nations.

A rereading of this essay today shows that Heisenberg, in an effort to transcend wartime difficulties, attributed to science and scientists a purity of purpose no longer admitted by a later postwar generation who see scientists as fallible fellow citizens, not superhumans. In trying to explain why scientific collaboration fails to prevent wars, Heisenberg gave as a reason that "only very few people in each country are really connected with science," as if scientists genuinely possess more political wisdom than other citizens, but are simply outnumbered and overwhelmed by misguided nonscientists.[57] This view was patently simplistic, but Heisenberg saw an unquestionable moral value in choosing science in 1946 as a convenient and feasible first avenue for the reestablishment of bonds among nations. It is no surprise that he exaggerated the purity of science at a moment when so little was available upon which to rebuild intellectual life and international contacts.

The Last Editions of Wandlungen

After including the above-described essay on science as a means of international understanding in the seventh edition (1947) of *Wandlungen*, Heisenberg added only two new essays to the volume until its tenth edition in 1973. (He also later included his older Nobel Prize lecture of 1933.) These two essays were entitled "Fundamental Problems of Present-day Atomic Physics" (1948) and "Planck's Discovery and Fundamental Philosophical Questions of Atomic Theory" (1958).

In the 1948 essay he emphasized that the real goal of atomic theory is, like the goal of ancient Greek physical theories, the explanation of the world in terms of "a homogeneous substance (*Stoff*) based on one unified principle."[58] He expressed concern that this goal might be lost sight of even by some physicists because of the increasing multiplicity of forms entering physics through the discovery of more and more elementary particles. He reminded his readers that all of these "particles" are really different manifestations of one entity—energy—and observed that it was not inconceivable that in the not-too-distant future there would be a single mathematical equation from which the different manifestations could be derived.

Since relativity theory had shown the equivalence of matter and energy, Heisenberg recognized that this underlying unity might be described as a new form of materialism of the sort defended in the Soviet Union by Fock, but, true to his own culture, he resisted this conclusion. He cautioned that, at any rate, the "materialism implied would in no way denote the anti-spiritual tendency we commonly attach to this word."[59] Indeed, a defense of "spirit" and even of "soul" was important to Heisenberg in 1948 and he stated that physics was incapable of explaining such phenomena, and even incapable of explaining "life."

In the last chapter of *Wandlungen*, written in 1958, Heisenberg inquired into the philosophical significance of Planck's discovery of the quantum of action in 1900 and the subsequent development of quantum mechanics. Among the several interesting philosophical implications of quantum mechanics, the one that most attracted Heisenberg was the new relevance it gave to Plato's view that nature should be described in terms of mathematical forms. Heisenberg had often emphasized in the past the importance to him of Platonic philosophy, even in his earliest days, but his stress upon Plato in his public pronouncements definitely grew in his post-World War II writings.[60]

As a recognition of the increasing mathematization of physics and its distance from simple materialistic views, Heisenberg's emphasis on Plato was quite understandable. In his particular case, however, the attraction of Platonic views was probably something more than a mere intellectual choice. The purity of the Platonic ideal of a mathematical vision of the world was also a welcome relief from a social reality and a life history that contained more than its share of unpleasant political events and esthetic disappointments. Heisenberg himself acknowledged the importance of these considerations. In the last paragraph of this final chapter of *Wandlungen*, written twenty-five years after the first chapter, Heisenberg revealed how pleasant it was for him to turn from the grubby reality of contemporary life to the refuge of mathematical physics:

> The 100th birthday of Max Planck [1958] comes at a time which, if one compares it with earlier epochs, makes a very chaotic impression in many areas—politics, art, value standards. It is therefore calming to think of a personality as harmonious as that of Max Planck and to know that at least

in one area, the one to which Planck devoted his life work, nothing chaotic is to be found; rather, we find here a simplicity and a transparent clarity just as definite as at the time of Plato or Kepler or Newton.[61]

The preference for an abstract and elegant mathematical world best represented by the Platonic philosophy was a constant in Heisenberg's life, although he deemphasized this view in the mid-1930s when abstract mathematization was under criticism in National Socialist Germany, and he stressed it more and more in the last years of his life. Underneath these small changes, the continuity of this view in his thoughts is evident, and its importance to him not only as a scientific philosophy for finding order in apparent chaos, but also as a refuge from unpleasant realities, can be traced back to his youth. Once again, Heisenberg himself has given the best description of this early influence. In 1919, after Germany's defeat in World War I, the seventeen-year-old Heisenberg was assigned to a cavalry rifle group to help squelch a communist revolution. Heisenberg described the events as follows:

In the spring of 1919, Munich was in a state of utter confusion. On the streets people were shooting at one another, and no one could tell precisely who the contestants were. . . . Pillage and robbery (I was burgled myself) caused the term "Soviet Republic" to become a synonym of lawlessness, and when, at long last, a new Bavarian government was formed outside Munich, and sent its troops into the city, we were all hoping for a speedy return to more orderly conditions. . . . Quite often it happened that, after spending the whole night on guard in the telephone exchange, I was free for a day, and in order to catch up with my neglected school work I would retire to the roof of the Training College with a Greek school edition of Plato's Dialogues. There, lying in the wide gutter, and warmed by the rays of the early morning sun, I could pursue my studies in peace, and from time to time watch the quickening life in the Ludwigstrasse below.

One such morning, when the light of the sun was already flooding across the university buildings and the fountain, I came to the *Timaeus*, or rather to those passages in which Plato discusses the smallest particles of matter. . . . I was enthralled by the idea that the smallest particles of matter must reduce to some mathematical form. . . . Meanwhile, my uneasiness continued, though perhaps it was only part of the general disquiet that had seized all German youth. I kept wondering why a great philosopher like Plato should have thought he could recognize order in natural phenomena when we ourselves could not. . . . We had all of us grown up in a world that seemed well-ordered enough. And our parents had taught us the bourgeois virtues underpinning that order. . . . [But] the old structure of Europe had been shattered by our defeat. That, too, was nothing special. Wherever there are wars there must also be losers. But did that mean that all the old structures had to be discarded? Was it not far better to build a new and more solid order on the old? Or were those right who, in the streets of Munich, had sacrificed their lives to prevent a restoration of the old ways, but for all mankind, even though the majority of mankind might have no wish to build such an order?[62]

Heisenberg answered these questions for himself by deciding that it was better to build a new and more solid order on the old, not by trying to construct something entirely new, and he remained loyal to this principle in later years in both politics and science. Against this background, the elements of continuity in scientific thought which he emphasized in the various chapters of *Wandlungen*, written at very different times, seem fitting with his whole approach to life. At certain early points he carried this philosophy of continuity to such an extreme that, as we have seen, he even denied that quantum mechanics represented a revolution in physics, when surely if any event in physics deserves revolutionary status, it is the advent of quantum mechanics.[63] It also should be noted that Heisenberg chose carefully the aspects of the past he wished to retain, disregarding many which did not fit his taste. His support of continuity was, therefore, somewhat arbitrary. For example, he selected *which* of the many elements in classical Greek philosophy he wished to continue into the realm of modern physics (e.g., not Democritus' materialism, but Plato's idealism), and his choices were meaningful to the cultural currents of his society.

The desire for mathematical order, and the Platonic philosophy that underlay it, was probably the most important feature in Heisenberg's thought. Compared to this aspect of his philosophical interpretation of nature, several of the other features of his thought which appeared in certain of his works—the positivism of the late twenties, the concessions to National Socialist ideology in the thirties—were temporary episodes of small significance in his total activity as an authoritative interpreter of modern physics for the educated lay person. To the end, he was "enthralled by the idea that the smallest particles of matter must reduce to mathematical form," and he transmitted the pleasure of that Platonic vision to his readers.

Heisenberg's Later Popular Writings

I have concentrated in this analysis of Heisenberg's views of the relationship of science to larger issues on only one of his publications, his *Wandlungen in den Grundlagen der Naturwissenschaft.* I believe that a good case can be made that this publication, in all its forms, is the best single indicator of the evolution of his philosophical views and of the role he played as an authoritative interpreter of science to the general public, and to the German public in particular. It is the only one of his popular books to appear during the formative period in the interpretation of the new physics—before World War II—and it is also the only one of his books that he continually revised and supplemented, a process that lasted throughout a major part of his professional life. It therefore is a particularly revealing record of the evolution of his views, and the relationship of those views to society at the time each new chapter was written. The book was widely read at a time when the new physics was still an object of great curiosity to the public. By the time of the appearance of his later popular books (after 1955), Heisenberg was a "grand old man" of physics, and his philosophic viewpoints had been rather fully developed in his earlier articles, the most significant of which had gone into the *Wandlungen.*

If we briefly look at the later books, we will see that they were further developments of the basic ideas contained in the *Wandlungen.* The four publications of the most general interest were his *Das Naturbild der heutigen Physik* (1955); *Physik und Philosophie* (1959); *Der Teil und das Ganze* (1969); and *Schritte über Grenzen* (1971).

These books were written at a time when the achievements and validity of the new physics were not to be doubted, and when the society in which Heisenberg lived was not under the degree of social stress often present when the various chapters of the *Wandlungen* had been written. They contain most of the same ideas expressed in the *Wandlungen,* but the tasks of defending science and of healing intellectual rifts were no longer quite so urgent. Therefore, the viewpoints internal to Heisenberg's interpretation of science are often stated less cautiously than before.

His emphasis on Platonism, as earlier noted, continued to grow.[64] He no longer hesitated to call quantum mechanics a "revolution" in science, as he had in the thirties, but he still warned his readers that in natural science "only a person who tries to alter as *little* as possible can achieve success, because he thereby makes evident what the facts compel."[65] He no longer spoke as appreciatively of Goethe's view of science as he had in 1941, although he continued to refer frequently to the great thinker's artistic vision.[66] He spoke more often of religion than he had done earlier, and he called for a restrictionist recognition of the validity of "the two kinds of truth" which religion and science can bring.[67] And the traits of positivism that were so visible in the young Heisenberg's writings had not only disappeared, but, indeed, had been replaced by explicit criticism of this philosophic view: "The positivists have a simple solution: the world must be divided into that which we can say clearly and the rest, which we had better pass over in silence. But can anyone conceive of a more pointless philosophy, seeing that what we can say clearly amounts to next to nothing? If we omitted all that is unclear, we would probably be left with completely uninteresting and trivial tautologies."[68]

The point in illustrating these shifts is not to find inconsistencies in Heisenberg's thought. Indeed, if we survey the whole of Heisenberg's life we find a coherent evolution that can be understood in terms of his changing role of, at first, innovator in science, and later on, interpreter of science for the public and defender of his culture. The evolution from narrow scientific issues to broader intellectual and cultural ones is a feature of many other authoritative interpreters of science, researchers who earned their credentials by achievements in science itself and who then, by dint of that status, were able to command appreciative audiences on general questions. Younger scientists in all countries often discount the significance of their elders who "moved on into philosophical questions," but this transition is probably not only inevitable, but also indispensable, for it helps integrate the divisions of intellectual life, often grievously split between the scientific and the philosophic or literary. Heisenberg believed more strongly than many scientists that the harmony of society depended in an important way on the reconciliation of older cultural traditions with the disruptions in intellectual conceptions brought about by science, and a large part of his efforts in his nonscientific writings was devoted to showing that a complete break with the past was both impossible and, if attempted, potentially disastrous.

He rejected the view that a scientist should divorce himself from his political and cultural traditions, even if aspects of those traditions were clearly obsolescent, false or dogmatic. In what is surely a revealing description of his own attitude toward political authority under different regimes, Heisenberg wrote in 1958 that the scientist must adjust to the political order because:

> . . . we belong to a community or a society. This community is kept together by common ideas, by a common scale of ethical values, or by a common language in which one speaks about the general problems of life. The common ideas may be supported by the authority of a church, a party or the state and, even if this is not the case, it may be difficult to go away from the common ideas without getting into conflict with the community. Yet the results of scientific thinking may contradict some of the common ideas. Certainly it would be unwise to demand that the scientist should generally not be a loyal member of his community, that he should be deprived of the happiness that may come from belonging to a commmunity, and it would be equally unwise to desire that the common ideas of society which from the scientific point of view are always simplifications should change instantaneously with the progress of scientific knowledge, that they should be as variable as scientific theories must necessarily be. Therefore, at this point we come back even in our time to the old position of "twofold truth" that has filled the history of Christian religion throughout the later Middle Ages. . . . If in our time political doctrines and social activities take the part of positive religion in some countries, the problem is still essentially the same. The scientist's first claim will always be intellectual honesty, while the community will frequently ask of the scientist that—in view of the variability of science—he at least wait a few decades before expressing in public his dissenting opinions. There is probably no simple solution to this problem, if tolerance alone is not sufficient; but some consolation may come from the fact that it is certainly an old problem belonging to human life.[69]

In this fascinating and deeply conservative passage we see more clearly than perhaps anywhere else the task which Heisenberg set himself in his conciliatory role of integrating science with society. Heisenberg had lived under too many different regimes of radically different character to take the simple view that the scientist must "always and immediately tell the truth and disregard the consequences." He had seen Imperial Germany, Weimar Germany, National Socialist Germany, the Western Federal Republic of Germany, and the Marxist German Democratic Republic. Heisenberg's retrospective attitude toward these regimes was not to support some and condemn the others, but to assert that every community must have a common set of ideas and to recognize that these common ideas may be supported by the authority of "a church, a party, or the state." The obligation of the scientist to be a loyal member of his community was even higher than his obligation to science, and hence the occasion might arise when the public asks the scientist "to at least wait for a few decades before expressing in public his dissenting opinions." Heisenberg had lived according to these principles. One wonders if he ever recalled Aristotle's saying in the *Politics* that a man can be a good (loyal) citizen of a bad state—hence, a good citizen can be a bad man.

Heisenberg certainly realized that playing his conciliatory role meant making compromises, and he made many of them. Yet he was deeply committed to his society, his nation, and his cultural traditions; therefore, he believed that his compromises were justified. Thus, we see how clearly Heisenberg reflected the characteristics of his time and of his culture in his attempts to find continuity and harmony in intellectual life and to be a loyal member of his community.

Werner Heisenberg: Expansionist or Restrictionist?

Of all the scientists discussed so far, Heisenberg is the most difficult to classify in terms of expansionism and restrictionism because of the fact that he expressed rather different views about the relationship of science to values during the different periods of his life. The young positivistic Heisenberg cannot be classified as either a restrictionist or an expansionist, since he rejected the issue of the relationship of science to values as a meaningless question, one which in principle cannot be answered. Later, however, he took up the question quite explicitly in numerous essays. When he felt that science was under the greatest social pressure, as it was during the National Socialist period, he retreated to a restrictionist position, as many scientists do when they avoid political difficulties by saying that they will stick to science and let political and religious leaders handle ideological questions. Heisenberg's 1941 essay on Goethe was a masterpiece of restrictionist writing in which Goethe's intuitionist approach was given priority over the subjective worlds of aesthetics and values, and Newton's mathematical physics was defended as supreme in the objective world. Yet even in this essay Heisenberg revealed his belief that the division between objective and subjective was not absolute, and he admitted his hope for a day when the two worlds could be united. Heisenberg was thus never a true restrictionist; once he became interested in the place of science in general culture he saw that science was too important to be confined forever to a small realm. He knew that its concepts and findings influenced moral and aesthetic behavior. But he observed in 1941 that it was "premature" to hope for an "early return to a more direct and unified attitude to nature." Such an understanding could come only later.[70]

After the war, when the pressure upon science lightened, Heisenberg was more willing to extract social insights from the activity of science. In his first postwar speech he spoke of the "international understanding" to which science leads (although evidence of science bringing peace is singularly absent in history), of the "beauty" of science, and of its "incorruptible method," which always distinguishes the false from the true.

In his later years Heisenberg often supported a mild form of restrictionism, stating that science "deals with the objective material world," while "religion, on the other hand, deals with the world of values. It considers what ought to be, or what we ought to do, not what is."[71] Yet Heisenberg was clearly uncomfortable with this traditional bifurcation of the world even though it served him well. He admitted his nostalgia for the view of the medieval and ancient philosophers who believed that the adjectives "true" and "good" were intrinisically linked. He openly expressed his hope for a method of cognition that would allow one to see reality as the "unum, bonum, verum," the "one, good, true."[72] In other words, he yearned for a uniting of the factual and the normative, a breakdown of restrictionism. And he also often indicated his realization that modern physics was *already* affecting man's values.

At one point in *Physics and Philosophy*, he observed, "One sees that—after the experience of modern physics—our attitude towards concepts like mind or the human soul or life or God will be different from that of the nineteenth century."[73] However, he never sketched out clearly the paths and mechanisms of this influence, and, on occasion, as we have seen, fell back on the traditional restrictionist "is-ought" dichotomy.

Werner Heisenberg reflected in his long life the most intense scientific and political controversies of our century. He never found a thoroughly satisfying way of relating the worlds of science and values, but he was not alone in that difficulty, and few scientists have been put to such tests as he was in trying to establish a relationship between these realms.

chapter five

BERGSON, MONOD, AND FRANCE

EACH NATION has produced its own authoritative popularizers of science, and each of these writers has reacted not only to developments in science but also to those in his culture. Although Eddington, Fock, and Heisenberg are not the only choices of authoritative popularizers of science I could have made for the three countries concerned, they are among a small handful of those who were influential enough to be eligible. In France, two similarly significant writers on science in this century have been Henri Bergson and Jacques Monod. I will discuss their views on the relationship of science and values in this chapter.

The most influential interpretation of nature in France in the early years of this century was that of Henri Bergson (1859–1941), whose best-known book, *Creative Evolution*, was translated into all the major languages of the world. Bergson's philosophy of nature resonated with ideological currents of his period. The farther we advance in time from the first decades of the century the more obvious becomes the influence of intellectual fashions in his work and—even more—the importance of these trends in promoting the great eagerness with which his opinions were seized upon by society. It is with a touch of embarrassment now that we read the encomiums on Bergson of some of his contemporaries. Thus, William James acclaimed,

> Oh, my Bergson, you are a magician and your book is a marvel, a real wonder . . . a pure classic in point of form . . . such a flavor of persistent euphony, as of a rich river that never foamed or ran thin, but steadily and firmly proceeded with its banks full to the brim.[1]

And Édouard Le Roy began his 1912 *Une Philosophie Nouvelle: Henri Bergson* with the announcement:

> Without any doubt the work of M. Henri Bergson will be considered by common consent in the future to be among the most characteristic, the most fruitful, and the most glorious of our epoch. It marks a date which history will not forget. It opens a phase of metaphysical thought. . . . The revolution which it produces is equal in importance to the Kantian or even the Socratic revolution.[2]

Few people would describe Bergson in these terms today, but it would be a mistake to consider Bergson's philosophy of nature no longer interesting or relevant. The Bergson who was lionized by the press and the public was not the same as the one who wrestled with the epistemology of change and indetermination in *Introduction to Metaphysics* and *Matter and Memory*, works much less well known to general readers than *Creative Evolution*. As one critical admirer commented, "The Bergsonism of pure instinct, Bergsonism as the enemy of thought, dada-Bergsonism, is a caricature about as exact as the Socrates of the *Clouds* when he measures a flea's jump"[3]

Born in 1859 of a Polish Jewish father and an Irish mother, Bergson early demonstrated aptitude in both the sciences and the humanities. At the Lycée Condorcet and the Ecole Normale Supérieure in Paris he excelled in mathematics, while at the same time becoming thoroughly familiar with classical studies. His professional career as a philosopher began with teacher's posts in several lycées in Angers and Clermont-Ferrand. Following the publication of his doctoral dissertation in 1889 under the title *Essai sur les données immediates de la conscience*, Bergson returned to Paris, where at first he held a teaching post at the Lycée Henry IV. Then in 1900, in recognition of his rapidly growing stature, he was appointed professor of philosophy at the Collège de France. Bergson gradually became interested in public and international affairs, serving as the first president of the Commission for Intellectual Cooperation and in several other diplomatic roles. He received many honors, including the 1928 Nobel Prize in literature. In the last decades of his life Bergson stated that personally and morally he adhered to Catholicism, but he simultaneously maintained his identity as a Jew. As Europe was swept by the anti-

Semitism of the Hitler period, Bergson spoke less publicly of his attraction to Catholicism, and more of his Jewish identity, wishing to take his stand with those who were persecuted. When, near the end of his life, the Germans occupied France, he insisted on registering as a Jew, even though he was offered an exemption from the Nazi-imposed system.

Bergson's Form of Restrictionism

Throughout his life Bergson was absorbed with the question of the relationship of science to personal and moral life, that issue that defines the expansionist-restrictionist dichotomy. He had grown up in a time when positivism and scientific skepticism raged in French intellectual circles, and as a young man was thoroughly familiar with the works of Taine, Renan, and Spencer. He particularly admired Spencer's interpretation of the biological origins of the human psyche in evolutionary stages. In his late twenties, however, Bergson turned against these earlier models and developed a form of restrictionism which attempted to conquer science by mastering it and revealing its inadequacies. In that way he hoped to show that there are two different ways of knowing, the one based on the analytical methods of science assuming discontinuities in nature, the other relying on an intuition that is immediate and continuous. Bergson chose this path to defend ethical freedom and traditional values from what he saw as the assaults of positivistic science. This form of restrictionism, based on sharply contrasting but complementary paths to truth, was a familiar one in the history of Western thought, but Bergson by his brilliance, style, and insights gave it a new and extraordinary appeal. Furthermore, his knowledge of mathematics and attention to modern science lent authority to his writings. Thomas Hanna said of *Creative Evolution* that it became "one of the rarities of philosophical literature, a smash. If not actually read by everyone, it was, like the Bible, known by everyone and quoted by all."[4]

The appeal of Bergson's interpretation of science rested not only on his talents but also on the thirsts of his readers. In the first decades of this century a revulsion against nineteenth-century materialism swept France, as it did much of Western Europe.

Bergson's antideterministic message fitted well with the currents of thought running in French society, just as Eddington's did several decades later in England. French Catholicism spawned an unorthodox modernist movement that tried to defend religion by taking science into account, rather than ignoring or defying it as in the past. Bergson's ideas about the relationship of values and human freedom to science provided inspiration to members of this movement, for it recognized science while restricting its ambitions. Particularly through the person of Édouard Le Roy, Bergson's views were made more patently religious and in turn were influential in forming the thought of Teilhard de Chardin, another French writer on the relationship of science and values who will be discussed in a later chapter in this book.[5] But Bergson's appeal went far beyond the religious realm. He became important to France even in a patriotic sense during World War I, when he enthusiastically contrasted the French "spirit" or "élan" to German "materialism" and "militarism."

THE CONCEPT OF DURATION

The key philosophical concept upon which Bergson built his restrictionism was that of "duration" (*la durée*) by which he meant qualitatively experienced time during a continuous period in which something "new" appears: "The more we study the nature of time, the more we shall comprehend that duration means invention, the creation of forms, the continual elaboration of the absolutely new."[6] These novel forms were by definition unpredictable (else they would not be truly new), so Bergson's duration was distinctly different from the scientist's time t, used in describing predictable events. Duration was irreversible, continuous, and uniquely characteristic of the organic world, while scientific time was reversible, discontinuous (instantaneous time, dt, was literally a *denial* of duration to Bergson, and a thoroughly artificial concept), homogeneous, and fashioned to fit the world of inorganic solids.

The distinction that Bergson made between mathematical time and real time can be best illustrated by examining the conceptual difficulties which arise when one considers mathematical treatments of dynamic processes. Best known are Zeno's paradoxes, such as the one arising when one discusses the flight of an arrow.

If at each moment the arrow is at a certain spot, then no number of "spots" added together can amount to movement, since the arrow is conceived as being motionless at each spot. On the other hand, if one begins by conceiving the arrow in terms of its flight, then no comment can be made about where the arrow "is" at any one moment, since conceptually this would amount to arresting its flight. In this discussion, Bergson stood with those who believed that only change or movement is "real," and that therefore the mathematical treatment of moving objects by means of the calculus, in which the flight is divided into infinitesimal increments of time and space, is an "artificial" operation. According to Bergson, "the arrow never *is* in any point in its course."[7] But positive science, which is "concerned only with the ends of intervals and not with the intervals themselves," treats the arrow in this "unnatural" fashion, regarding time as if it could be infinitely divided. Scientific time, then, was a less valid concept than the real time experienced by conscious beings, where succession in one direction is an undeniable and irreducible fact. "If I want to mix a glass of sugar and water," said Bergson, "I must, willy-nilly, wait until the sugar melts. This little fact is big with meaning. For here the time I have to wait . . . is no longer something *thought*, it is something *lived*. It is no longer a relation, it is an absolute."[8]

In this quotation Bergson revealed the emphasis on subjective, direct, and individual experience that underlay his comparison of duration as an "absolute" quantity to scientific time as a "relative" one. In emphasizing the significance of the paradoxes inherent in scientific treatments of dynamic processes, Bergson recognized that nature must be studied not only in terms of its discontinuous aspects but of its continuous ones as well. But instead of calling for a combination of the two approaches as the only way to gain a complete understanding of the phenomenon of movement, he assigned a higher value to continuous duration as a "natural" concept, while discontinuous scientific time became "artificial." In so doing, he demoted the place of science.

Relying on his concept of duration, Bergson advanced a critique of "radical mechanism" and "radical finalism" as world views. By radical mechanism he meant the belief that all the natural world is potentially explicable by rigid mathematical laws. It was an opinion that had gained considerable currency in the nineteenth century, when Newtonian physics was often seen as lying at the bottom of all phenomena. Bergson contrasted radical mechanism with radical finalism, which he described as the doctrine that all natural events are proceeding in accordance with a plan or a program previously arranged. This plan is a goal which the universe is in the process of fulfilling.

Bergson pointed out that radical mechanism and radical finalism are essentially the same in that both imply that "all is given." Radical mechanism tells us that all can be calculated in advance in accordance with mathematical laws; radical finalism tells us that all is postulated in a master plan, which may not be mathematically calculable, but which is nonetheless a priori. Both radical mechanism and radical finalism, therefore, deny Bergson's "duration," in that nothing totally new, in principle unpredictable, is in the process of being created. Both disregard the action of time, since according to these views, the final state of the universe *could* be instantaneously unfurled without disturbing either the mathematical formulas or the master plan. Duration, on the contrary, "gnaws on things and leaves on them the mark of its teeth."[9] The *process* of real time was therefore essential to the configuration and essence of the things of the universe, causing them to follow nonrepetitive patterns.

BERGSON AND TELEOLOGY

Bergson was not nearly as suspicious of the teleology inherent in radical finalism as he was of the rigid causality of radical mechanism. He admitted, in fact, that his view of nature partook of finalism to a certain extent, and this finalism was built in to his conception of evolution as a result of an original impetus, or *élan vital.* According to Bergson, the impetus of life was "a need for creation."[10] Bergson's view contrasted, therefore, with that of most antievolutionists as well as that of Darwinian evolutionists; the former postulated "original forms," the latter "successive forms," while Bergson saw "original force" at the base of evolutionary transformism. The force of life strove for ever greater individuality and freedom, a struggle in which inert matter was the greatest obstacle. Not able entirely to overcome matter, the life force insinuated itself into inert matter and used it for its own purpose.

The fact that, according to Bergson, the *élan vital* had "purpose" reveals the teleological framework of his analysis. He spoke of the "effort" of consciousness to "raise matter" and of life's "search for individuality," thereby attributing a volitional quality to the life process.[11] But Bergson denied that his scheme was inherently teleological, because in his view the endpoints of biological evolution are unpredictable in the sense that evolution could go in any of a number of different directions.[12] Furthermore, Bergson did not attribute specific goals to the biological process, only indefinite "strivings" and "tendencies." Thus, Bergson's teleological flavor was always mild, although ever present and essential. Yet to a modern biologist, who sees purposive behavior in organisms as a result of previous evolutionary history, and not as the underlying cause of that evolution, Bergson's conception here seems both clearly teleological and scientifically wrongheaded.

Bergson advanced the view that the force immanent in life had originally possessed two different and divergent modes of knowing, the first the "instinctual" mode, the second the "intellectual." In man the intellectual mode had developed at the expense of the instinctual. The intellect was "fashioned by evolution during the course of progress; it is cut out of something larger"[13] Intellect is similar to a nucleus within a larger cognitive field of which instinct is the outer fringe. And Bergson believed that the crystallization of this nucleus in humans involved very real losses in the modes of knowing, since he considered intellect to be a partial and artificial cognitive mode. Intellect was, he thought, peculiarly adapted to the "unorganized" (inanimate) bodies of physics. Bergson described the character of a physical object as the image of a human's potential action upon it. As he wrote:

The distinct outlines which we see in an object, and which give it its individuality, are only the design of a certain kind of *influence* that we might exert on a certain point of space: it is the plan of our eventual actions that is sent back to our eyes, as though by a mirror, when we see the surfaces and edges of things. . . . The bodies we perceive are, so to speak, cut out of the stuff of nature by our *perception*, and the scissors follow, in some way, the marking of lines along which *action* might be taken.[14]

But since the goal of future action determines our intellectual faculty, it simultaneously *restricts* our vision, making it a partial vision. According to Bergson, only by returning to the original source of cognitive modes, the *élan vital*, and by recovering the instinctual mode, can we gain a full knowledge of reality: "While intelligence treats everything mechanically, instinct proceeds, so to speak, organically. If the consciousness that slumbers in it should awake, if it were wound up into knowledge instead of being wound off into action, if we could ask and it could reply, it would give up to us the most intimate secrets of life."[15]

A few recent interpreters of Bergson have attempted to show that he is not accurately described as antiintellectual or irrational in his emphasis on the instinctual mode of knowing, but merely aware of the deficiencies and conditioned nature of nineteenth-century mechanistic physics.[16] When he spoke of the limitations of the "intellect," they maintain, he actually meant the limitations of Newtonian physics. What he was calling for, they continue, was not a deemphasis of rationality, but a broadening of it based on acquaintance with realms of being in which mechanistic science was inadequate. These reinterpreters of Bergson have clearly exaggerated his faith in a new rationality and tended to regard with indulgence his romantic affection for the vitalistic and the irrational, but, at the same time, they have underscored one of the most fruitful concepts in Bergson's thought: his genetic epistemology.

GENETIC EPISTEMOLOGY

Essential to Bergson's study of man's way of knowing was the belief that evolutionary analysis should be applied not only to the physiology of animals but to their psychology as well. In the case of man, the ways in which we *think* are as surely conditioned by our evolutionary past as are the ways in which we appear. The joining of epistemology and biology (a more novel suggestion in Bergson's early years than it is today) emerges in Bergson's observation that

. . . *theory of knowledge* and *theory of life* seem to us inseparable. A theory of life that is not accompanied by a criticism of knowledge is obliged to accept, as they stand, the concepts which the understanding puts at its disposal: it can but enclose the facts, willing or not, in preexisting frames which it regards as ultimate. . . . It is necessary that these two inquiries, theory of knowledge and theory of life, should join each other, and, by a circular process, push each other on unceasingly.[17]

What does it mean to join a theory of knowledge to a theory of life? Bergson believed that the intellect in man had developed as "an appendage of the faculty of acting."[18] Hence, the intellect was a means of fitting the body to its environment, to represent external relations. It would, therefore, be reasonable to expect that the intellectual faculty would reflect in some important ways the conditions of its origins. The external environment of evolutionary man was, for example, only a narrow band within a much broader objective reality. The relations of external things within that narrow band might differ in important ways from the relations encountered as man began, at a later stage, to broaden the band of his observable environment.

This insight into a genetic or evolutionary epistemology can be applied to the development of scientific concepts. The environment within which evolutionary man developed was a Euclidean one. Even an animal, at the moment of pursuing prey, must take actions that inherently follow the rule that the shortest distance between two points is a straight line, or else it would be at a serious disadvantage in hunting.[19] Bergson's "logic of solid bodies" (which he equated with "intellect") was a logic that he saw as being conditioned by the environment in which man developed. A belief in a mechanical causality similarly arose out of the conditions of man's early existence. As man began to broaden his realm of observed experience, it became increasingly clear that many of his mental categories are "middle-band categories" situated between the microworld of the very small and the megaworld of the very large. The significance of the development of quantum mechanics and relativity physics, which study the micro- and megaworlds, may be that adaptation of the older categories was no longer possible, and entirely new ones became necessary. Bergson was not writing on genetic epistemology in response to these direct challenges from physics, but notice how appropriate his analysis was for an understanding of these developments:

> Our reason, incorrigibly presumptuous, imagines itself possessed, by right of birth or by rights of conquest, innate or acquired, of all the essential elements of the knowledge of truth. Even where it confesses that it does not know the object presented to it, it believes that its ignorance consists only in not knowing which one of its time-honored categories suits the new object. In what drawer, ready to open, shall we put it? In what garment, already cut out, shall we clothe it? Is it this, or that, or the other thing? And "this," and "that," and "the other thing," are always something already conceived, already known. The idea that for a new object we might have to create a new concept, perhaps a new method of thinking, is deeply repugnant to us.[20]

Scientists important in the development of quantum mechanics, such as Louis de Broglie, have praised Bergson for supplying a helpful framework for dealing with quantum phenomena.[21] This framework helps to explain why it is so difficult, even yet, for people to think in terms of wave-corpuscle duality or of a four-dimensional space-time continuum. These aspects of nature were never encountered in man's early history and it was not previously necessary for him to become adapted to them.

ROMANTIC NOSTALGIA IN BERGSON'S THOUGHT

Bergson used his insight into the conditioning of cognition to promote a romantic, irrational, and—yes, it must be admitted—antiintellectual philosophy. Instead of putting his emphasis on a broadening of intellect in the future through an attempted assimilation of the new realms outside the middle range to which man was accustomed, Bergson called for an attempt to capture from the divested regions of instinct powers of cognition that had been lost in the process of evolution. Throughout Bergson's writings there is a romantic nostalgia for the past, a sense of deprivation. Bergson yearned not for past history, as many romantics do, but for a lost way of knowing, that of instinct. As Bertrand Russell observed, "The division between intellect and instinct is fundamental in his philosophy, much of which is a kind of Sandford and Merton, with instinct as the good boy and intellect as the bad boy."[22] To Bergson, "The intellectual faculty . . . possesses naturally only an external and empty knowledge," whereas "instinct is sympathy" and "is molded on the very form of life."[23]

To Bergson, the original impetus of life was simply a given; he made no attempt to ask where it came from, and in that sense his philosophy was more akin to creationism than evolutionism, with the vitalistic life force the first creation. Furthermore, the "impetus is finite, and it has been given once for all." As life proceeds, the force is divided as wind divides when it encounters an obstruction. In direct contrast to naturalistic descriptions of evolution, he maintained that "Life does not proceed by association of elements, but by dissociation and division." Therefore, to Bergson, the world was harmonious but the "harmony is rather behind us than before."[24] Thus, the genetic epistemology which could have been used by Bergson with as much success in pointing toward an expanding rationality was, instead, employed in an effort to rescue lost elements of a fictitious past. Even if one could lend credence to Bergson's assertion of the lost powers of instinct, surely those powers would be even *more* conditioned by the middle-range environment of evolutionary history than those of intellect and would therefore likely be of negative value in handling the strange phenomena of the newly observed realms.

BERGSON ON RELATIVITY THEORY

Bergson's reliance on subjective cognition emerged prominently in his debates with physicists over relativity theory. The specific issue over which the differences occurred concerned the relativity of time measurement; the best illustration of this difference centers on "Langevin's Voyager." The French physicist Paul Langevin had asked what would happen, in accordance with relativity theory, if one assumes the following set of events: A person on earth named Peter launches a rocket at enormous speed on which another person named Paul is enclosed; after a long period of time, say, fifty years, the rocket returns to earth, and Peter and Paul compare their watches and their biological ages. Most qualified interpreters of relativity theory said that if the velocity of the rocket relative to the earth had been close to that of the velocity of light, there would be a very real difference in the two watches and the two ages. During the time that Peter aged fifty years, Paul might have aged only ten, depending on the relative velocity of the rocket.

This example was perceived by Bergson as a challenge to his opinion that organic time was an absolute while scientific time was only artificial and empty. In *Creative Evolution*, in fact, Bergson had entitled one section "Individuality and the Process of Growing Old," in which he had tried to show that the process of aging is a reflection of duration, and therefore expresses an absolute relationship. Aging was not simply a matter of organic destruction in terms of physics and chemistry; instead, "under these visible effects an inner cause lies hidden," a "continual recording of duration."[25]

In 1922 Bergson published what is almost universally regarded as his least successful book, *Duration and Simultaneity*, devoted to the significance of relativity theory to his doctrines, and in 1924 he engaged in a debate on the same topic in *Revue de philosophie* with André Metz.[26] In this debate Bergson was finally forced to fall back on the position that time is a purely psychical category. He bifurcated the cosmos into inner and outer worlds with psychical time expressing the natural relations of the inner world and scientific time expressing the "illusory" relations of the outer world.

Bergson wanted to hang on to an absolute—his inner real time or duration—and he therefore denied the time differentials that followed from the example of Langevin's Voyager. Yet relativity theory, despite Bergson's fears and many popular misinterpretations, did not postulate that "all is relative." As Bergson's opponent in the debates, André Metz, observed, in accordance with relativity theory "certain entities ('space' and 'time') considered until then to be objective properties of the external world, were henceforth to be considered as 'relative to the observer' . . . behind the properties which were becoming relative, others were rising up to restore the absolute, which science needs to establish its foundations."[27] The new absolute was the space-time continuum, but Bergson preferred his old absolute of duration, which was, however, too closely linked to ordinary time to withstand the assault of the new science.

BERGSON ON EVOLUTION

Bergson was deeply disturbed by the Darwinian view that evolution was the result of selection out of random variations rather than being the result of some kind of overall guiding principle of development. As he observed, "Where we fail to follow these biologists is in regarding the differences inherent in the germ as purely accidental and individual. We cannot help believing that these differences are the development of an impulsion which passes from germ to germ across the individuals, that they are therefore not pure accidents, and that they might well appear at the same time, in the same form, in all the representatives of the same species"[28] In seeking the principle which might cause uniform, nonaccidental variation, he appealed once again to the past, to the original impetus of life, which had a continuing "psychological" influence even after the evolutionary roads diverged: "If our hypothesis is justified, if the essential causes working along these diverse roads are of a psychological nature, they must keep something in common in spite of the divergence of their efforts, as school-fellows long separated keep the same memories of boyhood. . . . Something of the whole, therefore, must abide in the parts; and this common element will be evident to us in some way, perhaps by the presence of identical organs in very different organisms."[29]

In his conception of a guided evolution, Bergson was following a view similar to Lamarckism, still very powerful in France, but differing from it in several important ways. Lamarck had postulated an evolution based on the inherited effects of use and disuse of organs as a result of the efforts exerted by individual members of a species. The most celebrated example was, of course, that of the giraffe, which, according to Lamarck, stretched its neck to reach leaves on trees; this acquired characteristic was then supposedly inherited in subsequent generations. This opinion contradicts the neo-Darwinian view that giraffes with genetic endowments giving them longer necks had a greater survival capability. The gradual lengthening of the neck was a result of natural selection, not of an effort made by individual giraffes. Bergson, like Lamarck, also put emphasis on "effort," but he interpreted the phenomenon in his own way: "Neo-Lamarckism is . . . of all the later forms of evolutionism, the only one capable of admitting an internal and psychological principle of development. . . . And it is also the only evolutionism that seems to us to account for the building up of identical complex organs on independent lines of development. . . . But the question remains, whether the term 'effort' must not then be taken in a deeper sense, a sense even more psychological than any neo-Lamarckian supposes."[30] The inadequacy which Bergson found in Lamarckism was its emphasis on *individual* effort rather than a common striving of the species as a whole, based on the instinctual memory deriving from the original impulse.

BERGSON ON MORAL AND RELIGIOUS VALUES

Moral and religious values had been glimpsed in Bergson's works by many of his earliest readers, but it was not until 1932, twenty-five years after *Creative Evolution*, that Bergson explicitly addressed these issues. This 1932 book, *The Two Sources of Morality and Religion*, drew heavily upon his earlier ideas. It is best described as showing a deepening of earlier trends, not a deviation from them. Bergson still saw life as an effort to obtain certain goals by insinuation of the life force into brute matter; instinct and intelligence were still tools by which life worked toward those goals; and in social insects and human beings the greatest achievements of the life force had been realized. (The dual focus on ants and humans was characteristic of several significant interpretations of biology of the twentieth century, including Bergson, William Morton Wheeler, and E. O. Wilson, who will be discussed later.)

The title of *The Two Sources of Morality and Religion* already points toward its basic framework. According to Bergson, moral and religious systems have two origins: social pressure and the appeal of moral heroes and mystics. The first influence, social pressure, is a mixture of instinct and intelligence. The uniqueness of humans can be seen by comparing them to social insects, such as ants and bees, which are governed entirely by instinct. The societies of ants and bees are very complex, since multiple and differentiated sets of rules of behavior are being followed, but the observance of these rules is innate and mechanical. In human societies, on the contrary, a degree of freedom of behavior is present. In man nature has decreed cohesion by *moral obligation*, which is transmitted through social pressure as a constraining influence. Education, moral teaching, social approval and disapproval all help preserve society. Intellect now plays a certain limited role, ordering and systematizing individual rules of behavior, but underneath the layers of rationalization instinct continues also to be at work, protecting human beings from the dangerous "dissolution" that intellect along might bring.

If social pressure were the only source of morality and religion in man, Bergson continued, mankind would be permanently stalled at a certain "peak" of evolution, just as the social insects are stalled at their impressive but stagnant stage. Human societies that never break out of the social pressure stage remain, according to Bergson, "closed societies." All primitive societies are closed societies. Such a society is provincial in its views, always distrusting the outsider, incapable of progressing to a love of all mankind as opposed to love of kinsman or neighbor. The *élan vital,* however, is constantly seeking a new way of breaking out, of pushing man to a more refined stage of development; the push is toward the "open society" in which a higher, universal morality will prevail. What is the method by which the *élan vital* reaches toward this goal? Not that of intellect, reasoning, or science. These ways of apprehending reality are hopelessly limited, thought Bergson, by the fact that they are derived from the necessity of acting on the "exterior" of the things of the world. They can result only in more social pressure, but social pressure already reached the limits of its effectiveness in primitive society, where it was a powerful influence in the family and the tribe. What is needed to go further is not "pressure behind" but "appeal ahead," an influence that will draw mankind forward toward its universalistic possibilities.

The moral hero or mystic is the means by which the *élan vital* strains toward the new stage, and the appeal of the hero or mystic is the second source of morality and religion. The attraction of these individuals can cause a new morality to arise which is "open, dynamic, absolute and human."[31] What is the nature of this attraction? Emotion or feeling (*sensibilité*), replied Bergson, since neither instinct, habit nor intellect is powerful enough to have a direct action on will. The mystic who through appeal captures the will of mankind is a person who has broken out of the bonds of nature as it has existed until that time and has made direct contact with the creative *élan*. By both reaching back to the original life force while attracting the species forward he has provided the dynamism necessary for further evolution.

Bergson presented his views here in an explicitly religious way and it is clear that he had been influenced by reading the Christian mystics. He described the advent of Christ as the beginning of the passage from the closed to the open society.[32] He concluded his book with the famous description of the universe as a "machine for the making of gods."[33] Little wonder that some French Catholics began to see Bergson as a strong ally, and little wonder that he was so influential on the thought of Teilhard de Chardin, whom we will consider later.[34]

Sociopolitical Values in Bergson's Work

If one simply made a list of all the writers who have either condemned or praised Bergson, noting their political and social ties, and then tried to decide, on the basis of the list, what the sociopolitical implications of his doctrines were in the minds of his audience, one might conclude that Bergsonism could be equally well used toward any social or political goal whatsoever. He was castigated by many Marxists as a social reactionary, but he was praised by Georges Sorel, leader of French syndicalism, as a philosopher of revolution.[35] He was feared by some Catholics as an evolutionary monist, a denier of Thomas Aquinas, but he was praised by others as a person who was renewing the Catholic faith of a previously secularized world.[36] He was described by not a few critics as a person trying to prop up the old order in a period of decaying capitalism, but he was attacked in France by the right wing, and Walter Lippman once commented, "if I were interested in keeping churches, constitutions, and customs fixed so that they would not change, I should regard Bergson as the most dangerous man in the world."[37]

Obviously, there were elements of both conservatism and radicalism in Bergson's thought, and his most perceptive critics saw both sides. The Russian Marxist G. V. Plekhanov spoke in 1909 of the "materialistic ideas" inherent in Bergson's *Creative Evolution*. Plekhanov continued:

If Bergson had wanted to take these ideas to their logical conclusion, then, by virtue of his outstanding ability and his strong tendency to dialectical thinking, he would have illuminated the most important questions of the theory of knowledge with a bright light.[38]

On the other hand, the French Thomist Étienne Gilson wrote of his youth as "the age of Bergson," and further described it thus:

For the first time since Descartes and Malebranche, France then had the good fortune to possess that rare being, a great metaphysician. . . . [Bergson introduced] us to a new world as he himself was discovering it step-by-step. No words will adequately express the admiration, the gratitude, the affection we felt and still feel in our hearts for him.[39]

Yet the philosophical antipodes Plekhanov and Gilson were united in ultimately rejecting Bergson. To Plekhanov, Bergson "returned to an idealistic harbor after a few materialistic excursions,"[40] while Gilson described Bergson as "not a Christian," indeed, as a "pagan."[41] In a sense, both these critics were correct. That Bergson was a philosophical idealist is beyond doubt; that he refused to accept orthodox Christian theology is equally clear.

On an abstract level, one can see easily enough the contradictory aspects of Bergson's thought that attracted and repelled diverse members of his audience. Bergson was not a revolutionary—he highly prized the past and tradition, and his scheme of cosmic evolution put the greatest value on a reactionary reaching back to regain something that had been almost lost, the supposed power of intuition. On the other hand, Bergson was not a traditional conservative—he believed in constant mobility of all forms and orders, sought constantly the creation of novelty, and believed in the openness of the future.[42]

Yet we should not conclude from our recognition of the diverse implications of Bergson's thought that these tendencies all canceled each other, leaving the world with an ideologically neutral message. Bergson's views coincided more clearly with those of the people who wished, after a little repair work, to rescue the old assumptions of the European order than with the views of those who wanted to build something entirely new. He was much closer to Gilson's

Catholicism than he was to Plekhanov's Marxism. This tendency emerged particularly clearly in his late work, *The Two Sources*, but the antecedents of religious inspiration were present in *Creative Evolution*.

The American intellectual historian Arthur Lovejoy saw these trends almost twenty years before Bergson made them explicit. Writing in 1913, Lovejoy observed:

> When a man appeals from the understanding to intuition in ethical and religious matters, he is very likely to find that, at the end, his intuitive insight strangely coincides with the beliefs of his grandfathers. . . . The neo-romanticism of Bergson, while it has developed both a right and a left wing, inclines more strongly to the right than to the left. . . . The preponderant, though by no means the only, result of Bergsonism upon contemporary society has been to give new courage to obscurantism in all its forms and all its degrees.[43]

It is neither my goal, nor is it possible, for me to analyze here all the sociopolitical values or ideological implications that various interpreters have read into Bergson's doctrines. My concern is with the position that Bergson himself took on the relationship of science to values, which is in itself a question of considerable ideological impact.

Science was an important subject to Bergson, perhaps his greatest preoccupation. His position can be described as antiscientific, but only if one sees that Bergson attempted to turn science in on itself, to defeat it by mastering it, to restrict science by revealing the inadequacy of its methods, to rescue mystical and intuitive values by showing that their significance could be revealed by what he called "experience" and "the scientific method." Thus, Bergson denied that he was antiscientific at the same time that he did everything he could to beat back science's expanding powers.[44]

Bergson was not content merely to criticize the mechanism and scientism of the previous century. If that had been his only goal, his admirers would have had a point when they maintained that Bergson should be seen not as an opponent of science but as a promoter of the new twentieth-century science over its obsolescent nineteenth-century forms. Bergson, however, wanted to confine science in general, of whatever century, to demote its "way of knowing" while promoting another supposedly better extrascientific way of knowing. He asserted, "The intellect is characterized by a natural inability to comprehend life."[45] Like most other vitalists, he wanted to preserve a part of reality from the encroachments of science. This effort is pure restrictionism, but Bergson went beyond the classical form of restrictionism that would give both scientific and religious or mystical ways of knowing equal validity in their specific realms to an assignment of superior value to the nonscientific way of knowing. Intuition, not intellect, was the path toward this true "reality," for intuition was a nonverbal way of seizing the ineffable. Science, hobbled by the distorting effects of language and logic, could give only a pale substitute of reality.

In the face of the new challenges of science to ethics that came in later decades, Bergson's views were clearly inadequate. By taking the restrictionist approach of excluding the rational and the scientific from significant roles in influencing values, he was of little help to people who wrestled with questions of the impact of scientific theories and technological applications upon human values. Indeed, by asserting that the surest guides to the new "open society" were emotion, feeling, and the yielding to the appeal of the religious mystic, he opened the door to the possibility that people might give up critical analyses of ethical dilemmas and instead rely on blind intuition and emotion. Despite his well-known writings on evolution and his scientific tone, Bergson was engaged in a restrictionist flight from science.

Jacques Monod

It is interesting to compare the views of Bergson with those of one of France's subsequent authoritative popularizers of science, the Nobel laureate biologist Jacques Monod (1910–1976). Monod's best-known book *Chance and Necessity* never quite enjoyed the vogue of *Creative Evolution*, but it, too, was a best seller in France and abroad, and it attracted attention from a wide spectrum of readers. In the two generations that separated *L'évolution creatrice* (1907) from *Le Hasard et la Necessité* (1970) a great many changes occurred in both the science of biology and in popular attitudes toward it. In biology a whole new field, molecular biology, had been created.

Monod remarked that "To biologists of my generation fell the discovery of the virtual identity of cellular chemistry throughout the entire biosphere."[46] Monod believed that the rapidly tightening principles of molecular genetics provided the base for a new look at the philosophical implications of biology, and for consideration of the question of the relationship of science and values. The latter issue was one to which Monod devoted much attention, just as Bergson had done in his day.

In France Darwinism was met in the late nineteenth and early twentieth centuries with more coolness than anywhere else in Western Europe. The reasons for this attitude are complex, and can only be alluded to here: the existence in France prior to Darwin of a theory of organic change, *transformisme*, which allegedly explained evolution without natural selection in the Darwinian sense; a centralized educational and scientific bureaucracy which made intellectual change difficult; a certain French disdain for British "empiricism" and a readiness to explain Darwinism as an expression of peculiar British socioeconomic values; and a realization that Darwinism sorely lacked a true explanation of the causation of the phenomena it described.

Even when Bergson wrote his famous book on evolution, he stressed a number of modified Lamarckian elements, as we have seen, and he criticized the Darwinian concept of adaptation through natural selection as a "merely negative influence."[47] To Bergson, the progressive story of evolution could not be explained without reference to the effort of the positive life force to express itself through novelty.

The great acclaim in France for Monod's *Chance and Necessity* signified the widespread acceptance, at last, among educated lay readers of the Darwinian version of evolution. Monod defended the mechanism of natural selection in the bluntest terms, and called Bergson an "illustrious proponent of metaphysical vitalism." In his youth, Monod remembered, "no one stood a chance of passing his baccalaureate examination unless he had read *Creative Evolution*." Now, he observed with satisfaction, Bergson's philosophy "seems to have fallen into almost complete discredit."[48]

Monod differed with Bergson on cardinal points. Bergson's distinction between "inanimate matter" and the "life force" which tries to contend with it was to Monod an example of the old "illusion" of duality between mind and body.[49] Bergson's emphasis on evolution as the "principle of life" was to Monod an error: "For modern theory *evolution is not a property of living beings,* since it stems from the very *imperfections* of the conservative mechanism which indeed constitutes their unique principle." While Bergson refused to accept the idea that biological variations are random, preferring an overall guiding principle of development, Monod unmincingly announced, ". . . chance *alone* is at the source of every innovation, of all creation in the biosphere. Pure chance, absolutely free but blind, [is] at the very root of the stupendous edifice of evolution." And where Bergson saw intellect as only one path to knowledge (and not the most valuable one), Monod wrote that "objective knowledge is the *only* authentic source of truth."[50]

Monod based his analysis on the philosophical position of mechanistic reductionism. To him, "living beings are chemical machines." The idea of nonreductive levels of being in nature, popular in France among Marxists, was to Monod an unnecessary and useless assumption: "Nothing", he wrote, "warrants the supposition that the basic interactions are different in nature at different levels of interaction."[51]

In the last chapter of his book Monod turned to the question of the relationship of science and values. He maintained that "our societies are still trying to live by and to teach systems of values already blasted at the root by science itself." Both the Judeo-Christian tradition and Marxism were, to Monod, "animisms" which are no longer relevant or intellectually potent. Both these belief systems, said Monod:

> exist at odds with objective knowledge, face away from truth, and are strangers and fundamentally *hostile* to science, which they are pleased to make use of but for which they otherwise do not care. The divorce is so great, the lie so flagrant, that it afflicts and rends the conscience of anyone provided with some culture, a little intelligence, and spurred by that moral questioning which is the source of all creativity. It is an affliction, that is to say, for all those among mankind who bear or will come to bear the responsibility for the way in which society and culture shall evolve.[52]

Monod did not say whether he was referring to all forms of the Judeo-Christian tradition and of Marxism, or specifically to the orthodox and Stalinist forms common in France. One senses that he would have had little patience even for the more latitudinarian forms, but he did have a few kind words for these great traditions. Marxism, he acknowledged, has "on the face of it a more solid moral framework than the liberal societies boast, but perhaps more vulnerable by virtue of the very rigidity that has made its strength up until now." And Christianity, by making a distinction between the sacred and the profane, had prepared the way for the "principle of objectivity" which he saw as the basis of modern science.[53]

In stating that scientific knowledge had "devastated" traditional value systems, Monod was obviously making a strong expansionist claim for the power of science to impinge upon values. But while he granted science great destructive power over old value systems he was much more cautious about the constructive power of science to set up new values. He believed that there is only one true link between science and values that can be translated into an objective guideline, a rule of conduct. That link is the belief that "value is objective knowledge itself." Without making that ethical choice, Monod believed that science is impossible. And science, to Monod, was the first principle of a modern society, a principle that would lead to a "humane socialist society":

> Where then shall we find the source of truth and the moral inspiration for a really *scientific* socialist humanism, if not in the sources of science itself, in the ethic upon which knowledge is founded, and which by free choice makes knowledge the supreme value—the measure and warrant of all other values?[54]

Monod, then, countered Bergson's effort to restrict science to the status of one limited mode of cognition with his own claim that science was not only the sole source of knowledge but the "supreme value—the measure and warrant of all other values."

Monod accurately identified many of the obsolescent elements and metaphysical flights in Bergson's arguments but in the process he revealed logical gaps in his own. If science is to be the supreme value, and if science contains only one ethic, that of "objective truth," where did Monod obtain the ethical principles that assured him that a society that was both socialist and humane was a good society? "Humanity" and "socialism," in the senses in which Monod used these terms, are principles which many people would agree are "good," but that judgment does not flow out of a principle of "objective truth," the ethic Monod postulated as supreme. Might not "humanity," for most people, and "socialism," for many, be seen as higher goods than mere scientific knowledge? If so the system of priorities erected by Monod would be in trouble, at least in gaining wide acceptance. The arrogance of the scientist in assigning the highest value to his own profession would not escape the notice of his audience. Furthermore, a strong argument can be made that science, in the final analysis, is not an absolute value, but a social function that derives its value, both cognitive and practical, from its relationship to society.

The contrast between Bergson and Monod brings us to a crucial issue concerning restrictionism and expansionism which I shall raise again in the final chapter of this book. What alternative exists between a form of restrictionism like Bergson's which subordinates science to obsolescent and traditional systems of value beliefs—ideologies which are, indeed, vulnerable to scientific assault—and a form of expansionism like Monod's which dismisses older value beliefs as "disgusting farragos" and "animisms" while providing nothing to take their place other than an apotheosis of science itself?[55] I shall conclude in the final chapter that reductionists like Monod have presented us with a powerful research program but not with a system of values. Reductionism alone cannot solve the problem that Monod saw so clearly.

PART TWO

THE BIOLOGICAL SCIENCES

BIOLOGICAL KNOWLEDGE AND VALUES

chapter six

STUDYING ANIMALS TO LEARN ABOUT MAN

THE ADVANCES made in the twentieth century in the life sciences, and, particularly, in the study of man, were not tied to conceptual revolutions of the epistemological type in physics, but they were equally potent in their implications. The discernment of common regularities in all forms of life effected a major unification in twentieth century biology, as Monod emphasized in his description of how biology had been transformed during his lifetime. The rediscovery of Mendel's laws at the beginning of the century initiated a search for universal regularities and structures that continues to the present. The hereditary "factors" that Mendel discussed were soon defined in terms of "genes," which in turn were soon associated with material structures, the chromosomes. For a considerable period of time biologists were uncertain whether the material carrier of heredity was protein or nucleic acid, both of which are present in chromosomes, but the work of Avery in 1944 established nucleic acid as the likely candidate. The well-known description of the structure of DNA in 1953 by Watson and Crick was the culmination of this particular line of development and simultaneously the impetus to even more intensive research leading toward practical applications.

A comparison of the philosophical implications of the developments in physics and biology reveals that in a certain sense the two fields were going in opposite directions. Physics in the nineteenth century was mechanistic and synthetic, that is, it relied on perceptual models of explanation and it assumed that the same laws were valid throughout the whole field of physics. By the end of the first decades of the twentieth century, however, physics had given up its mechanistic commitment to perceptual models (theoretical physicists do not suggest that we try to picture in our minds four-dimensional space-time, nor, for that matter, a photon) and had also divided its earlier unified world into three connected sub-worlds: in the microworld of subatomic particles, the laws of quantum mechanics now governed; in the macroworld of everyday experience, Newtonian laws still were valid; while in the megaworld of cosmic distances and velocities, relativity physics was now the governing set of regularities.

In biology, in contrast, the emerging world was one of surprising unity, even simplicity. All living organisms, it turns out, from paramecia to human beings, possess a genetic structure of the same nucleic acids, the internal order of which, but not the constituents, differed from species to species. Furthermore, physics may have given up its mechanistic modes of description, but biologists were more and more frequently using mechanical terms such as "template" and "replicate." And, after all, the double helix had been modeled by Watson and Crick on a metal array of brackets affixed to a chemical laboratory bench. To be sure, the "mechanization" of the biological picture, like the "fragmentation" of the physical one, was only relative, since ultimately the laws of quantum mechanics were inherent in biology as well, but the fact remains that the greatest champions of mechanism after the mid-point of the twentieth century were molecular biologists, not physicists.

One of the effects of the unification of biology around a common genetic substance was the further narrowing of the gap between human beings and the rest of the organic world. Humans share DNA with the other animals, and the differences in structure between human DNA and that of humans' closest relatives among the primates appear almost insignificantly small. The same process of closing the gap between humans and the other animals was being hastened by developments in behavioral psychology, ethology, and animal behavior studies. Developments in behavioral psychology led increasingly to comparisons between the behavior of humans and animals. Success in conditioning animals led a number of behavioral psychologists to extend their conclusions to human beings. This extension implied an abandonment of the Cartesian division of fact and value, since scientific studies of human behavior necessarily include questions about the *origins* of the values possessed by people which affect their actions. Behaviorists and ethologists frequently differ fundamentally on the importance of heredity, but they agree in assuming that the origins of values can be discussed in factual, scientific terms.

Any phenomenon, such as consciousness, that has a naturalistic origin seems susceptible to a naturalistic explanation. Values themselves, to the behaviorists and ethologists, have such origins. In this way emerged the full range of problems that later received popular attention in Skinner's *Beyond Freedom and Dignity*. Although historians can discern the beginnings of this emergence in the last decades of the nineteenth century, it was not until the development of behaviorism, ethology, and sociobiology in later decades that the full meaning and difficulty of the problem became evident to a considerable portion of the educated population. It gradually became apparent that an absolute separation of human values from scientific explanations was no longer tenable. A new wave of expansionism, coming out of biology, was sweeping scholarship and popular literature.

The areas of biological and behavioral studies we shall examine in the next two chapters were all directed, ultimately, toward an understanding of human values, even though the actual subjects of scrutiny were often simpler organisms—rats, pigeons, ants, ducks, gulls, chimpanzees, baboons, and apes. Of course, the scientists studying these animals were also interested in learning as much as they could about the particular species under study, and they created an impressive body of literature on the behavior patterns of these species. But a striking feature of the great figures in these fields—J. B. Watson, B. F. Skinner, Konrad Lorenz, Niko Tinbergen, and E. O. Wilson—is that all of them assumed that the knowledge they were creating would illuminate the nature of man.

We shall see that these studies can be divided into different phases and emphases, even though the lines of division are not clean. At first, especially in America, it was assumed that "intelligence" and "learning" were independent subjects amenable to scientific study and that the basic principles of learning were the same in all animals. The best way, then, to find these principles was to study a relatively simple animal—a rat or a pigeon—under laboratory conditions. The emphasis here was on controlled environment; relatively little attention was paid, at least in the early years of comparative psychology, to genetic differences. This environmentalism fitted comfortably with American assumptions about the perfectibility of behavior if the right conditions were provided. The emphasis of the early behaviorists was not on what the animal "wanted" to do, but on what the experimenter could force the animal to do, however unrelated that behavior might be to the animal's normal environment.

Other scholars of animal behavior, particularly in Europe, paid more attention to genetic differences among various animal species and countered the American devotion to environmentalism with an emphasis on "instincts," a controversial word only vaguely defined in the early ethological literature. The ethologists also differed from the behaviorists in stressing that animals should be studied in their natural environments, not in laboratory settings. Only in that way, they thought, would the biologically based forms of behavior, the results of natural selection in native habitats, become visible to the observer.

In recent years a synthesis between the various approaches to the study of animal behavior has been taking place. With the rise of primatology and sociobiology the importance of genetics and the study of species in native surroundings have been recognized throughout the various fields of behavioral studies, but laboratory study of animals still retains value for certain problems, such as attempts at language acquisition by primates.

Analyses of the relevance of biology to human values involve three different levels: 1) biology, 2) social behavior, and 3) values. The most ambitious of the sociobiologists hope eventually to sketch a line of causation from the lowest level, the organization of DNA (e.g., a gene for "altruism"), to the highest, the value beliefs of human beings. At the present time this sort of ultimate reductionism is far away and there are many biologists who believe the goal is not a realistic one. Nonetheless, as the methodology of animal behavior and sociobiology has become more refined, the specialists in these fields have become more explicit about what they see as the relevance of their studies to human values.

Comparative Psychology

The field of "comparative psychology" may be said to date from the publication in 1864 of a book of the same name by Pierre Flourens, a protegé of the naturalist Georges Cuvier.[1] Flourens hoped that the field would combine neurophysiology of the Cartesian tradition with animal psychology as earlier represented by Réaumor, Leroy, and his mentor Cuvier. Like Cuvier, he emphasized the importance of laboratory analysis, a characteristic which remained strong in comparative psychology throughout its history. Also like Cuvier, he differed with the scientists who believed that the most important evidence for understanding the animal world could be gained from studying animals in their natural habitats, and considered laboratory study to be much more useful. Another school of thought in France at that time promoted field study; out of this group, centered around Geoffroy Saint-Hilaire and then his son Isidore, came early French work in the field of ethology.

Although comparative psychology continued to develop in France, England, Russia, and Italy in the last part of the nineteenth century, it found its real home in America. Starting with the publication in 1878 of John Bascom's *Comparative Psychology, or the Growth and Grades of Intelligence*, the field grew impressively in the United States. It soon was linked to the behaviorism of John B. Watson, in whom the environmentalist approach to both animal and human behavior reached its most extreme development, probably more extreme even than that of the Pavlovian school in Soviet Russia. Watson wrote his doctoral dissertation on an animal that was to become symbolic of an entire field, the white rat, and he assumed throughout his career that what one can learn from the study of animals such as the rat is highly relevant for the rest of the animal world as well, including humans. His object of study was the subject of "learning" itself, and he thought of that subject as a single, coherent body of principles that could be transferred to other circumstances than the one in which the tests were executed, and to other species. Watson's insistence upon disregarding "consciousness" and his consequent revolt against introspection in psychology brought a refreshing and energizing rigor to the field of psychology; as time went on, however, and the field of comparative psychology continued to develop on the basis of the assumptions Watson emphasized, its conceptual and empirical bases became more and more narrow. By the 1930s and 1940s many psychology departments in the United States were dominated by "rat labs" in which the same sorts of problems were investigated over and over again: animal intelligence, maze problems, learning processes, and stimulus-response experiments.

Underlying much of this work was a set of dubious assumptions that were identified particularly clearly in 1971 by Professor Robert Lockard of the University of Washington.[2] The comparative psychologists, he observed, believed that there is a phylogenetic scale, a linear arrangement along which animals such as chickens, rats, and monkeys could be arranged, with humans at the top. Differences among animals in terms of intelligence were only in degree, not in kind. Without this assumption, one could not hope to find out much about humans by studying rats or pigeons. Lockard also noted that the comparative psychologists considered learning to be the key to animal behavior, assuming that most behavior is acquired. Animals might have a few innate tendencies—hunger, thirst, sex drive, etc.—but these innate features were merely the unconditioned responses on which the richer behavior patterns were built.

One of the most influential efforts to break the orthodoxy of these assumptions was a now-famous speech (and later article) by Frank Beach to the American Psychological Association in 1949. Beach pointed out that from 1930 to 1948 more than half of the articles in the *Journal of Animal Behavior* (note the broad aspirations in the title) were devoted to one species, the Norway rat, even though the Norway rat "represents .001 percent of the types of living creatures that might be studied." He further noted that even *within* the problem of "rat behavior" American researchers had studied one issue to the exclusion of a host of possible alternatives. That one issue was "learning," and Beach attributed its emphasis to "the anthropocentric orientation of American psychology."[3] It was clear from Beach's analysis that if rats are not a valid basis on which to generalize about animal (including human) behavior, and if learning is not the single most important aspect of animal behavior, then American comparative psychology had worked itself into a dead end.

By the late forties information was already being produced that showed that no single animal species is valid as a basis for erecting generalizations about animal behavior; other information was coming to light that illustrated that learning plays a considerably smaller role in animal behavior than earlier had been thought to be the case. As a result, a strikingly different view of animal behavior from that of comparative psychology began to emerge. It became apparent that nonhuman animals display a "plurality of behaviors," and that most of these have genetic bases. Natural selection operates upon animal behavior in a way similar to its operation upon morphological features, and to be "behaviorally adapted" is as important a concept as to be "physically adapted." As Lockard observed:

> Pick any animal species at random, study its behavior in its normal habitat throughout its life cycle, and you will discover an intricate set of behaviors, many of them of almost incredible matching relationships to demands of the environment, like a lock and key. Study further and you will find that the social organization of the species has a genetic basis, and is an evolved adaptation to the particulars of the environment.[4]

This new approach, when fully absorbed, revealed the naiveté of the earlier approaches to animal psychology, which often implicitly assumed the existence of something called "intelligence" which could be assigned some single value on a linear scale stretching from simple organisms to humans. For example, one might ask, "Are birds intelligent enough to discriminate the shapes and other visual characteristics of objects?" within the framework of this linear assumption about intelligence without asking whether such discrimination is necessary for the particular bird. It turns out that some birds, for example, seagulls, cannot identify their own eggs and will as cheerfully sit upon a rubber ball or even a brickbat as an egg. Other birds, whose existence have been threatened by mimicry of their eggs by other parasitic species, are quite adept at identifying the shape and pattern of coloration of their eggs. An example is the hedge sparrow, preyed upon by the cuckoo. Natural selection played the crucial role in determining these varying responses. The hedge sparrow needed visual discrimination in order to survive, the seagull did not.

As knowledge developed along these lines, the old assumptions of comparative psychology were increasingly replaced by new ones that more and more took species differences into account. Among the new assumptions were the following: animals should be seen as specialized organisms contemporaneous with man, and as highly evolved in their own ways as man is in his; "intelligence" is thus not a single entity but an aggregate of special abilities best understood in terms of the ecological demands on any organism; and rats, as rodents, are so distant from humans in terms of evolutionary ancestry that they are of little use for understanding human behavior.[5]

The shift from the old assumptions of comparative psychology to the new ones of animal behavior amounted to a conceptual revolution for the fields involved. A constant element before and after the revolution, however, was the ultimate goal—that of the study of man. The comparative psychologists hoped to build the principles of psychology on the study of simple animals under controlled conditions; the animal behaviorists hoped to find these principles in more advanced organisms in natural conditions.

Before examining these later stages of animal behavior studies, however, we must turn to B. F. Skinner and Konrad Lorenz, two innovative representatives of older traditions.

B. F. Skinner

B. F. Skinner emerged from the schools of comparative psychology and classical behaviorism, but he was able to escape the demise of those approaches by introducing conceptual and experimental innovations. Although he did not pay much attention to heredity, he insisted that his approach was not antithetical to a genetic one. His fame as a representative of behaviorism was greatly enhanced by his skill as a writer for nonspecialists. His novel *Walden Two* became a best seller, somewhat belatedly, when it became a favorite item of reading for the counterculture of the United States in the nineteen-sixties. One suspects that the utopian and communitarian aspects of the novel were far more attractive to these readers than the determinist and reductionist philosophy that underlay it, a supposition strengthened by the fact that several utopian communities patterned after *Walden Two* soon departed from the authoritarian leadership described in the novel.[6]

Skinner posed with unusual clarity the question of the relationship of science to sociopolitical values, and he gave a strongly expansionist answer. He believed that such values can, in principle, be changed by "controllers" possessing the requisite scientific knowledge. And he sharply criticized some of the sociopolitical values widespread in Western society.

Skinner focussed on the conception of "freedom" in Western society. What do we mean, he asked, when we speak of being "free"? Skinner replied that the most common answer given in the Western "literature of freedom" was that to be free meant to be able to do what one wants to do, to be able to act without encountering restraints or prohibitions. He quoted Leibnitz that "liberty consists in the power to do what one wants to do" and Voltaire that "when I can do what I want to do, there is my liberty for me."[7] Out of such sources, said Skinner, the Western concept of "autonomous man" has been built. Within each person's physical body there allegedly resides, figuratively speaking, an inner autonomous being making free decisions.

Skinner believed that this literature of freedom was fundamentally mistaken. That literature always asks, "What prevents a person from doing what he wants to do?" It then usually suggests ways of overcoming the people or obstacles in the way of the desired action. Unfortunately, this traditional literature does not ask, continued Skinner, "What made the person want to do that particular thing in the first place?" If it turns out that the very desire to do a certain thing also has its causes, the person can hardly be described as a free agent. Skinner was convinced that such explanations can be found, although he said the explanations should be stated in terms of actions that are "functions" of environmental reinforcement, not in terms of "inner desires" acting as causes of other events.

Skinner conceived of human beings as reservoirs of possible behaviors (and here there was leeway for genetic influence) that are actualized in sculpted forms depending upon reinforcement. The acts that a person performs bring consequences, and if those consequences are rewarding to the person they will be reinforced; that is, the person will tend to repeat the antecedent act. Reinforcement is of two types: positive reinforcement, which is associated with direct reward, and negative reinforcement, which comes from the removal of an "aversive condition." An aversive condition, according to Skinner, is one that people try to avoid. An unrigorous follower of Skinner's doctrines might simply call "positive reinforcement" the arising of a "pleasant" event after a given act and "negative reinforcement" the avoiding of an "unpleasant" one, but Skinner always eschewed terms that refer to internal mental states, as the words "pleasant" and "unpleasant" obviously do. His aim was to describe behavior with the help only of externally observable information.

Negative reinforcement and punishment are entirely different phenomena in Skinner's scheme and should not be confused. Negative reinforcement comes when a person successfully avoids an aversive condition. If a child stops crying after its parent gives it a pacifier, the parent is negatively reinforced and is likely to repeat the action when the child cries again. Punishment, in contrast, is the imposition of an aversive condition by the controller. While punishment can control behavior, it is, in Skinner's view, a very inefficient means. Punishment is, for example, nonselective; the controller must punish all behavior except the desired behavior. Positive reinforcement, on the other hand, can be made selective with high precision.

As a result of these distinctions, Skinner criticized legal systems for being based on punishment; far better, he said, to reward "good" behavior then to punish "bad." Some of his followers believed Skinner's views pointed the way to a "non-punitive society."[8]

One of Skinner's significant achievements was to go far beyond the mere stimulus-response psychology of classical conditioning of the Pavlovian type. That form of conditioning emphasized highly specific physiological responses, such as salivation or eye-blinks, and has been condescendingly described by some people as "twitchism." Skinner used the term "operant conditioning" to describe his method, and contrasted his control of complex behavior (the operant) to Pavlov's control of the simple reflex (the respondent). Highly successful examples of operant conditioning demonstrated by Skinner were the complex and unusual maneuvers performed by conditioned pigeons (performing figure-eight maneuvers; guiding a missile toward a target; performing assembly-line functions).

The basic assumption behind operant conditioning was, then, the belief that an organism's behavior is shaped by the results of previous behavior. Skinner extended this assumption in full strength to man, and maintained that its eventual acceptance would mean a scientific revolution:

> Science has probably never demanded a more sweeping change in a traditional way of thinking about a subject. In the traditional picture a person perceives the world around him, selects features to be perceived, discriminates among them, judges them good or bad, changes them to make them better (or, if he is careless, worse), and may be held responsible for his action and justly rewarded or punished for its consequences. In the scientific picture a person is a member of a species shaped by evolutionary contingencies of survival, displaying behavioral processes which bring him under the control of the environment in which he lives, and largely under the control of a social environment which he and millions of others like him have constructed and maintained during the evolution of a culture. The direction of the controlling relation is reversed: a person does not act upon the world, the world acts upon him.[9]

The acceptance of this revolution is obstructed, said Skinner, by the literature of freedom and dignity, written by humanists who wish to flatter man by describing him as autonomous, when in fact he is under the control of reinforcing conditions. This literature, he continued, may have served some useful purposes in the past by helping humans to resist tyranny, but it has now outlived its usefulness. It now results in an uncontrolled individualism leading to the "catastrophes threatened by unchecked breeding, the unrestrained affluence which exhausts resources and pollutes the environment, and the imminence of nuclear war."[10]

Skinner's analysis of the question "What does it mean when we speak of human dignity?" followed from assumptions similar to those of his analysis of human freedom. He observed that people recognize a person's dignity or worth when they give him credit for what he has done. And the less obvious the reasons for a person's behavior, the more credit we are willing to give. The person who receives the greatest credit is the person who does a "good thing" for no compelling reason at all; thus, in the traditional view, the truly autonomous man is most worthy of credit, while the person who acts in an involuntary way deserves little praise.

Skinner emphasized that the supporters of this view of human dignity were automatic opponents of his behaviorism, for they made a virtue of ignorance of the sources of behavior. They said that so long as we do not know why a person did a certain thing, he deserves great praise (if we like what he did). But since Skinner's behaviorism was based on the attempt to identify the reinforcements that explain any given behavior, the success of his science would mean the removal of the reasons for the praise. Once we see that a person's behavior is determined by prior conditioning, it seems a bit pointless to express an appreciation for that behavior. And, indeed, in the society of Skinner's utopian novel *Walden Two* the very expression "thank you" had fallen into disuse. The residents had recognized the fallacy of the traditional "literature of dignity."

Skinner denied the customary modern separation of science and values, and he further believed that his operant conditioning could affect not only actions, but values. To the extent that behavioral science is concerned with operant reinforcement, he wrote, it is a "science of values."[11] His criticism of the Western concepts of "freedom" and "dignity" was already an attempt to alter certain values that he considered misguided. But his critique of values went far beyond personal preference; he believed that behaviorism could successfully expand into the province of values, just as it had into that of actions. Indeed, the first step, he believed, was to see that the categories of "actions" and "values" do not represent essentially different phenomena. Both can be discussed, he said, in factual terms.

Skinner noted that the traditional view is that values raise questions not about facts but about how people *feel* about facts.

The special character of values, so the argument runs, is that they deal not with what humans *can* do, but with what they *ought* to do. Science, we are told, is capable of handling the "can" questions, but not the "ought" ones.

"It would be a great mistake for the behavioral scientist to agree" with this standpoint, asserted Skinner. "How people feel about facts, or what it means to feel anything, is a question for which a science of behavior should have an answer. A fact is no doubt different from what a person feels about it, but the latter is a fact also."[12]

According to Skinner, the things people call "good" are those things that have acted as positive reinforcers. The "feelings" about which people talk are not essential matters; they are "by-products" of reinforcement. There is no important causal connection, he wrote, between the reinforcement that comes from a stimulus and the feelings which follow it. What counts, he said, are actions, not feelings.

Skinner then directly attacked the modern fact-value dichotomy by quoting the philosopher of science Karl Popper on the subject as follows and then giving his own critique. Popper wrote:

> In face of the sociological fact that most people adopt the form "Thou shalt not steal" it is still possible to decide to adopt either this norm or its opposite. . . . It is impossible to derive a sentence stating a norm or a decision from a sentence stating a fact.

According to Skinner, this position is invalid, and to him it represented "autonomous man playing his most awe-inspiring role." Skinner then gave what he considered an entirely factual account of the norm "Thou shalt not steal":

> Long before anyone formulated the "norm," people attacked those who stole from them. At some point stealing came to be called wrong and as such was punished even by those who had not been robbed. Someone familiar with these contingencies, possibly from having been exposed to them, could then advise another person: "Don't steal." If he had sufficient prestige or authority, he would not need to describe the contingencies further. The stronger form, "Thou shalt not steal," as one of the Ten Commandments, suggests supernatural sanctions. Relevant social contingencies are implied by "You ought not to steal," which could be translated, "If you tend to avoid punishment, avoid stealing," or "Stealing is wrong,

and wrong behavior is punished." Such a statement is no more normative than "If coffee keeps you awake when you want to go to sleep, don't drink it."[13]

Thus, Skinner believed he had reduced the norm "Thou shalt not steal" to a statement of contingencies equal in logic to the assertion "Don't drink coffee." Of course, the statement about coffee refers to "natural" contingencies (the link between caffeine and sleeplessness), while the one about stealing refers to "social" contingencies (the link between stealing and social disapproval or attack), but both refer to probable consequences of given actions. All other norms, such as "Tell the truth," could, according to Skinner, be similarly reduced to contingencies. Social norms can be given a general form, if one follows Skinner, that runs "If you are reinforced by the approval of your fellow man, you will be reinforced when you do such-and-such."

THE PROBLEM OF CIRCULARITY IN SKINNERIAN PSYCHOLOGY

Interesting as Skinner's analysis here is, it conceals a circularity of logic. The philosopher Max Black has noted that norms such as "Tell the truth" can be expanded, if one pursues Skinner's analysis, into the statement, "If you are reinforced by those who approve of truth-telling, you will be reinforced when you tell the truth." The answer is, "Of course," and the tautological nature of the analysis emerges.[14] Both honest people and liars would agree with this statement, and the honest people would go on telling the truth and the liars would go on telling lies. Skinner might temporarily save himself from tautology by saying that more people are truth-tellers than liars, and that therefore the danger of disapproval or attack is greater for the liar. But, if so, the tautology is only driven deeper, since one can then ask, "Why are more people truth-tellers than liars?" One is then back to the beginning of the circle of logic that started with the question, "What is the source of the norm 'Tell the truth'?"

A similar circularity of logic resides in Skinner's proposal to "design a culture" that will determine new values. Before criticizing that analysis, however, we should remind ourselves that circularity of logic is not a demonstration of falsity, but merely of inadequacy of purported proof. Indeed, however inadequate, methodologically faulty, and overblown is Skinner's expansionist approach to values, the traditional restrictionist responses to his views are even less convincing. Skinner raised the right questions, but the gap between his claims and his achievements—when one looks at humans, not pigeons—is enormous. Furthermore, he refused to face fully the dangers, political and intellectual, that a successful science of behavior would present.

Skinner called for the design of a culture that will determine new values. He urged mankind "to save itself," saying "the intentional design of a culture and the control of human behavior it implies are essential if the human species is to continue to develop."[15] But if we have become convinced Skinnerians, where, we may ask, do we get the motivation for change? If our values have been determined by our existing culture, where do we gain the new values by which even to think of designing a new culture? Once again, a circularity in Skinner's reasoning comes to the surface. Arnold Toynbee, while espousing a historical idealism even more unsubstantiated than Skinner's determinism, identified Skinner's flaw with precision:

> If a human being's genetic endowment and his social setting, between them, determine the whole of his behavior, how it is possible for a human being to have a policy? Yet Skinner does attribute policies to human beings. . . . Passages in the book [*Beyond Freedom and Dignity*] assume that some, at any rate, among the participants in a human society design cultural practices, create cultural environments, control the behavior of other human beings, and do these things by operating a technology of human behavior. These assumptions are surely incompatible with Skinner's fundamental thesis, which is that freedom is a delusion; that human beings have no power of taking the initiative.[16]

Max Black put the dilemma in which Skinner was caught more pithily: "The spectacle of a convinced determinist *urging* his readers to save the human race is bound to be somewhat comic."[17]

Skinner was aware of this problem, but knew that there was no escape from it. He demonstrated this awareness in *Walden Two*, when Frazier, the founder of the utopian community, admitted that its creation was more similar to Christian cosmogony then science.[18] In other words, the motivation for a change in values cannot be born within a mechanistic framework of conditioned values; it needs either an outside impulse or a different explanation.

Even if somehow that different set of values arises, however, the question of the origin of change in Skinner's scheme is still vexing. Let us assume that a new society has been established, with new values appropriately conditioned by the environment. Are not the inhabitants trapped in a static set of values? Carl Rogers, in his well-known debate of 1956 with Skinner, centered on this immobility of Skinner's system:

> If we choose some particular goal or series of goals for human beings and then set out on a large scale to control human behavior to the end of achieving those goals, we are locked in the rigidity of our initial choice, because such a scientific endeavor can never transcend itself to select new goals.[19]

We see that Skinner's attempt to overcome the ancient problem of free will was not successful. Instead of solving that issue, he was excruciatingly caught by it. While espousing a thoroughgoing determinism, he nonetheless exclaimed, "A scientific view of man offers exciting possibilities. We have not yet seen what man can make of man."[20] This is an unusual determinism. A sophisticated Marxist might have tried to identify a dialectic at work in the tension between determinism and voluntarism, but Skinner never attempted this approach. His mechanism and reductionism were too deeply rooted.

Under examination even Skinner's attacks upon "freedom" and "dignity" turn out to be criticisms only of some versions of freedom and dignity, namely, those that resist behaviorism. Skinner himself defended other, implicit concepts of freedom and dignity. On the next-to-last page of his most famous book he described the bright future he thought behaviorism could help bring about, a world of happiness, enjoyment, and dignity, one that "would promote a 'sense of freedom and dignity' by building a 'sense of confidence and worth.'"[21] By placing the troubling words in quotation marks he indicated that he was compromising with traditional language, but the very need to do so shows that Skinner, too, emphasized those characteristics of man that are most difficult to explain in terms of an expansionist methodology derived from the study of other animals.

The weakness of Skinner's philosophical analysis of the means of change within his system is connected with his political naiveté, with his lack of appreciation of the problem of power in human relationships. Skinner maintained that "the technology of behavior which emerges is ethically neutral. It can be used by villain or saint."[22] But what about the "contingencies" of which Skinner so often spoke? In a society where power is unequally distributed and where the powerful possess values with which Skinner disagrees (i.e., all existing societies), is it not likely that the new technology of behavior will be used in the service of goals which Skinner finds repugnant? Skinner's failure to confront this problem in his writings in the detail and with the care it demands was the single greatest omission in his works. Of course, Skinner was correct in saying that in principle a technology of behavior *could* be used toward humanitarian as well as inhumane goals. The important questions, though, are "Is it likely that it will be used primarily for humanitarian purposes?" And "How would we protect ourselves if it is not?" Skinner's talk of the need to have "countercontrol" by the "controlees" is not very reassuring. In the present world, where only a minority of governments could conceivably be called humane, and where none can be regarded as automatically and eternally so, an effective technology of behavior would be exceedingly dangerous. That fact is no reason to reject such a technology a priori. Indeed, one regrets that Skinner's methods for improving literacy, aiding the retarded, and for fulfilling other such goals are not even more successful than they, in fact, are (and some of the successes are striking). Nonetheless, the implicit dangers of behaviorism are grounds for regarding Skinner's political innocence as unhelpful, if not irresponsible; we need to look at behaviorism with all the critical awareness that we are learning to bring to bear upon other technologies.

A final characteristic of Skinner's approach is his tendency to look at the fear of control displayed by human beings as an irrationality. He observed:

> It is sometimes said that the scientific design of a culture is impossible because man will simply not accept the fact that he can be controlled. Even if it could be proved that human behavior is fully determined, said

Dostoevsky, a man "would still do something out of sheer perversity—and he would create destruction and chaos—just to gain his point. . . . And if all this could in turn be analyzed and prevented by predicting that it would occur, then man would deliberately go mad to prove his point."

There is a sense in which Dostoevsky may be right. A literature of freedom may inspire a sufficiently fanatical opposition to controlling practices to generate a neurotic if not psychotic response. There are signs of emotional instability in those who have been deeply affected by the literature.[23]

Why was Skinner not willing to entertain the possibility that this fear of control is a human characteristic as natural and inherent a feature as other human traits? If it is so widespread, it may not be an "emotional instability," but a characteristic necessary for human survival, and perhaps selected for in the process of evolution. Here we return to the importance of genetics, a topic that Skinner, expecially in his early period, gave little attention.

In some psychology laboratories where experiments are conducted with animals, a sign is displayed, "The animal is always right." In other words, if the results gained from an experiment seem perplexing to the experimenter, don't argue with the animal. Skinner was engaged in an argument with his animal. The human animal displays a remarkably consistent aversion to Skinner's definition of its nature. Yet Skinner insisted that this aversion is not an innate characteristic, but instead a by-product of a supposedly transient "literature of freedom."

Ethology

One of the areas of science that had the greatest impact on mid-century concepts of the place of man in nature was ethology, a discipline frequently defined as the comparative study of animal behavior. The major ethologists have made it quite clear that the extension of their study to human beings was one of their primary goals. This prospect understandably caused concern among a portion of the lay public and at least some scientists, those who wondered if the diminishing of the gap between human beings and the rest of the animal world could lead to policies and attitudes that impugn human dignity and support questionable ethical policies. The brilliance of the ethologists' achievements in understanding animal behavior gradually won over many initially critical readers. Whatever limitations and ideological biases may be inherent in the views of leading ethologists (and I will discuss several of these later), it is now clear that they have made significant contributions to our understanding of animal psychology and of hereditary factors in animal behavior, as well as raised questions about the possible existence of similar processes in human psychology and behavior—questions that deserve the most serious consideration. The Nobel Prizes awarded to three ethologists in 1973 were belated recognitions of these achievements.

The origins of ethology as a science are closely tied to late nineteenth-century biology. Darwin himself has been justly called a forerunner of ethology, since his interests went far beyond the comparison of physical, anatomical characteristics of animals to emotional and behavioral ones as well, as evidenced by his work *The Expression of Emotions in Man and Animals*. As Niko Tinbergen, a leader of the twentieth-century ethologists, later wrote, Darwin's "work contains much concrete material that we now call ethological."[24]

Darwin's successors did not follow his lead in the area of the study of animal behavior, in large part because his work pointed to more obvious and appropriate research in morphological studies: the further identification of the structural characteristics of species, the construction of phylogenetic schemata which portrayed the evolutionary histories of the plants and animals of the world, and the development of the fields of comparative anatomy and palaeontology. Those few scholars who returned now and then to behavioral studies within the framework of evolutionary theory (Morgan, Whitman, Jennings, Heinroth, Verwey, Huxley) failed to elicit a sustained response from their fellow scholars.[25]

Konrad Lorenz

The central figure in the development of modern ethology was the Austrian physician and zoologist Konrad Lorenz (1903–   ), a student of Heinroth. Not only did his ideas and research dominate ethology for many years, but the other prominent ethologists were often associated with him or his students. Niko

Tinbergen, the Dutch-English scholar who also played such an important role in ethology, was closely associated with Lorenz since the thirties and frequently cited Lorenz' influence on his work; Desmond Morris was Tinbergen's student, while Irenäus Eibl-Eibesfeldt was Lorenz' student.

These ethologists were products of the German-speaking scholarly world, either directly or indirectly. Central Europe was an area where Darwinian comparative studies had struck particularly deep roots in the late nineteenth century, and where evolutionary doctrines had been more enthusiastically extended to the study of man than perhaps anywhere else in Europe. Both Lorenz and Tinbergen later cited ideological reasons as additional factors for the greater hospitality of Central Europe to the hereditarian doctrines contained in ethology than the more democratic nation of the United States (and, to a lesser degree, Great Britain), with its preference for environmentalist views.[26] Whether these political factors were truly important in determining the locus for the origin of ethology is speculative, but the fact that the two leading exponents of the field believed them to have been important reveals in itself something of the ideological currents in the field.

Several American writers have accused Lorenz and other Europeans of disregarding early American work on animal behavior and thereby "usurping" the word ethology and giving it a particularly narrow meaning. Although John Stuart Mill used the term ethology in 1843, and Isidore Geoffrey St. Hilaire also did in 1859, according to Mary and Howard Evans it was really the American William Morton Wheeler (of whom more later) who "did much to establish the word *ethology* in its broader sense in a series of articles in 1902–1905."[27]

Between 1927 and 1972 Lorenz published approximately one hundred fifty major works, including a dozen books. His early publications were all on birds—jackdaws, sparrows, swallows, etc.—and he was particularly interested in categorizing their habits and the degree to which they could perform operations such as flying, pecking, feeding, grasping, and nest-building without apparent learning. The more he studied, the more extensive the list of instinctive behavior patterns (*Triebhandlungen*) in birds became. Again and again he emphasized how little plasticity and variability were present in such behavior. A great deal of the activities of the birds could be divided up into separate modal action patterns, each one performed as if by a programmed automaton, often in response to specific releasing stimuli. Lorenz also presented the extremely fruitful concept of "intercalation" (the intermixing of hereditary fixed action patterns with conditioned elements of behavior) such as occurs in the nest-building of jackdaws. He also advanced the concept of "imprinting," the process by which, at certain critical periods of a bird's life, specific behavior patterns can become fixed in response to certain stimuli. Thus, greylag goslings taken into human care after hatching become attached to a human being as a mother, following the person from place to place. A generation of Europeans and Americans became familiar with photographs of Lorenz being followed about his gardens and fields by a string of goslings.

In his earliest research works Lorenz made little effort to show that the results shed light on human behavior, although a few side comments indicated his awareness of the possibility of extrapolation. He observed, for example, that imprinting in birds seemed analogous to pathological fixations on the drive object in human psychology.[28] Writing in 1932 he modestly observed of his work on birds: "This provides us with a certain insight into the origin of such behavior as found in the highest mammals and man. We can only hope to gain an understanding of the more complex system by analyzing the simpler form, although much of the behavior of mammals, by virtue of its exceedingly complex structure, will never be open to analysis even with the new addition to the methods available to us."[29]

One of Lorenz' early articles on the influence of conspecifics on bird behavior attracted the attention of the young Dutch ethologist Niko Tinbergen, and in 1938 they jointly published a classic article on the eggrolling movements of the greylag goose. In this article the two authors showed that the behavior of the goose in recovering with its beak and neck an egg that has rolled out of the nest can be separated into a purely automatic movement ("instinctive behavior pattern") and a modifiable movement dependent on the movement of the object (egg). Thus, if the goose loses the egg at an early point in the process of rolling it back with its beak, it nonetheless continues the movement all the way in a purely mechanical fashion, oblivious to the actual presence or absence of the egg under its beak. However, the weaving of the goose's head back and forth depends on the shape of the egg; a rolling cylinder replacing the egg will not elicit the weaving response. Here seemed to be another persuasive example of Lorenz' concept of "intercalation."

On the basis of such careful and insightful research, Lorenz and his colleagues gradually built up an impressive body of ethological research. Niko Tinbergen contributed valuable researches on digger wasps, butterflies, and eventually on gulls, the species for which he is best known as a master ethologist. The German Karl von Frisch decoded the "dance" of honeybees, showing how the characteristics of the dance indicated the location of nectar-bearing flowers. Through the work of these scholars and others less well known, ethology gradually won its place among biologists. Ethology was, along with the development of genetics and molecular biology (with which it would later be linked), one of the breakthroughs in the biological sciences in this century; in the early years this achievement was largely overlooked by the geneticists, who were absorbed by their own exciting studies, and who often lumped ethology together with nineteenth-century descriptive biology. In the decades after World War II, however, recognition was won.

With growing recognition came controversy, a good part of it created by the ethologists themselves. After the war Lorenz wrote a number of popular books arguing the importance of instinct in human behavior. In these works we can often see the operation of political and ideological elements, as well as provocative high-order journalism.

Lorenz' best-known work was *On Aggression*, originally published in 1963 in German under the more revealing title, *The So-Called Evil: Toward a Natural History of Aggression*. Although Lorenz' *On Aggression* was read by hundreds of thousands of people, the popular image of the book is still somewhat erroneous. The meaning of the book is often thought to be the following: an animal with a hunting and conflict-ridden evolutionary history—man—is endowed with aggressive drives which are the result of natural selection in his earlier development and which are now passed on to him genetically. Despite his rational desires to overcome these violent instincts, no longer valuable to him in a state of civilization—indeed, maladaptive in a world of nuclear weapons—man cannot subdue his innate violent drives. Therefore, he remains a dangerous creature who may destroy himself. If he does, it will be a result of the evil legacy of his genetic past.

A careful reading of Lorenz' *On Aggression* will show, I believe, that this popular understanding is not loyal to the intention of the book, whose main message is far more pessimistic even than the popular interpretation of it. In Lorenz' opinion the main cause of man's predicament is not his instinctual drives—indeed, Lorenz deeply valued those instincts. The cause of man's present maladaption is civilization itself, which has developed so rapidly that the necessary instinctual controls have not been able to keep up with the changes in phenotypical behavior. Lorenz' criticism was not directed toward man's ancient animal past, but instead toward his recent human experience. The problem is not that man is instinctually a killer like the great carnivores, but that he developed murderous weapons more rapidly than he could develop by means of slow natural selection the instincts necessary for handling such weapons. The great carnivores that can "kill at one blow," such as the predator cats or the sharp-beaked birds of prey, possess these controlling instincts which man lacks and consequently rarely use their weapons on each other, or for purposes other than gaining food. Lorenz observed: "One can only deplore the fact that man has definitely not got a carnivorous mentality! All his trouble arises from his being a basically harmless, omnivorous creature, lacking in natural weapons with which to kill his prey, and, therefore, also devoid of the built-in safety devices which prevent 'professional' carnivores from abusing their killing power to destroy fellow members of their own species."[30] Man has many instincts, including an aggressive one, Lorenz believed, but not those appropriate for his new, frightening powers. To make the environmentalist assumption that man can be easily changed so that he may control these powers in new circumstances is, Lorenz maintained, a gross error.

Lying behind this interpretation of man's place in the animal world is a radical critique of Kantian ethics, somewhat similar to Skinner's; Lorenz noted that, according to Kantian ethics, the "goodness" of an act is a result of an unconditioned good will, a will that refuses to be subservient to factual circumstances. "If we were utterly logical Kantians," said Lorenz, "we would have to value most the man who instinctively dislikes us but who by responsible self-questioning is forced, much against his inclinations, to treat us kindly." Such a person would be most virtuous in a self-willed way. Yet this hope is fruitless, believed Lorenz, since it will eventually break down; to put a moral veto on human aggression "would be about as judicious as trying to counteract the increasing pressure in a continuously heated boiler by screwing down the safety valve more tightly."[31]

What can be done, then? Eliminate the aggressive drive by eugenic breeding or genetic engineering? Not at all, said Lorenz, since aggression is also the source of many of our most valuable characteristics (ambition, creativity, laughter). Furthermore, the very power of moral responsibility itself derives from the deep aggressive instincts, taking its energy, as does power steering in a car, from the "same kind of motivational sources as those which it has been created to control."[32] Thus, man's greatest hope to control his aggressive drives, according to Lorenz, comes not from his rational considerations but from his admittedly weak self-protective instincts. The hereditarian element is not something to be overcome, but something to be built upon and used, as the best available protection from disaster.

LORENZ AND POLITICAL VALUES

For all of the contributions that Lorenz made to studies of animal behavior, his extrapolated ideology contained some dubious and potentially harmful elements that were linked to his own political environment. The weakness in his argument is not primarily his emphasis on genetic elements in human behavior. Lorenz was probably correct in maintaining that democratic and egalitarian intellectuals, particularly in the United States, often exaggerated their environmentalist biases into a refusal to see that humans, as evolutionary products, are endowed genetically not only with obvious anatomical characteristics but less obvious behavioral capabilities or potentialities as well.

As we develop a deeper knowledge of "open" and "closed" behavior systems and evolutionary strategies, it becomes ever clearer that a direct opposition between "innate" and "acquired" behavior is false. What seems to be much more common in the biology of higher mammals is "innate ability" for learning in complicated fashions in which both heredity and environment play incredibly complex and subtle roles. All of this is a subject for further research, much of it going on at the present time. And it is not even clear that a strengthening of the argument for innateness would necessarily provide support for political conservatism, despite much educated opinion in that direction. Noam Chomsky, for one, has argued that man's innate factors may actually serve as a protection against authoritarian regimes, which cannot make of their citizens what they might wish, but instead must accept their innate human qualities.[33]

What was disturbing about Lorenz' views was the unrestrained degree of his expansionism, the extent to which he extrapolated a still only partially understood, quite ambiguous biological research program into a clear and consistent political and social vision. Furthermore, this vision, going far beyond the logical implications of animal behavior studies, was highly reminiscent of right-wing political views in Germany and Central Europe in the first half of the century, the time when Lorenz' world view was formed.[34]

If one reviews the political opinions expressed by Lorenz over the years, a pattern emerges. Writing in the National Socialist period Lorenz made an attempt to show that his studies into the innate releasing mechanisms and inborn schemata of animals were directly applicable to man, and that specific policy conclusions could be drawn from them. Underplaying the enormous differences between different species and widely separated phyletic levels, Lorenz maintained that civilization was having a disastrous effect on the genetic constitution of man, leading to a "degeneration" similar to that of domesticated animals. Lorenz here drew heavily upon the fears of genetic catastrophe that had played such an important role in eugenic thought in the twenties and thirties (not only in Germany but in many other countries as well). Lorenz maintained that degenerative mutations in domesticated animals lead to the loss of species-specific releaser mechanisms in mating patterns, with the result that the purity of the stock is not maintained. Only concrete policies by the state authorities could avoid this danger, he thought, and he praised the National Socialist regime for educating its youth to the importance of racial ideas and for establishing a race hygiene policy. In several articles in 1940 Lorenz adopted National Socialist terminology and helped create propaganda for the German state. In one article he emphasized the importance of biological instruction in the schools not only for the sake of science education but also because of its "general ideological significance."[35] In another, he described decadent art as an example of genetic degradation of human beings, and called for steps to reverse such trends:

> The selection for toughness, heroism, social utility . . . must be accomplished by some human institution if mankind, for want of selective factors, is not to perish because of domestication-induced degeneracy. The racial idea as the foundation of our state has already accomplished a great deal in this direction. The Nordic movement has been expressly directed against the "domestication" of man.[36]

We see that Lorenz used the assumptions of biological determinism to advance a wildly speculative and highly political vision of the disastrous effects of civilization that went far beyond any biological evidence and that supported inhumane state policies.

If this viewpoint of Lorenz' were merely an aberration linked to the exigencies of 1940, it would not deserve more than passing mention here. Let us turn, however, to a book Lorenz published in 1973, *Civilized Man's Eight Deadly Sins*. The consistency of the sociopolitical message between 1940 and 1973 is clear.

Civilized Man's Eight Deadly Sins is a deeply pessimistic book about the future of mankind in which Lorenz inveighed against urban civilization, against industry, against radical youth, against the "caricature of liberal democracy" that he saw leading to the decline of the West. An entire chapter of the book is devoted to "Genetic Decay," and he concluded "There is no doubt that through the decay of genetically anchored social behavior we are threatened by the apocalypse in a particularly horrible form." As evidence for the genetic decline, he cited the growth of crime and violence. He repeated his assertion of 1940 that the genetic decline that occurs in domesticated animals is similar to that allegedly occurring in civilized man:

> We know from our domestic animals, even from wild forms bred in captivity, how quickly social behavior patterns disintegrate when specific selection is missing. In several fish species, bred for a few generations by commercial dealers, the genetic pattern of brood tending is so disturbed that, among dozens of fishes, one barely finds a pair still capable of caring for their young. . . . Countless young people are hostile to modern society and to their parents. The fact that, in spite of this attitude, they still expect to be kept by this society and their parents shows their unreflecting infantilism.
>
> If the progressive infantilism and the increasing juvenile delinquency are, as I fear, signs of genetic decay, humanity as such is in grave danger.[37]

Between 1940 and 1973 Lorenz retained his basic assumptions, but he had changed his mind on the means to achieve this constant goal of halting genetic degeneration. In 1940 he thought interference by the state was necessary; in 1973 he relied on individual action as the only hope. He insisted on the difference, and lamented that "in speaking of human beings, even the words 'inferior' or 'valuable' cannot be used without arousing the suspicion that one is advocating the gas chamber." And he concluded his chapter on the danger of genetic decay with an underscoring of the importance of individual responsbility: "To prevent the genetic decline and fall of mankind, all we need to do is follow the advice . . . I quoted earlier. When you look for a wife or husband, do not forget the simple and obvious requirement: she must be *good*, and he no less."[38]

We now know enough about Lorenz to see that for all of the importance of his scientific work on animal behavior, he is a dubious source at best for prescriptions for human society. Nowhere in his discussion of social aberrations, crime, and psychopathology did he give full credit to possible nongenetic factors contributing to these phenomena, such as social and economic conditions. And his description of the coming genetic apocalypse would be rejected by many fully qualified geneticists.

But it is important to notice that biological explanations of social behavior are not inherently and necessarily fascist or repressive. We shall see in Chapter 8, where I discuss eugenics movements in Weimar Germany and Soviet Russia, that strikingly different conclusions can be drawn from biological theories of human behavior. Whether human beings have innate biological "aggressive" and "altruistic" drives are controversial questions with social implications, but we already know enough about the merits and demerits of various types of government to know that no answer to these biological questions could lead uniquely to a specific political ideology or form of government. The ways in which Lorenz and a number of other ethologists tried to answer these questions revealed something dangerous about their own ideologies, but that fact should not screen from us the growing importance of biology to an understanding of man. As one of Lorenz' most eloquent critics, Leon Eisenberg, remarked, "Let us agree: behavior, like structure, is under genetic control."[39]

The control that genetic structure exercises on human behavior is most striking for its latitude, for what it permits, rather than what it prohibits. As Stephen Gould has emphasized, instead of speaking of biological determinism in humans, one might better speak of "biological potentiality."[40] It is not too much to say that ethology explains best about man what is least distinctive about him. It uncovers the universals between man and the other animals, but, as a field of study based on comparison, it is least useful in explaining the distinctive. The autonomy of the social sciences and the humanities is not threatened by ethology, not because the distinctive aspects of man are outside the realm of scientific study, nor because man contains a mystical or religious essence, but simply because the level of complexity of human behavior can not yet be approached by the tools of ethology.

chapter seven

PRIMATOLOGY AND SOCIOBIOLOGY

IN THE late 1950s a new type of animal behavior studies emerged. Unlike the comparative psychologists, the new specialists put a strong emphasis on heredity. And unlike the ethologists of Lorenz' generation, they utilized the quantitative science of population genetics.

The new animal behaviorists shared with their predecessors a preoccupation with man. Instead of studying species as distantly related to man as the rat, however, they emphasized two other types of comparative studies: those of species closely related to man, the primates; and those of species existing in similar environments and subject to similar selection pressures, such as the pack-hunters.

Primatology and Related Studies

The study of the primates was the most promising avenue, and in the last thirty years primatology has grown enormously. According to one scholar who surveyed the research being conducted on primates, field studies in primatology, measured in cumulative man-months of research, grew over tenfold in just one decade, from 1955 to 1965.[1]

Predecessors such as Robert M. Yerkes, who studied primates in the twenties and thirties largely within the older comparative psychology tradition, laid the groundwork for the burgeoning field. Primatologists in the postwar period put great emphasis on field studies. Significant research on primates in recent years has included Schaller's study of mountain gorillas, Jane van Lawick-Goodall's of chimpanzees, Irven De Vore and S. L. Washburn's on baboons, and Phyllis Jay's on langur monkeys.[2] All of these were examinations of primates in their own habitats, studies made in the field under conditions that were as natural as possible for the subjects of study (and often rather unnatural for the human investigators!). In addition there have been laboratory studies of primates which have added to our knowledge of their natural behavior, such as the well-known studies by H. F. and M. K. Harlow on the effects of social and maternal deprivation on rhesus monkeys.[3] These researches have added tremendously to our knowledge of primates.

Citation of just a few of the findings will give some indication of the interest the studies contain for the general public:

Tool use: Chimpanzees use sticks to extract termites from holes and use leaves for absorption of water and occasionally for personal cleanliness.

Aggression: Primates frequently express aggression through standardized threat postures toward each other. Threats and dominance relations seem to be adaptations to avoid violence. Early studies seemed to indicate that intraspecific violence among adults was rare, but later ones showed that primates sometimes kill each other.

Dominance: Dominance hierarchies are common among primates, but vary considerably in the strength of their expressions, some primates being tense (e.g., rhesus monkeys, baboons), others more relaxed (langurs). Dominance relations are often expressed sexually, in both heterosexual and homosexual forms.

Territoriality: Groups of primates have home ranges, but some intermixing occurs. Vocal spacing mechanisms ("whoops," "hollers") are used by some primates to determine relative positions of groups.

Parent-offspring relations: An affectionate bond between mother and offspring appears to be important for the future development of the infant primate. Deprivation of such a bond in rhesus monkeys lead to severe maladjustment in the infant. Among many primates, the males are not involved in care of the infants; in others, they are. Juvenile females often "practice" the skills of caring for an infant before they become mothers themselves.

Grooming: Individual bonds are formed and dominance relations expressed through mutual grooming, which therefore seems to have both a social and a physical significance.

Innate Releasing Mechanisms: There is evidence for the existence of innate releasing mechanisms in primates. Rhesus monkeys raised in isolation react strongly to pictures of other monkeys in a threat posture and also to pictures of infant monkeys. Visual and olfactory stimuli related to sexual activity are probably also in this category.

Learning: Primates learn from the group. Experiments have shown that ape and monkey mothers raised in isolation do not know how to care for infants. In addition, there is some evidence for "cultural evolution" across several generations in terms of food habits.

Maximization of Individual Gene Contributions: Dominant males in at least one species have been observed to kill the offspring of other males of the same species. An evolutionary explanation of such behavior would be the effort of dominant males to maximize their contribution to the gene pool of the subsequent generation.

It is obvious that such findings are at once a significant step forward in our knowledge of primates and a potential source for speculative mischief and even socially dangerous conclusions. Already a rash of publications extrapolating such conclusions to human beings have appeared in the press. Examples of such speculations are: Male baboons have heightened shoulders which they display in threat postures; ergo, the padded suitcoats and military epaulettes of man must be lineal descendants. Many primates express dominance relations by the sexual mounting of subordinate males by dominant ones; therefore, the homosexual activity in many prisons in which meek males are exploited by aggressive ones must be an evolutionary continuation. A strong affectionate bond between mother and infant is essential for normal development of rhesus monkeys; therefore, working mothers are violating nature and damaging their children when they are absent from home. Primates often have home ranges; therefore, man is a territorial animal who defends both his turf and his possessions out of evolutionary urge. All primates that have been studied display aggression; therefore, man is naturally an aggressive animal who can never be expected to follow the precepts of cooperation and altruism, nor be persuaded by rational arguments for avoiding wars. Primates behave in such a way as to maximize their contribution to the gene pool; therefore, population restriction measures run counter to man's inner urges. All these arguments and many, many more of similar logic have appeared in recent years. All are based on the questionable assumption that the behavior of animal primates and that of human beings are controlled by the same factors.

Although the growing field of animal behavior studies has provided much of the essential background for the popular literature about the primeval roots of human behavior, one should also note that the same field is increasingly providing materials for the criticism of these same popularizations. Furthermore, criticism of rash speculations has caused primatologists to be somewhat more cautious in their claims.[4] The field of animal behavior has matured greatly in recent years. On the one hand animal behaviorists have convinced many of the social scientists who were strong environmentalists that genetics is immensely important to animal behavior and probably important to human behavior. On the other hand, as more and more diverse information has been gathered, many of the earlier generalizations have been recognized as gross simplifications. Indeed, it seems likely that the high-watermark in the rise of speculative works by researchers like Lorenz and Morris has already been passed.[5] The variety of patterns of behavior now known even among primates (to say nothing of other animals) on such topics as dominance, sexual dimorphism, aggression, and social organization is large enough to throw into doubt the earlier, grand speculations. For example, Lorenz' earlier belief that only man among the higher mammals kills fellow members of the species (allegedly because of the disparity between man's genetic and cultural evolutions) has been discredited by better knowledge of intraspecific violence among primates.

It is also useful to remember that one of the principles of animal behavior studies previously mentioned was that there are *two* types of comparative studies that may have some validity: those comparing closely related species and those comparing species in ecologically similar situations. Primate studies have focused primarily on the first type. Monkeys and apes are man's most closely related species in evolutionary terms, but it has been millions of years since man was a tree-living animal and since he faced an environment similar to that of the other presently living primates. During the extremely long recent period (since the late Tertiary) man lived in a hunting society where the environment and the selective pressures were quite different from those surrounding the present primates. (It is true that some nonhuman primates—e.g., baboons and chimpanzees—occasionally hunt and that gorillas spend most of their time on the ground, but their environments, needs, and customary feeding habits are strikingly different from those of primitive hominid hunting societies.)

The fact that man is the only primate that became adapted to a way of life based primarily on hunting is, by itself, reason for hesitation in extrapolating from monkeys or apes to man. The transition from the situation of a tree-living primate to that of a ground-living hunter brought with it changes in many, many characteristics. Dominance in apes and monkeys, for example, is often expressed in ways tightly connected to feeding habits. A subordinate monkey yields place to a dominant one in taking a fruit. Food gathering among such monkeys is highly individualistic, with low emphasis on cooperation. When early man entered a closely organized hunting stage it seems likely that many of the characteristics of earlier primates were softened (dominance, promiscuity, perhaps territoriality) and likely that other characteristics were heightened (cooperation, communication, family unit formation). These changes, enormous in evolutionary terms, built up flexibility and reservoirs of options, and decreased the validity of close analogies to past patterns. From the standpoint of this analysis, man can certainly not be described accurately as merely "the naked ape."

Several animal behaviorists have been so struck by the significance of the hunting and ground-living stage that they have turned to animals quite different from the primates in order to gain social insights on humans, the pack-hunters. William Etkin wrote, "If the general thesis of this book is correct and behavior systems are easily modifiable in evolution by ecological demands, then we should look more to the wolf than the macaque [monkey] as a basic model for understanding protocultural man."[6]

And a switch of attention to the pack-hunters does, indeed, reveal, at least at first glance, some interesting differences from the nonhuman primates. Wolves, for example, share their food, in contrast to primates. Wolves cooperate during the hunt, alternating in chasing prey and pressing the attack. They have dens where the young are left behind during the hunt (the young of monkeys and apes travel with the group). The wolf pack seems to be based on family units, with lifelong male-female bonds apparently the rule (monkeys and apes are sexually promiscuous). Females hunt with the pack when not burdened with advanced pregnancy or newborn young. Dominance relations seem weak. Adult wolves return to the den to give food to their young, sometimes bringing the food in chunks, sometimes regurgitating it. As one animal behaviorist commented, "This is especially important for the young and for lactating females. . . . Here is one origin of altruism. Giving part of a kill to others entails not only restraint but also planning ahead. . . . Dominance, which usually involves prior right to food as to other desirables, is now inappropriate."[7]

Yet it soon becomes clear that the behavior of wolves also serves poorly as a model for human evolution. What has been gained in ecological similarity has been lost in genetic distance. Benefiting from many of the physical and behavioral characteristics of their primate past, early humans were so remote from the other pack-hunters in characteristics such as tool-making dexterity, mode of locomotion, olfactory and visual capabilities, and antecedent behavioral patterns that, once again, close analogies become difficult. If some animal behaviorists have emphasized the importance of the study of wolves in informing us about early man's development, others have downplayed that relevance, as the following quotation illustrates:

The importance of seeing human hunting as a whole social pattern is well illustrated by the old idea, recently revived, that the way of life of our ancestors was similar to that of wolves, rather than that of apes or monkeys. But this completely misses the special nature of the human adaptation. Human females do not go out and hunt and then regurgitate to their young when they return. Human young do not stay in dens but are carried by their mothers. Male wolves do not kill with tools, butcher, and share with females who have been gathering. In an evolutionary sense the whole human pattern is new, and it is the success of this particularly human way that dominated human evolution and determined the relation of biology and culture for thousands of years. Judging from the archaeological record, it is probable that the major features of this human way, possibly even including the beginnings of language, had evolved by the time of *Homo erectus*.[8]

The fact of the matter is that there is no other species of animal that is close enough to man in *both* biological and ecological respects to provide a very trustworthy analogy for explaining man. Thus, the contribution of the more sophisticated specialists in animal behavior has been to emphasize both the similarities and differences of man from the other animals.

Frequently in the history of science the sudden appearance of a new field has led in an early stage to the overextension of generalizations from that field to other related ones. Although these efforts frequently look naive in later decades, the efforts were in many ways understandable and even justified, since only in that way can the benefits and limitations of the analogical transfers be revealed. A danger for society exists at these moments, however, for such speculations at a moment of weak scientific control sometimes give room for the expression of political, racial, sexual, or class biases and actions. A society that is not under heavy social stress and that values and protects free discussion can weather these trials and emerge much stronger for them. Animal behavior study is, seen as a whole, a very significant and informative scientific achievement, and one still developing at a rapid rate. The responsible position for the rest of society is to make its insights a part of general education while recognizing the natural limits for its generalizations. Broad sociopolitical conclusions based on animal behavior should be challenged on the logic of their transitions from nonhuman animals to man, and questioned on the possibility of bias.

Sociobiology

The most comprehensive and authoritative delineator of social and human significance in biology in the 1970s was the Harvard entomologist, E. O. Wilson, whose 1975 book *Sociobiology* provoked a great controversy that entered into the pages of many general-interest publications.[9] Wilson was praised and criticized, hailed and condemned. In some quarters the discussion became primarily political, and Wilson was castigated for advancing conservative, establishment views that potentially could hinder social reforms.

I will give my view of the validity of these criticisms below, but first I wish to approach the problem of interpreting Wilson's work from quite a different direction, as I think this avenue is the most fruitful one. Instead of initially concentrating on the place of this one book in an ideological context, I would like to consider the development of Wilson's views over time and some of the main interests and motives that brought him to his major topics. The later analysis will then be built on this foundation.

A helpful approach to understanding Wilson's work is to examine his three major works of the seventies together, the 1971 *The Insect Societies*, the 1975 *Sociobiology*, and the 1978 *On Human Nature*. The 1971 book is important for understanding the 1975 and 1978 publications, and, indeed, for appreciating the overall direction of Wilson's intellectual development. And Wilson himself spoke of the "momentum left from writing *The Insect Societies*" which helped him "to attempt a general synthesis."

Although the books differ greatly in the comprehensiveness of the material covered, they have in some respects organizational or "morphological" similarities. Each book discussed a certain topic in great detail and then in its last chapters issued a call for extension of the analysis into a much larger area. Thus, *The Insect Societies* was almost entirely a discussion of the social insects, but the last chapter addressed a much more ambitious task, "The Prospect for a Unified Sociobiology," and contained in embryo the plan for the 1975 book. *Sociobiology*, in turn, was almost completely a description and analysis of invertebrate and vertebrate zoology and behavior, but the last chapter was entitled,

"Man: From Sociobiology to Sociology." It contained an ambitious expansionist call for the "biologicization" of ethics and sociology.[10] *On Human Nature* dealt to a large degree with ethics and values, but the last chapters were on "religion" and "hope." If the logic of the trilogy were continued, the next book would be on religion. And, indeed, Wilson stated in the third book that "religion constitutes the greatest challenge to human sociobiology and its most exciting opportunity to progress as a truly original theoretical discipline."[11]

In the research that led to these books, Wilson grounded himself in theoretical conceptions about the significance of the animal world for larger, ultimately human problems, but the nature of these conceptions changed in an essential way during the work. Indeed, I will speak of two different models that served as frameworks for his work, although they can be placed on a conceptual continuum on which the discrete two models are merely convenient points of analysis. What did *not* change was Wilson's desire to have such a model as a goal for his work, that is, as an illustration of the relevance of his work for large human concerns.

Entomologists have often been considered by other scientists, even by other biologists, as somewhat narrow empiricists, people who collect and classify a staggering variety of insect forms, but who have not made many theoretical contributions to the science of biology as a whole. In this century geneticists and molecular biologists have usually considered themselves much closer to the theoretical core of biology.

It seems likely that this perception of entomologists has strengthened the desire of many of them, including several of the most talented, to illustrate the theoretical relevance of their work. Wilson's illustrious predecessor at Harvard, William Morton Wheeler, wrote fascinating articles and books that are best described as "biological philosophy" as well as many classical, rigorous studies of ants and social insects. One of Wheeler's philosophical ideas, to be discussed below, had a strong influence on Wilson, even though Wilson eventually abandoned it. But Wilson always emulated his predecessor in requring that his work have comprehensive, theoretical impact.

The very first sentence in Wilson's book *The Insect Societies* is the question, "Why do we study these insects?", and his answer refers to the principles of organization of social insects and the utility these principles may have outside the field of entomology:

> The biologist is invited to consider insect societies because they best exemplify the full sweep of ascending levels of organization, from molecule to society. Among the tens of thousands of species of wasps, ants, bees and termites, we witness the employment of social design to solve ecological problems dealt with by single organisms. The insect colony is often called a superorganism because it displays so many social phenomena that are analogous to the physiological properties of organs and tissues. Yet the holistic properties of the superorganism stem in a straightforward behavioral way from the relatively crude repertories of individual colony members, and they can be dissected and understood much more easily than the molecular basis of physiology.[12]

This quotation says a great deal about Wilson's early work; the words and phrases "ascending levels of organization," "superorganism," and "holistic properties" can be unpacked to reveal whole chapters of biological thought that were important to Wilson's background.

The single most important influence on Wilson's early research was probably Wheeler, who was also a specialist in ants, and also the author of several books on insect societies. Wheeler was a proponent of the concept of "emergent evolution," the belief that "the whole is not merely a sum, or resultant, but also an emergent novelty, or creative synthesis." Wheeler saw in insect societies the emergence of qualities that transcended the individual members of the society. Indeed, as time went on, Wheeler began to regard ant colonies as superorganisms that could be regarded as single, living entities in a real, not merely metaphorical or analogous, sense. In a well-known article written in 1911 Wheeler wrote of the ant colony as an organism, and by 1928 the social insect colony was a *super*organism. He wrote in the latter year: "We have seen that the insect colony or society may be regarded as a superorganism and hence as a living whole bent on preserving its moving equilibrium and integrity."[13]

Wheeler's views sound romantically speculative today. And

Wilson wrote in *The Insect Societies* that "seldom has so ambitious a scientific concept been so quickly and almost totally discarded." And yet Wilson also claimed that the superorganism concept had a "major impact on current research," that it was "the mirage that drew us on."[14] He noted that "the idea was inspirational and . . . as originally formulated by Wheeler had just the right amount of fact and fancy to generate a mystique. . . . It would be wrong to overlook the significant, albeit semiconscious role this idea has played."[15] One gains the clear impression that Wilson is admitting that much of his intellectual drive in the study of social insects had been after a goal that proved to be illusory, a mirage indeed.

Why was this mirage so attractive, and how did it appear? The appeal of the superorganism project was the potential it had for providing explanations of broad biological significance. If an ant colony was analogous or identical to whole biological organisms, as Wheeler claimed, then one might be able to get at the principles of organization of whole biological organisms by studying the principles of organization of ant colonies. One can take apart and examine the individual functioning parts of an ant colony much more successfully than one can dissemble the parts of a living organism. As Wheeler wrote:

> Another more general problem is suggested by the insect society, or colony as a whole, which . . . is so strikingly analogous . . . to any living organism as a whole, that the same very general laws must be involved. But the biologist, with his present methods, is powerless to offer any solution of the living organism as a whole. . . . We can only regard the organismal character of the colony as a whole as an expression of the fact that it is not equivalent to the sum of its individuals but represents a different and at present inexplicable "emergent level."[16]

Wilson later said that Wheeler's use of words such as "powerless" and "inexplicable" here represented a challenge which his generation accepted. They wished to test their analytical powers against this riddle. How was it possible to explain the apparent "emergent levels" of insect societies? Wilson and his colleagues took on that task and worked on it so successfully that they abolished the superorganism concept itself.

A central question for early students of insect societies was how to explain the organizational, homeostatic or self-regulatory powers of the communities they examined. Honeybees, for example, maintain a remarkably constant internal temperature in the hive, fanning their wings and regurgitating water drops on hot days, and clustering in swarms on cold ones. Ants occasionally move from one nest site to another, carrying their eggs and larvae with them. Many social insects, such as termites, build nests of intricate architecture. All this activity would seem, at first glance, to require some sort of conscious coordination, and, indeed, some early specialists even spoke of the "spirit of the hive." One such writer, Maurice Maeterlinck, described bee and ant colonies in detail and then asked, "What is it that governs here, that issues orders, that foresees the future?" A slight leap of anthropomorphic thought, and one can easily conceive an architect-termite, plans in hand, directing the construction of the complex system of passageways and ventilating shafts of a termite nest.

The results of the work of a generation of entomologists have revealed that the most fruitful approach to this problem was that of behavioral genetics, a path that eliminates the necessity of postulating anything approaching conscious coordination. Communication exists in complex forms in insect societies, but it is made up of individual responses based on genetic control. Pioneer ethologists, such as Konrad Lorenz, played important roles in illustrating to entomologists the importance of concepts such as fixed action patterns and innate releasing mechanisms (FAP, IRM) in explaining the behavior of organisms. As Wilson commented, "They [the ethologists] convinced us that behavior and social sturcture, like all other biological phenomena, can be studied as 'organs,' extensions of the genes that exist because of their superior adaptive value."[17]

Yet the behavior of insects was certainly not "fixed" in every instance, but seemed instead to be based on probabilities. Consider, for example, Wilson's description of "mass actions" among ants and bees:

> An important first rule concerning mass action is that it usually results from conflicting actions of many workers. The individual workers pay only limited attention to the behavior of nestmates near them, and they are largely unaware of the moment-by-moment condition of the colony

as a whole. Anyone who has watched an ant colony emigrating from one nest site to another has seen this principle vividly illustrated. As workers stream outward carrying eggs, larvae, and pupae in their mandibles, other workers are busy carrying them back again. Still other workers run back and forth carrying nothing. Individuals are guided by the odor trail, if one exists, and each inspects the nest site on its own. There is no sign of decision-making at a higher level. On the contrary, the choice of nest site is decided by a sort of plebiscite, in which the will of the majority of the workers finally comes to prevail by virtue of their superior combined efforts. . . . The same process occurs in the construction of comb cells by honey bees. In order to obtain pieces of wax for cells of their own, the workers regularly tear away walls that are in the process of being constructed by other nestmates. . . . Although these various antagonistic actions seem chaotic when viewed at close range, their final result is almost invariably a well-constructed nest that closely conforms to the plan exhibited throughout the species.[18]

The clue, then, to the understanding of many mass actions of insects that appear to be coordinated "from above" is the concept of "the emergence of statistical order from competing elements," an order which in the final analysis derives from genetic programs that have evolved during natural selection. Probabilities for certain behaviors under certain environmental conditions and stimuli grow or decrease as selection pressures change. In "cooperative labor" among insects the most important stimulus for the future behavior of the insects is "the product of work previously accomplished, rather than direct communication among nestmates."[19]

Once this approach to the study of insect societies had been formulated, the appeal of Wheeler's superorganism concept rapidly faded. Differences between insect societies and higher organisms, in which overall coordination on the basis of central nervous systems definitely does occur, became ever more apparent. Wilson further observed, "It is not necessary to invoke the concept in order to commence work on animal societies. The concept offers no techniques, measurements, or even definitions by which the intricate phenomena in genetics, behavior, and physiology can be unraveled."[20]

The decline of the superorganism concept was a result of the achievements of Wilson and his generation, and yet one has the clear sense that Wilson, at least, found the passing of that era somewhat saddening. He stated quite clearly that the concept "drew them on," gave them a goal. Furthermore, the concept of the superorganism bestowed upon entomology the promise of making a special contribution to theoretical biology, of revealing the unique regularities that govern the social insects and simultaneously whole organisms. The new affirmation of the overwhelming importance of the genetic base of insect behavior, and the assertion that mass effects can be explained as the probabilistic meshing of individual actions was a de facto denial of unique regularities. In the final analysis, the reductionist biologists, even the molecular geneticists, held the promise of the ultimate explanation of the social behavior of insects.

Wilson asked rhetorically in *The Insect Societies*, "What vision, if any, has replaced the superorganism concept?" What goal could draw him on now? The answers to these questions are, in my opinion, the essential factors for explaining his next ambitious venture, the writing of *Sociobiology*.

Wilson was certain that no new holistic conception would replace the old one. Rather, the new goal consisted of taking new, higher levels of biological complexity and trying to apply similar analysis there, i.e., "the continuing quest for precise evolutionary, that is, genetic, explanations of the origin of sociality and variations among the species in details of social structure."[21] This quest, said Wilson, is "the exciting modern substitute for the superorganism concept." If insect biology could not provide clues to the unique principles of whole organs or "higher" organisms, since upon examination it turned out there were no such unique principles, then perhaps the very demolition of the appearance of unique principles could serve as a model for an attack on the allegedly unique principles of "higher" social behavior, including human behavior.

WILSON'S MAIN WORK ON SOCIOBIOLOGY

Sociobiology is not a book devoted to the study of human beings, and, indeed, is much more cautious about extending biological speculations to humans than a host of earlier works by ethologists and ethology popularizers. True, one chapter out of twenty-seven does concern man, but the book can easily stand alone without that chapter. Wilson made some ambitious and foolish expansionist claims in chapters 1 and 27 about the eventual absorption of ethics and sociology by biology, but the significance of the book is something quite different: the attempt to formulate the main theoretical problems and analytical categories of a sociobiology based almost entirely on the nonhuman animals.

The most important concept behind *Sociobiology* was that of kin selection. As Arthur Caplan later observed, "It is the theoretical models of kin selection . . . that provided the theoretical spark for sociobiology."[22] The importance of kin selection was, first, that it provided a theoretical answer to the question of how complicated social behavior, especially cooperation and altruism, could evolve in Darwinian terms, and, second, it provided by extension a base for what in humans might be called moral behavior. To some biologists kin selection seemed a means for showing how normative behavior can develop out of material evolution.

The original problem was that classical Darwinian evolution was heavily based on the individual. Any individual of a species that sacrificed itself for another, and therefore had fewer progeny, seemed to be a loser in the Darwinian struggle. Surely its inheritance would not spread in the population. Yet it was known in Darwin's time that altruistic acts occur in the animal world. Bees protect the hive by stinging intruders and killing themselves in the process. In fact, among social insects the Darwinian ultimate in cooperation and altruism occurs, since the specialized castes of workers are completely sterile. How can this phenomenon be explained in strict Darwinian terms?

As is often the case in the history of science, the first great figure in the field sensed the later difficulty more fully than many of his immediate followers. Darwin called the problem of social insects "one special difficulty, which at first appeared to me insuperable, and actually fatal to my whole theory." Darwin pointed toward a possible solution by saying that natural selection may occasionally act on levels higher than the individual, such as the family. If sterile worker-insects increased the fitness of the family as a whole, their sterility would be an adaptation of biological significance that might be preserved by natural selection.

With the development of population genetics in the twentieth century Darwin's suggestion could be examined in a more rigorous way. In the 1960s William D. Hamilton developed a formal and mathematical statement of kin selection based on the concept of *inclusive fitness*. He defined inclusive fitness as the sum of an individual's own fitness plus the sum of all the effects it causes to the fitnesses of its kin. A reciprocal relationship thus arose which can be stated in the following way: if the reduction of gene contribution to succeeding generations by the sacrificing individual is more than compensated for by increased similar gene contributions of its relatives, a self-sacrificing act can be considered biologically adaptive.

Wilson placed Hamilton's idea of inclusive fitness at the base of his sociobiological scheme. Wilson gave the following example of a simplified network consisting only of an individual and his brother, and proposed that it was legitimate to speak of "altruistic genes" (meaning genes for altruism):

> If the individual is altruistic he will perform some sacrifice for the benefit of the brother. He may surrender needed food or shelter, or defer in the choice of a mate, or place himself between his brother and danger. The important result, from a purely evolutionary point of view, is loss of genetic fitness—a reduced mean life span, or fewer offspring, or both—which leads to less representation of the altruist's personal genes in the next generation. But at least half of the brother's genes are identical to those of the altruist by virtue of common descent. Suppose, in the extreme case, that the altruist leaves no offspring. If his altruistic act more than doubles the brother's personal representation in the next generation, it will *ipso facto* increase the one-half of the genes identical to those of the altruist, and the altruist will actually have gained representation in the next generation. Many of the genes shared by such brothers will be the ones that encode the tendency toward altruistic behavior. The inclusive fitness, in this case determined solely by the brother's contribution, will be great enough to cause the spread of the altruistic genes through the population, and hence the evolution of altruistic behavior.[23]

On the basis of this sort of analysis, Wilson hypothesized that in humans social and moral values may be influenced by genes. In addition to "altruistic genes," he spoke of possible "conformer genes," and he said that "human beings are absurdly easy to indoctrinate—they seek it."[24] He thought that religious belief and other types of conformity further the welfare of their practitioners and may well have acquired a genetic basis. He thought it likely that many other forms of human social behavior are genetically influenced, and he named as possibilities the cooperative division of labor between males and females, incest taboos, sexual activity, language acquisition, artistic impulses, maternal care, aggression, dominance, territoriality, and group cohesiveness.

We should remind ourselves that in speculations of this sort nearly any kind of human conduct can be "explained" by postulating a genetic foundation. But when one does so, one may be committing the "dormitive virtue in opium" fallacy, that is, one may convert what should be an empirical investigation into a dance around tautologies. The verification procedures applied to such hypotheses may, if one is not careful, contain the original hypothesis in disguised form.

Wilson was aware of these dangers and he did not draw many conclusions from his speculations about the genetic bases of human values. Instead he postulated possibilities. He pointed in the direction of genetic explanations of human behavior, but he cited few specific examples of universal culture traits that he believed were genetically determined. Instead, he emphasized how labile human traits are, and how flexible human behavior is. He criticized the ethologists who drew parallels between animal rituals, such as the mating "ceremonies" of birds, and similar human rituals. He even observed:

> During the past ten thousand years or longer, man as a whole has been so successful in dominating his environment that almost any kind of culture can succeed for a while, so long as it has a modest degree of consistency and does not shut off reproduction altogether.[25]

Wilson later said that he was surprised by the controversy which his *Sociobiology* caused, and we have no reason to doubt him. The academic reviews were largely laudatory, but he was sharply attacked by a group of Cambridge, Massachusetts, academics in the "Sociobiology Study Group," who maintained that Wilson was a biological determinist whose arguments amounted to a justification of the existing social order by implying that it was an inevitable genetic product.

Wilson's most vocal critics often distorted his viewpoints and certainly overreacted to a book that was much less biologically determinist than many previous writings on man's animal origins. Yet the criticism probably had a beneficial effect, for it forced Wilson to face more directly than he had earlier the political and moral implications of his work. If one compares Wilson's 1975 *Sociobiology* with his 1978 *On Human Nature* several important differences become apparent, and it is highly likely that the criticism he received after publication of *Sociobiology* was a factor accentuating the changes. In *Sociobiology* Wilson speaks little of the moral goals to which man's future evolution should be, in his opinion, directed. Instead, he spoke of the past biological determinants of that behavior. In the last sentences of *Sociobiology* he quoted Camus on the alienation of man by his loss of illusions and "the hope of a promised land;" Wilson continued that "when we have progressed enough to explain ourselves in these mechanistic terms . . . the result might be hard to accept." [26] In *On Human Nature*, on the other hand, the pessimism of Camus is abandoned for a call for a "mythopoeic drive" that "can be harnessed to learning and the rational search for human progress."[27]

ON HUMAN NATURE

In the introduction to *On Human Nature* Wilson announced that "The time has at last arrived to close the famous gap between the two cultures." But Wilson was not merely calling for a bridging of the educational gap between scientists and humanists, as C. P. Snow had done several decades earlier, but for an assimilation of the social sciences and humanities by the natural sciences, at least temporarily. This assimilation would be based, philosophically, on a rejection of Cartesian dualism. According to Wilson, "The mind will be more precisely explained as an epiphenomenon of the neuronal machinery of the brain." Once this has been done, scientists will have penetrated to the origin of ethical principles and value judgments. As Wilson maintained:

> Innate censors and motivators exist in the brain that deeply and unconsciously affect our ethical premises; from these roots, morality evolved as instinct. If that perception is correct, science may soon be in a position to investigate the very origin and meaning of human values, from which all ethical pronouncements and much of political practice flow.[28]

Thus, according to Wilson, natural science was now in a position to begin a direct study of what had been, up to this time, primarily the subject of the humanities and social sciences, namely, the study of human nature.

So far, a reader of Wilson might think that there was not very much new or subtle about his viewpoint. His opinions seem to be one more rewording of what is by now an old theme in Western writing, the thesis of man-the-machine, tentatively advanced even in Greek times by Empedocles, Democritus, and Epicurus, ambitiously proclaimed in the eighteenth century by La Mettrie and Holbach, and reappearing frequently in the nineteenth and twentieth centuries in the writings of Vogt, Moleschott, Büchner, Pavlov, Loeb, and Skinner. But Wilson should not be lumped with the reductionists and vulgar materialists so easily. If we read him carefully we will see that he considered reductionism valuable but inadequate as an approach to human nature, that he rejected the simplicity of Lorenz' and Skinner's differing deterministic approaches, and that, finally, he did not believe that natural science would ever exhaustively explain the social sciences and humanities. This last point is particularly important, and must be examined in more detail. Wilson is an expansionist—indeed, one of the most ambitious expansionists of the current generation—but his approach is so flexible and subtle that it differs strikingly from the majority of past interpreters basing themselves on biological determinism.

First, let us establish the ambitions of his expansionism and we will then look at some of the factors in his approach which limit that expansionism and tend to render it potentially less destructive or damaging. Wilson believed that every scientific field is "both an antidiscipline to another field, and, in turn, has its own antidiscipline." This conception was based on an implicit hierarchy of fields of knowledge, very close to the levels-of-reality conceptions of holistic materialists. The lower antidiscipline of chemistry is physics, while chemistry is an antidiscipline to the chemical aspects of biology. Wilson thought that biology is now the antidiscipline of the social sciences like psychology, anthropology, economics and sociology. These social sciences are vulnerable to biology because their explanations of human behavior have not yet taken biology into account. And the dimensions of their vulnerability are illustrated by Wilson's proud pronouncement:

> The question of interest is no longer whether human social behavior is genetically determined; it is to what extent. The accumulated evidence for a large hereditary component is more detailed and compelling than most persons, including even geneticists, realize. I will go further: it already is decisive.[29]

This is a very ambitious expansionism indeed, and we are justified in suspecting exaggeration. But let us look further. How far does Wilson think biology can go in advancing the sort of explanations of human behavior that were previously assigned to the social sciences and humanities? And here we find some surprises. Wilson differed from many previous mechanists by maintaining that the natural sciences can only make a contribution toward an understanding of man; they cannot preempt the social sciences and humanities because the latter will always retain richer modes of explanation. This aspect of Wilson's thought is often overlooked by many of his critics, but it emerges clearly in the following statement in his *On Human Nature*:

> Biology is the key to human nature, and social scientists cannot afford to ignore its rapidly tightening principles. But the social sciences are potentially far richer in content. Eventually they will absorb the relevant ideas of biology and go on to beggar them. The proper study of man is, for reasons that now transcend anthropocentrism, man.[30]

Wilson was an expansionist, but a disciplined one. He asked social scientists to take biology into account so they could more effectively get on with the task of "beggaring" the ideas of biology.

At the root of Wilson's conception here is vestigial antireductionism, one that may well go back to the superorganism ideas of William Morton Wheeler. Wilson believed in the strength of reductionism as an analytical method but he considered it by itself insufficient for a complete understanding of complex social phenomena. Referring to his hierarchy of "antidisciplines" and "disciplines" he observed, "The laws of a subject are necessary to the discipline above it, they challenge and force a mentally more efficient restructuring, but they are not sufficient for the purposes of the discipline." Wilson was too much of an evolutionary biologist to be a simplistic reductionist, and as he addressed man more directly than he had in the past his antireductionism emerged more clearly. Wilson remarked,

> Raw reduction is only half of the scientific process. The remainder consists of the reconstruction of complexity by an expanding synthesis under the control of laws newly demonstrated by analysis. This reconstitution reveals the existence of novel, emergent phenomena. When the observer shifts his attention from one level of organization to the next, as from physics to chemistry or from chemistry to biology, he expects to find obedience to all the laws of the levels below. But to reconstitute the upper levels of organization requires specifying the arrangements of the lower units and this in turn generates richness and the basis of new and unexpected principles.[31]

The terms Wilson used here—"novel, emergent phenomena"; "level of organization"; "new and unexpected principles"—illustrate both his opposition to a pure reductionism and explain why he believes that the social sciences will be able to "beggar" biology. And this antireductionism also explains the voluntarism and ethical elements he built into the final pages of *On Human Nature*.

In the last chapter, entitled "Hope," Wilson spoke of the logical "circularity" that imprisons the person who wishes to explain human behavior in terms of determinism and who, at the same time, believes in the importance of ethical systems.[32] This was the same circularity that trapped Skinner so completely that he never escaped, a plight that cause Max Black to joke about the comic dilemma of a "convinced determinist *urging* his readers to save the human race."[33] Wilson skirted a similar circularity (a genetic rather than Skinner's environmental one), which he described in the following way: "We are forced to choose among the elements of human nature by reference to value systems which these same elements created in an evolutionary age now long vanished."[34] In other words, strict determinists have no way of establishing a leading edge, a leg up on determinism whereby they can decide which of the determined characteristics they wish to retain and which they wish to change.

But Wilson was committed to making such decisions on the basis of a new ethics, an emergent, novel, human guide to action. He spoke of using knowledge of biology to select a more "deeply understood and enduring code of moral values."[35] He then listed three "values" that he believed should stand at the base of any new moral code. They were:

1. The value of the long-term survival of the human gene pool.
2. The value of diversity in that gene pool.
3. The value of universal human rights.

A difference in kind exists between the first two of these values and the third. The first two, the survival and diversity of the gene pool, would be valuable characteristics for all species in changing environments; the last, universal human rights, is uniquely human. The leap from the first two of these values to the third is the most difficult of all transitions to explain within the framework of a purely biological scheme. Not only is commitment to universal rights uniquely human, but it is an extremely recent arrival in human history and even now is probably more frequently violated than observed. The principle is a product of the last centuries of civilization, and its fate is not assured. If it is a desirable adaptation in terms of inclusive fitness, one wonders why it did not come earlier and why it still spreads with such difficulty. Many Social Darwinists of the nineteenth and early twentieth centuries (see, for example, the discussions of eugenics in Chapter 8) saw it as maladaptive. We would be more likely to say now that the biological adaptive value of commitment to universal human rights is an open question. We are on safer grounds to support human rights for ethical rather than biological reasons.

Wilson attempted to give the principle of universal human rights a biological explanation, but his abbreviated statement—less than a page—was very weak and did not sustain his argument. Perhaps he included it largely to try to show his critics that he, too, was a libertarian. Contrary to almost all other libertarians, he tried to justify human rights through science. He maintained that since humans are mammals, not insects, they must give "grudging cooperation" to other members of their society for reasons of kin selection. Furthermore,

> We will accede to universal rights because power is too fluid in advanced technological societies to circumvent this mammalian imperative; the long-

term consequences of inequity will always be visibly dangerous to its temporary beneficiaries. I suggest that this is the true reason for the universal rights movement and that an understanding of its raw biological causation will be more compelling in the end than any rationalizations contrived by culture to reinforce and euphemize it.[36]

The justification of an ethical system on biological grounds that goes beyond talk of rudimentary altruism and kin selection is the most difficult operation within sociobiology, and we must observe that Wilson did not do a very convincing job of it. Nonetheless, his "third value" spared him from the worst excesses of Social Darwinism and brought him to a recognizably ethical and humanistic position. But his attempt to show that he derived his belief in human rights from biology was inadequate. Thus, Wilson *did* mix ideology and science, but his ideology was not the conservative one assigned him by his critics. It was a mildly liberal and democratic one, the sort of ideology likely for a bright and sensitive young man raised in the South who became much more liberal than most of his boyhood friends, only to move to an academic center in the North where the new youth were more radical than he. True, his vision contains such large elements of biological determinism that one wonders where he might come down in a pinch, say in a moment of uncertainty over the results of study of human differentiation. But just as Wilson says that the genes influence human behavior "on a very long leash" we can say that Wilson's commitment to universal human rights influences his ethical conclusions from sociobiology on a similar long leash.

It thus makes sense for Wilson to say, "We can hope to decide more judiciously which of the elements of human nature to cultivate and which to subvert, which to take open pleasure with and which to handle with care."[37] A rigid determinist could not logically make such a statement. It is the ethics of human rights which provides Wilson with criteria by which to judge the biologically influenced characteristics of human behavior.

Within this framework Wilson can call sexual differentiation "inconvenient and senseless" in humans, and talk about taking social steps to overcome what he sees as vestigial genetic differences between the sexes, so long as the social costs are not too great. His message here is obviously double-edged, and his vociferous critics will see only that Wilson believes that males and females are genetically different, not that he thinks that at the present stage of human development these differences are perhaps unfortunate, that they are small enough to be overcome by socialization, and that such a step "could result in a much more harmonious and productive society."[38]

Whether Wilson is right about the degree of genetic influence over human behavior is an empirical question. It would be no surprise if, like many previous biological determinists, he has exaggerated or simplified that influence. But compared to past biological determinists—the eugenists and Social Darwinists of the last one hundred years—he is advancing a fairly mild social message.

Yet the most serious and significant test of the sociopolitical impact of Wilson's views may still be in the future. The political uses of a scientific doctrine do not derive primarily from the theoretical framework of the doctrine, nor from its author's personal views, but from the social and political setting in which the doctrine appears. Wilson's new form of biological determinism may be seized upon by conservative political forces in society and used as one of their banners, as a scientific sanction for their reactionary social vision. This happened with eugenics in Germany in the 1920s, as we will see in the next chapter.

In the late 1970s Wilson's writings were picked up by the new right movement in politics in France and England.[39] If that movement should continue to grow the need for Wilson to state his opinion on its relevance to his views would increase in step. After all, Wilson has written many times that his biological theories can be applied to society. He therefore has an obligation to tell us which of the applications being made by others is, in his opinion, legitimate. If he does not denounce the conversion of his biological views into planks in an ideological platform the impression will grow that he is not really opposed to its use for right-wing, antifeminist, and inhumane purposes.

So far the squabbles over sociobiology have not reached this threatening political stage, and they may not. And so far we should give Wilson credit for his differences from earlier conservative political ideologists relying on biology for support. Notice, for example, how different Wilson is from Lorenz. Both Wilson and Lorenz are expansionists, explaining human behavior and values in terms of biological science. But Wilson looks to the future, saying "At some time in the future we will have to decide how human we wish to remain—in this ultimate biological sense—because we must consciously choose among the alternative emotional guides we have inherited."[40] Wilson takes consolation in the more just society that advancing technical civilization might bring. Lorenz, on the contrary, blames advancing civilization for all ills, and looks to the biological past for redemption. To try to overcome biological imperatives is, for Lorenz, like trying to solve the problem of an overheated steam boiler by tying down the safety valves. He lacks Wilson's concept of emergent ethics culminating in universal human rights. Thus, he would adjust to the old boiler, while Wilson would build a new one.

BIOLOGICAL KNOWLEDGE AND SOCIETY

chapter eight

EUGENICS: WEIMAR GERMANY AND SOVIET RUSSIA

ONE OF the precursors of the later debates about ethical issues in biological research was the great "nature-nurture" discussion of the pre-World War II period. The two areas where this controversy took the most extreme forms were Germany and the Soviet Union in the 1930s and 1940s. Here the potential political ramifications of the debate were realized in dramatic and tragic forms. The differing fates of biology in these two countries are profound case studies of the interaction of science and political values in the twentieth century.

The essential features of National Socialist attitudes toward race and heredity and the contrasting Soviet view favoring environmentalism (with Lysenko as the leading advocate) are well known to most educated people today. More unfamiliar is the story of the gradual development of attitudes toward heredity in the 1920s in these two countries. Since this early period offered more possibilities for alternatives and much greater intellectual interest, this chapter will focus on Germany and Russia in the 1920s.

In the 1920s, National Socialism had not yet come to power in Germany and scientists of greatly differing political views, including Marxists, socialists of various sorts, Catholics, liberals, and conservatives, discussed human genetics. In Soviet Russia in the same years no one had yet heard of Lysenko, and the spectrum of debate about human genetics was surprisingly broad: there were Marxists who were also eugenists and saw no contradiction in their respective positions, and there were anti-Marxists who were Lamarckians. In both Germany and Russia in the 1920s, then, single-minded political concerns had not yet overruled scientific judgment in discussions about human heredity and its social implications. How did it happen that in both countries this early period, with its complexities and uncertainty about the political value implications of theories of human heredity, gave way to a period in which Mendelian eugenic theories were usually linked to conservative views of society and Lamarckian theories were usually linked to left-wing socialist views of society? Was this expansionist process entirely a social and political phenomenon, essentially distinct from the scientific theories under discussion, or was there something intellectually inherent in each of the competing theories of heredity that supported a particular political ideology? If the answer to the latter question is affirmative, what do these inherent factors say about the allegedly value-free nature of science?

Background to the Debate

In Weimar Germany and Soviet Russia the peak of the debates over the social and political implications of theories of human genetics occurred in the mid-1920s. Scientists in each of these countries were aware of ongoing debate in the other. In both instances, however, the debates occurred within unique contexts formed by the recent histories of the two nations. In Germany, as in many other countries, a great debate had occurred over Darwinism and Social Darwinism in the last part of the nineteenth century.[1] Many of the views expressed there were similar to those uttered elsewhere, but several features were stronger in Germany and deserve particular attention. One characteristic of German Social Darwinism was a gradual decline in the last decades of the century in the optimistic version and a growth of the pessimistic one. The optimistic version is that interpretation of Social Darwinism that sees in the economic competition of capitalism an analogue to biological competition and postulates a direct benefit for the most creative and successful entrepreneurs and an indirect benefit for society as a whole. This form of Social Darwinism is so well known that it need not be discussed here; it was particularly popular in England and the United States. The reasons for its relative decline in Germany at the end of the century are complex. The general tendency toward cultural pessimism and discontent with rapid industrialization among educated classes often noted by historians undoubtedly played a role. Yet other influences are also evident: 1) the socialist critique of unfettered capitalism; 2) the popularity, even among the petty bourgeoisie and workers, of Social Darwinist views in the form presented by Ernst Haeckel and the monists; and 3) the strength of theories of an independent, immutable germ plasm, particularly the one promoted by August Weissmann. Socialism helped to convince a large portion of the German population, including many who were not socialists, that unrestrained capitalism was not appropriate to Germany. By the end of the century Germany had the largest socialist party in the world and the most extensive set of social welfare laws mitigating the effect of capitalist competition. Monism prepared a broad segment of the German population for biological arguments applied to society. And germ-plasm theory gave heuristic clarity—especially important to lay audiences—to the concept that each individual is a genetic custodian with the responsibility for preserving his or her own genetic "package" for the future. If those with the "best" genetic heritages handed them on to the next generation less frequently than others, the overall result for a population would be genetic degeneration.

The transition from optimistic to pessimistic forms of Social Darwinism can be seen in the contrast between two early German Social Darwinists, Otto Ammon and Wilhelm Schallmayer. In 1895, Ammon, an engineer and private scholar, published *The Social Order and Its Natural Foundations*, in which he maintained that natural selection among humans was the key to social progress. He tried to humanize this conclusion and to reconcile his Christian ethics with economic competitiveness by asserting that there are two kinds of natural selection: the one individualistic and egoistic, the other altruistic and social. Ammon looked upon this automatic and natural process with awe, seeing in it a secret mechanism beyond man's complete comprehension. Nature was superior to man and therefore any attempt by man to help the process along through eugenics was certain to go awry. "We must not attempt to breed men selectively as if they were cattle!" he angrily exclaimed. He considered it ridiculous to believe that "mere humans, especially small-spirited ones, who have never understood the iron law of heredity, might interfere with bold hands in order to improve it and remake it from the ground up."[2]

In contrast to Ammon, Schallmayer exhibited a pessimistic interpretation of Social Darwinism. Schallmayer in 1900 won first prize in a writing competition established by the German industrialist Alfred Friedrich Krupp on the topic, "What can we learn from the theory of evolution about internal political development and state legislation?"[3] In his award-winning book, Schallmayer observed that "the possibility of genetic progress stands face-to-face with the possibility—in case of inadequate selection or direct germinal impairment—of genetic regression, which can go so far in a relatively short period of time that it can pull down political power with it."[4] According to Schallmayer, those interpreters of Darwin who saw only progress in his teachings were simply misreading him, since the story of evolution is replete with episodes of extinction. Schallmayer feared that man's emphasis on individual, rather than group, interests was bringing him into similar danger. By developing a system of ethics based on the rights of the individual, civilization had actually done a disservice to the human race, since man was no longer willing to sacrifice the individual for the sake of community welfare. To Schallmayer, Weissmann's view of autonomous germ plasm graphically illustrated that what mattered from the standpoint of modern biology was that individuals secure the preservation of the species through proper transmission of their genetic heritage (*Erbgut*), not that man make more comfortable either his own temporal existence or that of his fellow man. In fact, more comforts and better social conditions might actually be detrimental to the future of the human race. Degeneration (*Entartung*) was more of a danger among civilized peoples than among primitive ones because social measures such as those already prevalent in Germany interfered with the "natural" process of selection.[5] Conquering disease through medicine, the care of the incapable and the unfit by social security, and the spread of contraception among the educated all had the effect of lowering the quality of the genetic heritage passed on to successive generations.

Although he seems to prefigure the extremes to which eugenic and racist doctrines later deteriorated in Germany, Schallmayer actually considered himself a radical democrat and in subsequent years stubbornly opposed the National Socialist interpretation of race. He agreed that life selection (the elimination of the unfit) occurs in nature, but he considered it totally impermissible among humans. Rather, he hoped for a system of voluntary eugenics, in which individuals would, through education, make careful decisions about the desirability of having large numbers of children. He insisted upon humane methods, and he saw no correlation between a biological interpretation of society and right-wing politics. Indeed, he considered eugenics a progressive doctrine, as did several political groups in other countries during the first years of the movement.[6]

The degree to which early German eugenists at first separated themselves from particular political viewpoints can be seen in the initial writings of Alfred Ploetz, a founder of the leading German eugenics journal and a central figure in the German eugenics movement of the 1920s. As a young man Ploetz' interest in the genetic future of man was stronger than his commitment to any particular social or economic order, and he criticized both capitalism and socialism for the contradictions which he saw between them and the ideals of eugenics. Ploetz, therefore, was no businessman or academic scholar defending economic competition on the basis of Darwinism, a model long familiar to us; rather, he wanted to find a third way, a eugenic society that would put the good of the future above the comforts of the present. Socialism presented the prospect of a humanitarian society through the alleviation of economic suffering. But socialism thoroughly contradicted the principles of eugenics. If the "weak" were protected through social legislation guaranteeing employment, health and life support, then they would propagate their kind in increasing numbers. But he considered capitalism no better. It caused great human suffering—Ploetz cited Karl Marx, Karl Kautsky, August Bebel, and other socialists to that effect. And, furthermore, by giving advantages to the hereditary wealthy in the struggle for existence, it hindered free natural selection.[7]

One might erroneously suppose from this analysis that the most appealing form of society to Ploetz was a somewhat softened capitalism, in which individual competition would still play the major role in determining success in any one generation, but in which inherited wealth and privilege would be prohibited. In such a society each generation would have to prove itself anew; the losers would fall to the bottom of the heap, where they would be protected from the most extreme forms of suffering. Ploetz, however, was both too complex and too humane to propose such a society. He seemed to favor all aspects of socialism except its alleged genetic consequences, and the specter of open competition within each generation was offensive to him. Therefore, he struggled to find a way of combining the humanitarian and egalitarian ideals of socialism with the eugenic goal of a constantly improving biological base for society, since the potential unification of the principles of eugenics with "humane socialist principles is of the utmost significance for the further evolutionary development of human society."[8]

Ploetz believed that he had found a way to reconcile his contradictory demands. He proposed that all people be protected by welfare legislation, as the socialists demanded, from the raw struggle for survival inherent in pure capitalism; but at the same time he wanted to ensure that each generation consisted of genetically healthy individuals by having each married couple select the "best" of their own germ cells for fertilization. Ploetz noted that the germ cells, or gametes, of each couple vary greatly in their genetic constitution; from the standpoint of eugenics, the children needed to be at least genetically equal to their parents. By "shoving selection back" to the prefertilization stage he hoped to achieve eugenic goals without tampering with individual human rights or interfering with natural legal parenthood. "The more we can prevent the production of inferior variations," he wrote, "the less we need the struggle for existence to eliminate them. We would not need it at all if in each generation we were able to ensure that the totality of variations was qualitatively somewhat superior, on the average, to that of the parents."[9] Ploetz's suggestion was naive and entirely impractical from the standpoint of the biological science and technology of his time. No one in the early years of this century had the faintest idea how to give an overall evaluation of the genetic quality of different gametes (nor can it be done now). In stating the problem of a clear and pressing threat of degeneration on the basis of quite inconclusive evidence and proposing as a solution a scheme that was wildly futuristic, Ploetz actually solved nothing. In addition, he left the road clear for less humane schemers. Furthermore, Ploetz continued to support the German eugenic movement as it deteriorated, abandoning his earlier humane principles.

Nonetheless, Ploetz made an important point which is more interesting to us now than it was at his time. The parents who today decide to abort a fetus found by amniocentesis to be suffering from Down's syndrome (Mongolism) are following Ploetz's policy of "shoving selection back" to the prebirth stage in the name of humane principles, although not to the gametes themselves. And a few recent efforts have been made even to extend selection back to the gametes, as Ploetz suggested, not only for the determination of sex but also for other specific genetic reasons. Controversial as some of these proposals are, Ploetz was prescient in seeing that value consequences of applying the science of human heredity depend more on *when* and *how* it is applied than on the question of what is inherent in the science itself. It would, of course, be a mistake to draw a close parallel between Ploetz's goals and those of contemporary genetic counseling based on amniocentesis. Ploetz worried primarily about avoiding an overall decline in the quality of future generations viewed collectively, a worry that we now know he grossly exaggerated; he was not primarily concerned with avoiding a genetic flaw and the consequent suffering for one child and its family. Ploetz's significance, then, lies not in the similarity of his situation to ours, but in his early effort to analyze the differing social value consequences of different approaches for applying the science of human heredity.

The Controversy in Weimar Germany

In the mid-1920s academic geneticists, anthropologists, and eugenists in Germany disagreed sharply even over the terminology of human heredity and race. Although these disputes seem somewhat scholastic at first glance, they formed the essential theoretical background for the subsequent degradation of the science of human heredity in Germany. What occurred in this period was the gradual crystallization of political value links to specific biological interpretations.

One of the issues that divided German specialists in human heredity was that of the appropriate German word to be used to translate the English term "eugenics"—a word devised by Francis Galton in 1883. At first the analogous German word *Erbhygiene* was preferred; later, the more direct *Eugenik.* The socialist Alfred Grotjähn, who strongly favored eugenic measures to avoid what he saw as imminent degeneration, employed the term *Fortpflanzungshygiene,* a word he thought accurately described his concerns, while showing that he rejected the racist interpretation of eugenics.[10] By the 1920s the term *Rassenhygiene* had gained in popularity, although its exact meaning remained in dispute.[11] Did "race hygiene" apply to one human race? If so, some people could maintain that in itself the term did not connote racial differentiation. Or did it mean "races" and a consequent hierarchy of racial worth—the definition ever more widespread in right-wing political literature? Some German eugenists, including Wilhelm Schallmayer, preferred *Rassehygiene*; the omitted "n" indicated that they were speaking of only one race, the human species. Others said the word *Rassenhygiene* was acceptable because the middle "n" was not a plural but a variation of the genitive singular.[12] Still others insisted that the "n" was, indeed, a plural form and that a differentiation of races on the basis of genetic "worth" was an entirely scientific goal.[13]

The word eugenics today is usually considered a pejorative term and is even, on occasion, erroneously equated with National Socialist doctrines, but in Weimar Germany "eugenics" was thought of by proto-National Socialist publications and organizations as a kind of leftist deviation. Race hygienists carefully followed the eugenics movements in other countries and frequently pointed out that the leaders of these movements often failed to understand the racial significance of eugenics. Whether one used the term *Rassenhygiene* or *Eugenik* became in the late 1920s a kind of political flag, often with the more right-wing members of the movement favoring the first term, the more left-wing members the latter. In the last years of the Weimar period, alarmed by the growing racist sentiment among specialists in human heredity and their increasing alliance with the political right wing, a coalition of centrist, Catholic, and social democratic biologists and anthropologists mounted a counterattack within the German Society for Race Hygiene with the goal of changing its name to the German Society for Eugenics. They received important support from Eugen Fischer, a prominent anthropologist, who, according to an historian of German anthropology, was still trying to reconcile his concept of human genetics with his Catholic faith. In later years Fischer ignored the pope's encyclical condemning eugenics, abandoned his religious hesitations, and joined fully with the race hygiene movement.[14] For the moment, however, Fischer aided the rebels combatting race hygiene. The campaign did not succeed, since it resulted in a weak compromise on the title "German Society for Race Hygiene (Eugenics)." A few of the rebels believed they had shown that *Rassenhygiene* was simply a German word for eugenics with no necessary racist connotations.

If Catholics frequently had reservations about eugenics, socialists and communists objected to the racist doctrines contained in the race hygiene movement, even if they accepted eugenics. The objections of the Catholics and the leftists had, however, quite different roots. The Church maintained a restrictionist position, keeping ethics and biology apart. Many leftists, on the other hand, supported the expansionist view that science can be applied to society and its norms, but they refused to accept the particular expansionist interpretation of the political right wing, one which supported racism. The leftists hoped that biology's social implications could be read in a different way. The position of German socialists and communists toward eugenics in the twenties was complex and interesting. Contrary to what we might assume, in the early and middle years of the decade the leftist press did not define the debate in terms of nature versus nurture where nature was politically conservative. The leftists noted that many of the eugenists opposed such traditionally conservative institutions of society as monarchy and nobility on the grounds that they were genetically regressive, having no correlation between influence and natural ability. Most Marxists prided themselves on their scientific outlook and so regarded knowledge of genetics as desirable. Furthermore, they, like the eugenists, definitely favored societal reform. Therefore, many socialists and communists supported eugenics.

Eugenic considerations, and the type of social analysis used by eugenists, early entered into the discussions of the Social Democrats on the issue of abortion. In the 1920s the proposal to permit legal abortion created great controversy in Germany and Austria, and socialists seriously divided over the issue. They universally condemned the existing system, in which abortions were illegal but widespread: the rich obtained adequate services while the poor were condemned to frightful conditions. Furthermore, the Social Democrats usually favored freeing abortions from religious and absolutist restrictions, but they questioned the degree to which they, as socialists, should regard abortion as an individual decision, as if the fetus were private property, and the extent to which abortion had to be a societal decision. Karl Kautsky, Jr., wrote that, after long pondering this question, he had decided his earlier support of the individual decision was "unsocialist"; the stake of society in the number and quality of children to be born was too high to allow pregnant mothers unrestricted right to control their own bodies. He noted that in 1924 the Organization of Social Democratic Physicians in Vienna had recommended against making abortion a completely free individual decision and had, instead, called for group decisions to be made jointly for each abortion by physicians and public committees, which would base their decisions on "medical, social, and eugenic" indicators.[15]

The principle of responsibility to society as a whole lay at the basis of many early eugenic and race hygiene programs. The Democratic Socialist writers of the 1920s often applauded this principle even when it was advanced by racist authors. In 1923, writing in one of the oldest socialist journals, the *Socialist Monthly*, George Chaym observed, "Socialism certainly does not take a negative position toward race hygiene (*Rassenhygiene*), insofar as race hygiene concerns theoretical and practical measures for the improvement of the race or avoiding its debasement (*Verschlechterung*)." Chaym continued that, correctly understood, race hygiene "unconditionally belongs in socialism's work program."[16] The most prestigious journal of German social democracy in the 1920s, *Die Gesellschaft*, contains numerous discussions of race hygiene and eugenics. Up to the end of the Weimar period the journal continued to support eugenics when it was divorced from extremist and racist interpretations.[17] Even the word *Rassenhygiene* continued to be used by many leftist writers. In 1926 Oda Olberg, in a review of *Biological Problems of Race Hygiene and Cultured Peoples*, called for "a more careful consideration of race hygiene requirements by socialists and other people in Social Democratic-oriented endeavors." In another issue the same author observed, "I am not able to understand why the question of the protection of our racial value (*Rassenwerte*) should be looked at through one or another particular pair of partisan, political, religious, or philosophical spectacles. This problem stands higher than any kind of party politics."[18]

In June 1926 Kautsky wrote a long review of Olberg's own book, *Degeneration and Its Cultural Conditioning*, a volume that contained a particularly clear exposition of the view that civilized man was genetically declining because social welfare measures softened the effects of natural selection. Olberg could not be dismissed as a right-wing racist fanatic; she was a loyal Social Democrat who had often written for *Die Gesellschaft* in the past. Furthermore, she stoutly defended democratic socialism in her work, including its social welfare program; she claimed that only when class differences had been eliminated from society could the environmental and genetic causes of social ills be separated. Once that was accomplished, she was convinced that socialism would have to adopt a eugenic program. Kautsky praised Olberg for raising these issues, but he was clearly worried about her analysis. For all of his own reverence for science, he saw that Olberg exaggerated its significance for politics. He questioned her main thesis—the belief that an increase in culture automatically brings with it an increase in genetic degeneration. He was bothered by her lumping together so many different phenomena as examples of degeneration—proliferation of crime, decrease in nursing by mothers, increase in mental illnesses, and so on. And he expressed fears about what the acceptance of such views by Social Democrats would do to their political programs. But he agreed with Olberg that eugenic concerns are important to society and should form a part of the political consciousness of Social Democrats.[19]

The first major criticism of the race hygiene movement to appear in *Die Gesellschaft* was an article in February 1927, by Hugo Iltis, a biographer of Gregor Mendel. In this article, which revealed a great deal about socialist attitudes toward eugenics, Iltis was directly critical of the majority of German race hygenists. He said they subverted science for politics and noted with regret the presence among them of a number of prominent academic geneticists. He criticized the typological description of races, which, displacing the populational view, attributed mental and ethical qualities to individual races. He also saw clearly the dimensions of the threat; indeed, he said that "delusion about race is on the point of conquering German science." And he predicted that "the time will come when the race hatred and race obscurantism of our day will be what witchcraft and cannibalism are for us now: the sad vestiges of depraved barbarism." Surprisingly, however, not only did Iltis continue to support eugenics, but he even accepted the term *Rassenhygiene*. He was convinced that the main problem of eugenics was its drawing of premature and inhumane conclusions on the basis of fragmentary data. In the future, he maintained, after "decades and perhaps centuries of hard work," we will create "the foundations of a genetic race hygiene," one which will serve the whole human race.[20]

Iltis saw in Lamarckism a way of softening the hard facts of human genetics, and in that way he helped to forge the links between leftist politics and Lamarckism that were growing in the 1920s. He believed that only through the inheritance of acquired characteristics would it be possible to show that everything the socialists were working for—better education, better living conditions—would have a beneficial genetic effect. As he observed, "once Lamarckism has found its great methodizers and as much time is given to its study as is given today to Mendelism, it will be possible to find the means by which not only the individual human races but the human race as a whole can be raised to a higher level, and its physical, ethical, and emotional qualities can be advanced."[21] Thus, Iltis held out an unattainable, scientifically unjustified ideal: a Lamarckian eugenics directed toward the genetic betterment of mankind, a science that would grant a place to both Mendelian and Lamarckian inheritance.[22]

Not all left-wing German eugenists supported Lamarckism, but the conservative race hygenists were delighted when they could hang the label of "Lamarckism" on their opponents; critics like Iltis became easy game. One of the leaders of the German race hygienists, Fritz Lenz, who applauded Hitler even before he came to power, retaliated for the right-wing movement by dismissing the most prominent socialist and religious objectors to race hygiene as hopeless Lamarckians trying to dodge the facts of modern biology by creating the fiction of the inheritance of acquired characteristics.[23] Lenz was an academically qualified geneticist, and on the question of Lamarckism he exploited his advantage to the hilt. He pointed out that Paul Kammerer, a dedicated socialist and admirer of the Soviet Union, was trying to prove the inheritance of acquired characteristics with his experiments on the midwife toad. The experiments were exposed as fraudulent, after which Kammerer committed suicide. In the Soviet Union, Kammerer was made a hero, the subject of a popular film.

Lenz described all of this in his journal *Archive for Race and Social Biology*, and then turned the tables on his left-wing critics. "They accuse us," he said, "of inserting our own values into biology, when actually they are the ones guilty of maintaining that biology contains humane principles, those of Lamarckism." "*We* know, on the contrary, that science is value-free," he continued. "We must follow the facts of human heredity wherever they take us, and those facts tell us that man will genetically degenerate unless the strong and the fit are given advantages in propagation." He declared that the inheritance of acquired characteristics was a myth and that Mendelian laws alone governed human heredity.[24]

Lenz here was hypocritically adopting a restrictionist position because it was to his advantage. By defending the position that science was value-free, he took a traditional academic stance that allowed him to criticize the Lamarckians for confusing science and politics. The Lamarckians, Lenz charged, supported an erroneous biological theory because it permitted them to draw political conclusions favoring egalitarianism and socialism. The Mendelians, however, Lenz continued, rejected such expansionism, and accepted science for what it was, without political fears. But despite Lenz' claims, *he* too was actually an expansionist. In his fervent support of racial eugenics he drew political conclusions on the basis of his understanding of biology, and the conclusions were those favored by Hitler's party.

Perhaps the most outspoken attack on the German race hygiene and eugenics movement to appear in the entire Weimar period was an article in 1928 in the communist journal *Under the Banner of Marxism.* In it Max Levien presented a classical Marxist critique, simplistic but trenchant, and he avoided the Lamarckian trap. The central thesis of his long polemic postulated that in the main, German race hygiene was a bourgeois science that served the ruling class of German capitalists. The leading race hygienists preached a false theory of genetic degeneration because, by portraying such a threat, they hoped to justify imperial expansions against "inferior races and nations," particularly the Eastern Slavs. Both of these goals were parts of their program of stemming the revolution already visible in the East. As did the democratic socialists, the communist Levien believed that eugenics was a progressive, useful science. Once the revolution was victorious, eugenics could serve the proletariat as it had once served the capitalists. Levien said that German Marxists must not let the race hygienists do damage to "serious scientific efforts to create a people's eugenics (*Volkseugenik*) that considers the whole population of the earth." Once race hatred had been deprived of its necessary base in the capitalist order, Levien asserted, man "will become a creative shaper of his own species; this advancement will bring about an upward surge in mankind's intellectual achievements and the power of science will thus enhance the rational, practical use of the laws of genetics for the development of an undreamt-of higher kind of man."[25] Levien's communist eugenics was to be part of an overall *Volkshygiene* program that included measures for both genetic and environmental improvement.

Levien's argument was of especial relevance to the Soviet Union. There the successful revolution he dreamed about had already occurred. The journal in which he wrote was the Western communist theoretical publication most widely circulated among Marxist intellectuals in the Soviet Union. Indeed, many of the articles were written by Soviet citizens. Would Soviet Marxists, freed of the capitalist order, follow Levien's suggestion by developing a socialist eugenics?

The Controversy in Soviet Russia

Observers of its early history are frequently surprised to learn that Soviet Russia in the 1920s possessed a strong eugenics movement. One might have expected revolutionary Russia, which prided itself on opposition to capitalist culture and aristocratic privilege, to have stood aside from the movement for "race betterment" that swept the world in those years and led to the establishment of eugenics societies in dozens of countries. To arrive at this conclusion, however, is to carry back into the third decade of this century ideas both about eugenics and Soviet views of man which took clear form only in later years. Here, as in Germany, there was a gradual expansionist formation of value links to scientific ideas, but in a very different political setting.

Scholars who have investigated the history of Darwinism in Russia in the late nineteenth century have noted the unusual enthusiasm with which it was received by intellectuals there. One historian wrote, "Unlike its reception in the West, Darwinism met almost no opposition in Russia from either the scientists or the social thinkers."[26] One reason for this receptivity was that by the last decades of tsarism the intelligentsia of Russia usually defined itself in terms opposite to those of the Church and monarchy; intellectuals seized upon scientific explanations of man's origins as natural allies in their struggle against these authorities. In addition, the intelligentsia wanted to close the gap between the culture of Russia and that of Western Europe; in their effort to do so, they often embraced rather uncritically the latest scientific trends coming from the West.

By the first decades of the twentieth century, however, biologists in Russia had moved far beyond mere imitation of Western science and had formed a center of outstanding genetics research, entirely in step with the new trends and in some respects even leading the way. Around the figure of S. S. Chetverikov a school of population genetics was established which is not fully appreciated even yet by historians of biology, largely because it disappeared in later years when Lysenko took over Soviet genetics.[27] In addition to Chetverikov, prominent early Soviet biologists included N. K. Kol'tsov, A. S. Serebrovskii, N. I. Vavilov, and Iu. A. Filipchenko. Several of these distinguished scientists were also heavily involved in the eugenics movement.[28] One of the outstanding younger Russian biologists was Theodosius Dobzhansky, who later emigrated to the United States, where he became world famous. Dobzhansky's career began with a position in the Bureau of Eugenics of the Soviet Academy of Sciences.

The Russian eugenics movement was limited almost entirely to the years between 1921 and 1930, and this period can itself be divided into two phases, with the division around 1925. In the first phase, Russian eugenics developed along lines quite similar to those in a number of other countries. The German example was probably the most significant, largely because of the number of German monographs on the subject and the closeness of Russian and German scientific relations. In the second phase, after 1925, Soviet eugenists made an effort to create a unique socialist eugenics of their own, an effort which met increasing opposition. Lysenko and the form of Stalinist genetics which later became notorious throughout the world were not involved in either of these phases (these developments came only in the late 1930s and 1940s); nonetheless, the controversies over eugenics in the 1920s were important elements in the foundation of later Soviet attitudes toward genetics.[29]

The two most important organizations in the Soviet eugenics movement were the Russian Eugenics Society and the Bureau of Eugenics of the Academy of Sciences, both created in 1921.[30] Each published a journal, the first entitled the *Russian Eugenics Journal*, the second (in its initial form) *Bulletin of the Bureau of Eugenics*. The latter was more academic than the former and also proved to be considerably more sensitive to the political opposition that soon developed to the eugenics movement. The members of the Bureau of Eugenics were professional geneticists, interested in studying human heredity from a scientific standpoint and anxious to protect this study from political considerations, while the members of the Russian Eugenics Society were a heterogeneous collection of scholars and lay people. The popular aspirations of the society were illustrated by its call at the end of the first year

> for cooperative work on eugenics by scientific and social workers in all specialties. As an expression of the most vital cultural problems, as a question of the eventual fate of culture, the eugenic idea must sooner or later gather under its banner all intellectually aware human beings. We must hurry, while there is still time, while it is still possible to ward off many of the misfortunes that are threatening cultured society, misfortunes that are the result of the one-sided development of modern civilization.[31]

The historian who today leafs through the pages of the *Russian Eugenics Journal*, knowing well the class antagonisms and radical currents still waiting to be expressed in Soviet society, is struck by the naiveté and blindness to political complications shown by the early leaders of the movement. One of the early concerns of the journal's authors in the years immediately after the revolution was the genealogy of outstanding and aristocratic Russian families; investigations, complete with family tables, were made of princely families of exemplary achievements, as well of all the members of the Academy of Sciences in the previous century. Several writers expressed dismay about the dysgenic effects of the Russian Revolution. The emigration of the nobility and of other upper-class families as a result of the revolution was seen as a serious loss to the genetic reserves of Russia, requiring eugenic correction. More predictable, perhaps, was concern about the enormous Russian losses during World War I, far heavier in absolute (though not in relative) terms than those of any other nation.[32] One Soviet eugenist, V. V. Bunak, summed up the effects of eight years of war, revolution, civil war, and famine in the statement that the total genetic impact "might exceed that of the West" in the same period. Thus, the deep eugenic gloom of postwar Germany was reflected among Russian eugenists, although there was some hope for backward Russia in that "the more cultured a country, the more the biological danger of war."[33]

The favored word in Soviet Russia for the subject was *evgenika* (eugenics), although the term *rasovaia gigiena* (race hygiene) was also used and racial interests were widespread. The Russian Eugenics Society soon established a special commission for the study of the "Jewish Race," following a major interest of the

German movement.[34] In one study the commission concluded that Jews were in no way inferior to other ethnic groups. Despite the conclusion, the Russian eugenics movement explicitly thought in terms of racial differences, psychological as well as physiological, and followed the Germans in accepting such terms as the biologically meaningless "Jewish race." The *Russian Eugenics Journal* frequently contained reviews of German books that were establishing the guidelines of race hygiene in terms of a static or typological definition of race (such as the famous textbook on human heredity by Baur, Fischer, and Lenz), and the reviewers were usually impressed by the new doctrines.[35] The first issue of the Russian journal contained reviews of fourteen German books on human heredity and eugenics and no others. Kol'tsov and Filipchenko also wrote several articles for the main German eugenics journal, using the terms *Die rassenhygienische Bewegung* (The Race Hygiene Movement) and *Die rassenhygienische Literatur* (Race Hygiene Literature) to describe Russian trends.[36]

The leaders of the Russian eugenics movement were anxious for contacts with other national eugenics societies and, indeed, worked for full international recognition. The fact that Soviet Russia was a revolutionary state that attacked the practices and world views of leading nations in most areas of international political and cultural relations seems not to have deterred the early activities of the Soviet eugenists. They established contacts with the Eugenic Education Society in England, the Eugenic Record Office in the United States, and the German Society for Race and Social Biology. In November 1921 the Russian Eugenics Society was recognized as a full member of the International Eugenics Union with headquarters in London. But in the international arena, Bolshevik Russia (as well as the former enemy nation, Germany) was ostracized by the former Allies at many international scholarly congresses after World War I, and this policy extended to eugenics, despite the formal contacts. Thus, neither Soviet Russia nor Germany was invited to the Second International Congress of Eugenics in New York in 1921, a slight that deeply annoyed the Russian eugenists.[37] This ostracism, however, had the effect of deepening scholarly connections between the two pariah nations. The president of the Russian Eugenics Society, the biologist N. K. Kol'tsov, did manage to obtain an invitation to the Third International Congress of Eugenics in Milan in 1924, where he lamented that the influence of the Catholic Church was so prevalent that the participants were one-sidedly cautious in their discussions of practical eugenic measures. He also noted that, if the attitudes of Catholic Italians were one extreme, those of the Americans with their system of sterilization laws, were another. No Americans or Germans were present at the Congress and, according to Kol'tsov, not one participant spoke in favor of the American system of sterilization.[38]

The eagerness of Russian eugenists for international contacts and recognition was also evident in the pages of the *Russian Eugenics Journal*, which regularly printed news of the other societies as well as summaries of the publications of the other eugenics journals; the *German Archive for Race and Social Biology* received the greatest attention.[39] When the German Society for Race Hygiene passed a series of resolutions in 1922 calling for greater attention by legislators to eugenics, the *Russian Eugenics Journal* printed the resolutions in full without comment.[40] Similar attention was given to proposals of the English and Swedish eugenics societies.[41] In 1924 the journal gave publicity to a German physician who appealed to all of his "colleague-physicians both in the cities and in the countryside to search for defective persons, taking advantage of the help of local authorities, teachers, philanthropic institutions, medical personnel, and to sterilize as many of them as possible."[42]

Although the Russian eugenists shared many assumptions of the international eugenics movement in general and of the German movement in particular, they should not be seen as proto-National Socialists. Substantive links between the Russian eugenics movement and incipient fascism were completely lacking. Similar eugenics organizations existed in most Western nations at that time and, in fact, were far stronger in England and the United States than they were in Russia.

Several early Soviet eugenists showed awareness that their analysis and program were potentially inflammatory from a political standpoint. In the early spring of 1925 the Russian Eugenics Society debated eugenics proposals that had been advanced in other countries, and some Soviet participants objected to the "coarseness" and ill-defined quality of a number of these proposals. The Leningrad branch of the society dissociated itself from plans to prohibit racially mixed marriages, and they voted in favor of sterilization "only on the basis of a decision of a special council of competent individuals and with the permission of the individual or his legal representatives and only on eugenic, not social, grounds."[43] They did, however, vote in favor of obligatory isolation from the opposite sex of "certain categories of the mentally ill and habitual criminals whose reproduction would be dangerous for society on social and eugenic grounds." But the real significance of the Russian eugenics movement lay not in its relationship to Western eugenic proposals, interesting as those connections were, but in its place in the debates over the nature of man that were beginning to take place in Soviet Russia itself. How was the eugenics movement perceived by Soviet intellectuals outside the community of eugenists? What was the relation of the doctrines of eugenics to the ideas of Russian socialism and communism?

The great debate over the issue was slow in developing. The commissar of public health, Nikolai Semashko, had given his approval to the eugenics movement; the Commissariat of Internal Affairs (a police organization) formally accepted the charter of the Russian Eugenics Society; and the Russian Eugenics Society received a small state subsidy.[44] These first official acts of recognition were probably not too significant in themselves, since in the early years the concept of eugenics was so new to most people in Soviet Russia, the essential issues so unexplored, that a sophisticated understanding of the movement by bureaucrats was hardly possible. To the extent that eugenics was understood, it was thought to be the science for the collective improvement of mankind, and as such it was an activity that the young Soviet government automatically found interesting. In 1925 the debate that had begun to simmer in lecture halls and in local publications spilled out into major Soviet intellectual journals. The activities of the Russian Eugenics Society and the Bureau of Eugenics—up to this time primarily the concerns of a rather small circle of geneticists, physicians, public health specialists, and university students studying the same fields—now became an object of public attention.

In one of the first comprehensive and critical analyses of the eugenics movement, Vasilii Slepkov, in his article "Human Heredity and Selection: On the Theoretical Premises of Eugenics" (which appeared in April 1925 in the major Bolshevik journal of Marxist theory), pointed out that the eugenists were emphasizing biological determinants of human behavior to the total neglect of socioeconomic determinants. Since most eugenists were biologists with little knowledge of the social sciences and less of Marxism, they had "absolutized" the influence of heredity to the detriment of the influence of the environment and thereby converted genetics into a "universalist" interpretation in which all of human history was merely a story of the replacement of one genotype by another. They had accepted the views of exponents of the Nordic race, who explained European history in terms of the domination by superior genotypes who became the European nobility. This conviction necessarily led to cultural pessimism, since the eugenists believed that the nobility was no longer reproducing at a rate sufficient to ensure its survival. Slepkov pointed out that this view totally ignored the principles of Marxism, which demonstrated that social conditions determine consciousness. Slepkov quoted Friedrich Engels on the importance of labor in the evolution of primates, Georgii Plekhanov's point that conservative thinkers had always explained human behavior on the basis of innate qualities in order to avoid social analysis leading to revolutionary conclusions, and Karl Marx' concept that "people are a product of conditions and education and, consequently, changing people are a product of changing circumstances and different education."[45]

A thief is not a biological type, created by heredity, said Slepkov, but a "social man" created by his environment, poverty, and unemployment. By emphasizing the hereditary differences of individuals to the complete neglect of social influences, the eugenists were inexcusably distorting the general understanding of human nature. Slepkov did not deny, however, that individuals differ genetically. In order to explain those differences he resorted to a theoretical position that was to have a long, and eventually tragic, influence on Soviet biology: that of the inheritance of acquired characteristics. He believed that Pavlov's research had demonstrated the influence of the social environment on human behavior. And he pointed out that Pavlov believed that conditioned reflexes acquired during the lifetime of an organism could become, in some cases, permanently hereditable. Slepkov also referred to the research of Kammerer on salamanders as additional evidence of the inheritance of acquired characteristics, ignoring the fact that Kammerer's work was rejected by the great majority of biologists. Kammerer's political radicalism, however, was well known and Slepkov used it to raise the credibility of his proposed expansionist correlation of Lamarckism and political reformism.

As criticism of their naive and one-sided biologism, much of Slepkov's castigation of the eugenists was appropriate; but, by placing so much emphasis on the inheritance of acquired characteristics, Slepkov praised antique biology in order to humanize modern biology. A more judicious criticism would have been to dissect the dubious assumptions of eugenists that society was threatened by a biological degeneration that demanded social and legislative solutions. Throughout the twenties, however, the inheritance of acquired characteristics remained only a remote candidate to replace classical genetics. The geneticists and eugenists still had very powerful arguments at their disposal. Even in political terms, in the Soviet Union during the mid-1920s few assumed that Marxist socialism and eugenics were incompatible or believed that Marxism and Lamarckism were uniquely compatible.

Not only the eugenists with their arrogant programs for the biological reform of society but also the more sober and scientific geneticists interested primarily in animal and plant heredity were beginning, nevertheless, to meet stiff opposition from radical students in Soviet universities. One biologist with a political commitment to the new regime, B. M. Zavadovskii, wrote that each year, when he gave lectures at the Sverdlov Communist University, the radical students reacted with hostility to his discussion of heredity; they called genetics "a bourgeois science," which contained implications that were unacceptable to the proletariat. Voices in favor of Lamarckism were becoming "louder and stronger," he wrote in 1925, as was the belief that Mendelian genetics contradicted Marxism and the social policy of the Communist Party. "This point of view is receiving support in the psychology of the masses," he commented, "whose first reaction to genetics is negative." Zavadovskii recognized that among educated Marxist scientists and social theorists the attitude toward genetics was much more positive than among students and lay people, since serious Marxists were anxious not to contradict the findings of science and many fully recognized the scientific baselessness of Lamarckism. Yet Lamarckism continued to grow, especially among pedagogues, who believed that education was much more important than heredity. Zavadovskii feared that, in dispensing with the erroneous views of the eugenists, Marxists might "throw out the baby with the bath water" and eliminate the science of genetics as well. Marxists, he said, have been "frightened" by the conclusions of bourgeois eugenists.[46]

By 1925, then, the first crisis in the debate between Mendelian or classical genetics and Lamarckism was coming to a head in the Soviet Union, and the eugenists were in the middle of it. The eugenists began to defend both genetics and eugenics simultaneously, and they considered the Lamarckians to be one of the most important groups among their opponents. As early as 1924 the most prominent Russian eugenist, Kol'tsov, had published an article, "New Attempts to Prove the Inheritance of Acquired Characteristics," in which he had dismissed such efforts as "the replacement of science by faith."[47] Several of the eugenists realized that the criticisms advanced against them were so serious that, unless some way could be found to reconcile their understanding of human heredity with Marxist aspiration, genetics itself was in grave danger. Iurii Filipchenko, one of the brightest of the eugenists, found an argument in favor of genetics that carried great weight, and for a time some observers thought he had outmaneuvered the radical critics of genetics. Let us assume for a moment, said Filipchenko, that the Lamarckians are correct and the Mendelians incorrect, even though scientific evidence at the moment all points in favor of the classical Mendelian theory. What would be the result for the proletariat, for the lower classes in general, and for the cause of social revolution if acquired characteristics were inherited? Most people seem to believe that such a theory points to rapid social reform, since by giving individuals a better environment, not only would those individuals benefit during their lifetimes, but supposedly the good effects of that better environment would be passed on to subsequent generations, holding out the possibility of rapid improvements built on a successively improving genetic base. The view of an unchangeable genotype contained in classical genetics, on the other hand, seemed to set iron limits to such improvement and therefore seems inherently conservative in its social implications.[48]

This view was superficial and false, said Filipchenko, because it assumed that only "good" environments have hereditable effects, while a consistent interpretation of the inheritance of acquired characteristics would show that "bad" environments also have effects. Therefore, all socially or physically deprived groups, races, and classes of people, such as the proletariat and peasantry and the nonwhite races, would have inherited the debilitating effects of having lived for centuries under deprived conditions. Far from promising rapid social reform, the inheritance of acquired characteristics would mean that the upper classes are not only socially and economically advantaged, but also genetically privileged as well, as a result of centuries of living in a beneficial environment. Thus the proletariat in Soviet Russia would never be capable of running the state; it was genetically lamed by the inheritance of the effects of its poverty. If the classical geneticists were correct, on the other hand, said Filipchenko, then combinations of genes were distributed throughout the lower classes that would give the individuals possessing them all the possibilities for being great scientists, musicians, artists, or whatever they might wish to be, if only the exploitative, debilitating environment were eliminated.

The logic of Filipchenko's argument seems to have caused critics of the geneticists to hesitate. Several radical journals ran articles that maintained that only the inheritance of acquired characteristics, not Mendelian genetics, was counterrevolutionary. One author stated that the international bourgeoisie constantly renewed efforts to establish the inheritance of acquired characteristics in order to show its own genetic superiority, but the proletariat was learning that science spoke against them. Another author, writing in *The Red Journal for All People*, said that every social reformer should read Filipchenko's argument in order to be armed for the political struggle.[49]

Filipchenko's argument was not persuasive enough, however, to stem the popular belief that the inheritance of acquired characteristics was more congenial to the idea of creating a new society and a new culture than was Mendelian genetics. First of all, the majority of radical reformers, in a time of social revolution, either never listened to or never understood Filipchenko's fairly sophisticated argument. Simple arguments won out over subtle ones. Many commentators agreed with Zavadovskii that the first reaction of young Soviet students, on hearing of the dispute between classical genetics and Lamarckism, favored Lamarckism because of its apparent possibilities for the rapid transformation of man. In other countries as well the nature-nurture argument has often been equated, quite exaggeratedly, with a political split between the right and the left. And even at the source there seemed to be some support for this view, for Lamarck had himself risen to high position after the French Revolution. Events in Germany in the late 1920s seemed to confirm the reverse side of the historical analogy: conservative race hygienists followed strict Mendelian genetics. Reasons for hesitating to make this neat equation—that Lamarckism has many conflicting interpretations of which a number are not closely linked to reformism or materialism, that Filipchenko's argument was logically stronger than those of his critics, that the whole question of directly linking politics to biology is intellectually dubious—all of these subtle points faded before that first and simple assumption of lay people that genetics seemed to set absolute limits to human improvement while the inheritance of acquired characteristics seemed to open limitless possibilities.

Even those relatively few scientifically educated Soviet participants in the debate who listened to Filipchenko's view and understood it were not all convinced, some for fairly good reasons. After all, if Filipchenko were correct about the debilitating effects of an assumed inheritance of acquired characteristics on the lower classes, a great deal would depend on how long it would take to erase those effects—many generations, or one or two? Furthermore, the most that Filipchenko could promise on the basis of his classical genetics was that *some* members of the proletariat could excel because of their fortuitous possession of the right genes. The eventual result would be a class society based on innate, unchangeable ability, a meritocracy. The inheritance of acquired characteristics held a more radical, democratic prospect: *all* members of the lower classes and their progeny might advance equally because of their social environment, and, given sufficient time, there was no reason for the former lower class to be at any genetic disadvantage to the former upper class.

Several Marxist scholars tried to develop sophisticated arguments to meet Filipchenko's intellectual challenge. Volotskoi, who had abandoned his earlier commitment to eugenics based on classical genetics and moved toward a Lamarckian eugenics, held that Filipchenko's mistake lay in believing that the genetic effects of an assumed inheritance of acquired characteristics would be unilaterally harmful to those who came from a background of exploitation and unilaterally useful to those who came from a privileged background. Marxists understood, said Volotskoi, that the division of labor that led to societies based on slaveowners and slaves, lords and peasants, or capitalists and workers has harmful *and* beneficial effects for both groups. In a host-parasite relationship in botany or biology, the parasite often degenerates because it obtains so much of its sustenance from the host; and the slaveowners, nobles, and capitalists of history have similarly degenerated while benefiting from the lower classes. On each side of exploitative relations, said Volotskoi, there is a balance sheet of plusses and minuses: the proletariat suffers from poverty, lack of culture, and unsanitary conditions, but it benefits from the physical labor and hard work necessary for survival; the capitalists benefit economically from their privileged position, but they do not have to labor and become slothful and corrupt. Therefore, Volotskoi continued, Filipchenko was incorrect in believing that an assumed inheritance of acquired characteristics would work only to the benefit of the upper classes; its effect would be mixed. The solution, he maintained, was to abolish the division of labor and conditions of privilege and suffering, for then human beings could be perfected through better social environment.[50]

Yet, in trying to refute Filipchenko, Volotskoi enlarged an assumption that was increasingly criticized by other Soviet Marxist theoreticians: the expansionist belief that biological laws could be easily extended to human beings. It would be better, the theoreticians maintained, to divide human from animal or plant heredity and to recognize the nonreductive and qualitatively distinct nature of man. In that way many difficult problems could be avoided. Furthermore, dialectical materialism, the Soviet Marxist philosophy of nature, incorporated this theory of nonreductive levels into its overall perspective. Volotskoi confirmed the suspicions of the Marxist theoreticians by marching right ahead with the idea of eugenics, much as he decried "bourgeois eugenists."[51] Indeed, in his promotion of a socialist eugenics based on Lamarckian biology, he went so far as to suggest mandatory sterilization and thus took a minority position.

Among scientists, these debates essentially resulted in a draw on the issue of whether or not the inheritance of acquired characteristics was more consonant with Marxism than was classical genetics. On a popular level, however, the belief among lay people and students that genetics was a bourgeois science continued to be strong and even grew as social conflict increased at the end of the decade with the commencement of the five-year plans. The obvious losers in the debate were the Russian eugenists who defined their field in terms of the Western eugenics movement. Not were they only supporters of classical genetics, with all of the controversy that field attracted in the Soviet Union, but they went far beyond these theoretical principles to extrapolations of biology to society. In these extrapolations they included a host of assumptions about the future of society, the nature of races and classes, and the relative influence of nature and nurture that did, indeed, conflict with prevalent views among the politically active elements of Soviet society. As a result of their possession of these distinct and controversial views and of their links to ideologically hostile foreign scholarship, the Russian eugenists became more and more easily defined as "bourgeois specialists," an appellation that increasingly became akin to a criminal charge.

The eugenists recognized that they must either abandon their concerns or radically change their activities in order to demonstrate that they actually had the interests of Soviet socialism, not Western capitalism, at heart. The scientists in the Bureau of Eugenics at the Academy of Sciences chose the prudent path of abandoning the field. Between 1925 and 1928 they shifted from a concern with human heredity to a concern with the genetics of plants and animals. The change of emphasis can easily be seen in the successive titles of their leading journal:

1922–1925	Bulletin of the Bureau of Eugenics
1925–1927	Bulletin of the Bureau of Genetics and Eugenics
1928	Bulletin of the Bureau of Genetics

The last articles in this journal concerned predominantly with human genetics appeared in 1925. Eugenists in the Academy of Sciences became plant and animal geneticists primarily interested in the most exciting experiments in the field of their time, the study of *Drosophila melanogaster*, the fruit fly. But their eugenics past was to be held against them in the troublesome 1930s.

The Russian Eugenics Society chose a more heroic and foolhardy path. The editors of the journal decided to change their emphases, to show how eugenics could be fitted to the purposes of social revolution and Marxism. Turning away from their genealogical studies of the nobility, they began to make studies of the reformers and revolutionaries of Russian history, beginning with the Decembrists of 1825 but continuing right up to contemporary Communist Party leaders, whom they noted were not reproducing at an adequate rate.[52] Kol'tsov did a study of the genetic backgrounds of the talented young proletarians being promoted up through the ranks of the universities and institutions of the Soviet Union.[53] Other eugenists began developing extended justifications for a unification of the goals of Soviet socialism and eugenics. But the whole effort became increasingly artificial and strained.[54] The eugenists with their record of connections with the international movement—including German race hygiene with its ever-clearer links to National Socialism—could never justify themselves by trying to be more radical than their critics, much as they were by then attempting. A scholar like Filipchenko could win individual debates through the strength of his arguments about the social implications of an assumed inheritance of acquired characteristics, but by style and background he was always a middle-class *intelligent* to his critics.

By 1930 the eugenics movement in the Soviet Union was finished. These were years in which political controls in the Soviet Union were imposed on many scholarly institutions, and collectivization was violently enforced in the countryside. The Russian Eugenics Society was closed and its publications suspended. By 1931, when the *Large Soviet Encyclopedia* published the volume of its first edition with an article on "Eugenics," the field was simply condemned as a "bourgeois doctrine."[55] All efforts to create a unique, socialist eugenics were ridiculed. By this moment in Soviet history the logical possibility of linking the Marxist desire to transform man with the eugenic desire to improve him had irretrievably disappeared.

The end of eugenics meant the end of discussions of human heredity, but it did not mean immediate victory for the doctrine of the inheritance of acquired characteristics. Debates on that issue had been indecisive, and the communist philosophers and political leaders did not wish to fall unwittingly into contradiction with science. The terms of the de facto intellectual truce were pragmatic: let the partisans of classical genetics and of the inheritance of acquired characteristics resolve the issue in practice, in the applied science of agronomy—not in the dangerous field of human heredity where the Germans in the 1930s were winning such notoriety, but in agriculture where the Soviet Union needed help. This truce set the stage for the rise of a clever schemer who could capitalize on the inability of geneticists in the early 1930s to produce agronomic miracles because of the immaturity of their science (hybrid corn and similar genetic innovations were applications of the 1940s), while he, surrounded by political supporters, could expose the eugenic past of a number of the prominent geneticists who tried to show the fallacies of his own innovations. Political loyalties, historical associations, and dedication to practical support of Soviet agricultural policies became far more important in determining the results of the discussions than logical rigor or scientific validity.

THEORIES OF HUMAN HEREDITY AND POLITICAL VALUES

By the early 1930s the process of the gradual crystallization of value links to conflicting concepts of heredity—links that were at first by no means clear or inevitable—was far advanced in Germany and Russia. Eventually the two societies went in opposite directions in their interpretations of the nature-nurture controversy. The genetic doctrines and practices of Nazi Germany postulated the existence of genetic differences among races and other categories of humans that have much greater importance in determining the worth and performance of individual members of those groups than anything that can be done during an individual lifetime. Thus, according to this assumption, "nature" obviously far outweighs "nurture." In the same years in which these doctrines became widespread in Germany, an opposite and equally unsubstantiated theory of genetics gained support and, finally, official approval in the Soviet Union. Lysenko's theory of agronomy, based on the inheritance of acquired characteristics, assigned an overwhelming weight to the nurture side of the equation. His naively optimistic theory of inheritance granted a much greater degree of control than the results of research in biology had indicated was possible.

These two chronologically simultaneous episodes with contrasting results are probably the best large-scale "test cases" that we are going to find in the flux of history for the question of whether or not different values are inherent in different theories of heredity. The known scientific facts of biology and the tested, verified theories of heredity were the same in both countries. Yet each, on the basis of different value systems, embraced different theories of heredity, and each labored mightily in later years to show the harmony between its biological science and its social morality. What do these examples tell us about the relationship between science and sociopolitical values?

Granted, these two episodes are not ultimately the ideal types for asking such a question, since the sets of variables in the two countries were, not surprisingly, different. For example, the official ideologies which later developed out of the controversies of the 1920s described here were not true opposites—that is, they were not mirror images of each other. For that to be the case, two different theories of *human* heredity would have had to be fitted to two different political ideologies. That logical circle was never quite closed, since Lysenko's theory of inheritance was never explicitly applied to man, only to plants and animals. The concept of a Soviet eugenics—of Mendelian, Lamarckian, or whatever kind—was condemned by Russian political leaders after the early

1930s. In a sense, however, that very condemnation is the exception that proves the importance of the interaction of scientific theories and political values. The Germans in the 1930s used eugenics in such a horrifying way that the entire concept of eugenics was discredited even in those countries where both the prevailing value goals and the particular theory of heredity contrasted sharply with the goals and theory of the Germans. The world thus never witnessed the ironic and bizarre scene of two different societies aiming toward two different sets of value goals by applying two different (both incorrect) theories of human heredity. But the potential for this scenario was present.

From the perspective of the present, it may appear that there is a natural alliance between eugenics and conservative, even fascist, sentiments. That link was not logically preordained, however, and was not perceived in the early 1920s by large numbers of radical critics of society. Marxists and socialists of many types then supported eugenics, as did liberals, progressives, and conservatives. In the early 1920s scientists and publicists who were interested in eugenics covered a rather wide range of political sentiment in Weimar Germany and Soviet Russia. In these early years eugenics was a faddish doctrine that was often considered progressive, the latest application of science for the benefit of man. If sometimes supported by aristocratic devotees of genealogical tables or middle-class members of social clubs, it also on occasion had support from committed socialists who believed that cultivation of true talent, rather than mere economic privilege, would destroy class society as it had been known. Both Weimar Germany and Soviet Russia were revolutionary states that had replaced recent monarchies. In the 1920s both were experiencing eclectic periods when the full political implications of the latest scientific hypotheses had still not been formulated.

Were the passing of this early heterodox period and the emergence of a high correlation between theories of heredity and political world-view phenomena that touched science itself, or were they only epiphenomena of social and political turmoil? Do different theories of heredity intrinsically contain different value implications? Is expansionism here inevitable and unilateral? The answers to these questions depend in part on what the main rival theories may be. If the contending theories are defined in terms of their post-1933 vulgarized and absolutized forms (that is, a National Socialist theory categorizing humans according to their "overall" genetic value—with Jews, gypsies, and certain types of "social misfits" at the bottom of the heap—opposed to a Soviet theory that dogmatically attributed genetic progress to improvements in man's social environment), then the scientific "theories" in themselves contained by definition clear value components. Obviously these absolutized explanations, encased in official ideology and supported by the respective governments, can hardly be classified as "science." Yet this fact should not be used as a convenient exit from the dilemma of the relationship of science to values. This chapter has intentionally been restricted to the period before 1930, when professional geneticists and eugenists in both Germany and Russia usually defined the contrasting theories in more academic terms. They often saw the difference between vulgarized extrapolations and core theories. Even Fritz Lenz spoke out in the 1920s against National Socialist "excesses" which, he believed, harmed the eugenics movement.[56]

During the German Weimar and Soviet New Economic Policy (NEP) periods, the rival core theories, stated in their scientific forms, did not explicitly contain value statements. The rival theories were those of Mendelian genetics and Lamarckian genetics, viewed as scientific alternatives. To many people they seemed to contain strikingly different value implications. One reason for the different value connotations stems from the state of genetics in the 1920s as a pure and an applied science. Whether or not science *qua* science contains values, most agree that technology does have impact on values. Therefore, the significance for values of a particular science will vary as the associated technology for that science develops. One interesting aspect of the development of human genetics in the 1920s is that not only was human heredity understood in an exceedingly inexact fashion, allowing much room for speculation based on social and class motives, but also there was no technology available for controlling the genetic constitution of organisms except by selection on the basis of phenotypes. Such selection was offensive to existing values when applied to people, and therefore humane individuals wished for an alternative even if they accepted the goal of more control over human genetics.

The relatively few geneticists who did not fall prey to the political ideologies of the time were left with a very undramatic argument, which could not carry their audiences in times of great social stress when the politically active were striving for answers to societal problems. The rigorous geneticists were saying, in effect: Yes, the science of genetics is in principle applicable to man. Genetics is, however, so immature as a science that any effort to apply it to man now would have disastrous and unpredictable effects. Therefore, do not try to apply human genetics as a science, but do continue to believe us when we tell you that science is ultimately the best hope for man.

In revolutionary situations popular audiences are not likely to find such an argument persuasive. Both Germany and Russia were undergoing social and political upheavals, and the audiences turned toward more radical, although quite different, answers to the problem of heredity, answers that contained possibilities for application. The German academic eugenists (as opposed to the race hygiene anthropologists) like Fritz Lenz were closer to science (as it then existed) when they attacked Lamarckists for their sentimental biology than were the Soviet Marxists who rejected genetics along with eugenics. The tragedy is that those scientists who, contrary to Lenz, remained loyal to genetics as a scientific theory, while rejecting the growing inhumane and antirational extrapolations of German race hygiene, were unable to find appreciative audiences in either Germany or Russia. The reason for their failure no doubt lies largely in the social, political, and economic strains both societies underwent. To give the details of those strains would require a long digression into the social and political histories of Weimar Germany and Soviet Russia. But the existence of these strains was only a necessary, not sufficient, condition for the emergence of radically different attitudes toward heredity. Also involved was a science of heredity so immature that it allowed room for wide speculation while offering few applications that were both reliable and ethical.

Particularly worthy of notice is the argument of the young Alfred Ploetz, still in his humanistic phase, because he emphasized that, if the time ever came when selection could be shoved back before birth, it might be possible to develop eugenic measures that were less offensive to traditional values than many of the proposals advanced in his day. Today, Ploetz's observation has more meaning—at least for specific genetic defects—than it did when he wrote. The differential impact on values of various technologies thus serves as a further warning that we should not attribute values easily to scientific knowledge itself; the more important value determinant is likely to be the technology derived from that knowledge.

We are now approaching the core issue in the question of the relationship of science to values as revealed in our study of Germany and Russia in the 1920s. We have already separated the question of political and social motivations from that of scientific inquiry to the maximum degree possible (it is never entirely possible) by focusing on the more academic debates of the 1920s, not on the ideological exhortations of the 1930s. We have separately considered the question of technology, trying to show where it had impact on value that is distinguishable from that of scientific theory. As the layers of this problem are peeled away, the central question—one that relates to scientific theory alone—emerges: Was there something inherent in a hypothetical Lamarckian theory of heredity that made it a substantial buttress for egalitarian political values, and something within the Mendelian theory of heredity that lent support to elitist views of society? Iurii Filipchenko tried to come to grips with this problem in 1926 when he maintained, with some success, that the Lamarckian theory of heredity, followed to its conclusions, would postulate the genetic impairment of those social classes that left-wing social reformers wanted to help. The existence of the deprived classes in poor environmental conditions for many generations would supposedly have resulted in the acquisition of hereditary characteristics that would genetically impair them.

This argument was also used by the American geneticist and socialist Herman J. Muller in his criticism of Lamarckism. (Muller and Filipchenko knew each other and the connection may not be coincidental.) Yet the argument by itself means very little. First of all, it is not completely clear that in a hypothetical case deprived environments would lead uniquely to genetic damage; some people might maintain the reverse, seeing a sharp selection of the "fittest." Since Lamarckism is a doctrine without empirical support, there is no way of knowing how a hypothetical Lamarckism would fit with normal Darwinian selection. Furthermore, no one knows—nor can anyone know in the absence of facts or verifiable hypotheses—how many generations would be required for a hypothetical Lamarckism to overcome the effects of earlier genetic impairment—one generation or many?

Two assumptions were carried by most Lamarckists into the debates over human heredity in the Soviet Union of the 1920s: either deprived environments do *not* result in impaired heredity, or, if deprived environments *do* result in impaired heredity, this impairment can be easily erased in a generation or two in the future through Lamarckian inheritance in improved conditions. If one makes these assumptions, then Lamarckian views truly fitted more comfortably with social reformism than did Mendelian genetics, and this "science" contains real value implications. The Lamarckists could then promise the quick advent of a bright genetic future for all members of the proletariat (or any other group) while the Mendelians could promise such a future only to the fortunate few. This formulation of the argument was the generally accepted version in the Soviet Union by the 1930s. Filipchenko lost the debate, the Lamarckists eventually won. But it should be clear that the chains of assumptions here took both into unreal worlds.

On examining the relative value ingredients supposedly inherent in different theories of heredity, we see that Filipchenko was not the only person to maintain that positive, humane values are inherent in theories of heredity that postulate a relatively stable genetic material and that negative values lurk in the supposedly humane Lamarckian view of a malleable genetic base. Noam Chomsky has taken the view that man's relative immalleability is a protection for him against potential tyrants:

> One can easily see why reformers and revolutionaries should become radical environmentalists, and there is no doubt that concepts of immutable human nature can be and have been employed to erect barriers against social change and to defend established privilege.

> But a deeper look will show that the concept of the "empty organism," plastic and unstructured, apart from being false, also serves naturally as the support for the most reactionary social doctrines. If people are, in fact, malleable and plastic beings with no essential psychological nature, then why should they not be controlled and coerced by those who claim authority, special knowledge, and a unique insight into what is best for those less enlightened?[57]

This statement contains not only an observation on aspiring genetic manipulators of man, but also the concern of a more recent generation about the control of human behavior. From Chomsky's standpoint, both the eugenic selectionists of Germany and the Russian "makers of a new Soviet man" represented threats to human freedom of a fundamental sort. Yet Chomsky's observation, valuable though it is, presents only one more insight and does not attempt to give an ultimate analysis of whether putative theories about the mutability of man, genetic or behavioral, contain inherently positive or negative value. He admits that the theories can work both ways, with hypothetical immutability "employed to erect barriers against social change and to defend established privilege" and hypothetical mutability used "as the support for the most reactionary social doctrines." *Which* way the theories would work in a given historical situation depends on the values of the political and scientific authorities who would employ the theories and the associated technology.

In my opinion the present state of our knowledge of human heredity does not allow an abstract answer—that is, an answer apart from reference to available technologies and existing sets of social forces—to the question of whether available theories of the nature of man have in themselves positive or negative value content. In the end, heredity and the environment form our values in ways that can be scientifically studied, and in that theoretical sense the expansionists are correct. But until we have a much more complete knowledge of human heredity, its relationship to neurophysiology, and the influence of the environment on behavior, we can not give a satisfactory answer to the question of the relationship of heredity to values. In the interim, we have to deal with different sets of incomplete information available at different historical and social moments.

But in given historical situations, such as those in Weimar Germany and Soviet Russia in the 1920s, rival scientific theories about human nature have invariably arisen within the context of given social and political circumstances, of competing political ideologies and economic motives, and of existing systems of technological capabilities. Within those frameworks of external factors, rival scientific theories *do* have differential value implications; but they derive their value meaning much more from their relationships to these external considerations than from anything inherent in the science. In terms of impact, then, we are not asking the right question when we ask, "Do scientific theories contain in themselves social or political values?" Instead, we should pay more attention to what can be termed the "second-order" links between science and values—those contingent upon existing political and social situations, current technological capabilities, and the persuasiveness of current ideologies, however flawed in an intellectual sense they may be.

If we look back at Germany and Russia in the 1920s with these overall extrascientific considerations in mind, it is clearly far from accidental that the societies eventually ended up on the opposite sides of the nature-nurture dispute with which we now associate them. Arguments for radical egalitarianism *could* be made more easily and more persuasively in accordance with Lamarckian environmentalism, and arguments for hierarchies of social and racial values *could* be placed more comfortably within the Mendelian system. The more obvious points of debate leaned in these directions, while the more subtle ones were not heard at moments of social and political trial. In this global, approximate, and intellectually flawed sense, then, these rival scientific theories did contribute heavily to arguments about values. These value implications derived from the relationships of the theories to their social and economic milieux, but that is normally the situation with science.

Since I have maintained that the theories of human heredity current in Soviet Russia and Weimar Germany in the 1920s contained in themselves much less inherent value component than is often thought to be the case, my viewpoint might easily be considered as a buttress for the argument that science is really value-free. My interpretation might even be used to attempt to free scientists from a sense of responsibility to society. My reply to this misunderstanding of my analysis would be to point out that the value-free interpretation of these theories of human heredity is persuasive only if the links of science and society are severed and if science is viewed in abstract isolation from its setting. In fact, every scientific theory and every technological innovation *always* exists in a social and political setting, and the impacts of these combinations upon the formation of values can be massive.

Nuclear science, seen in isolation, is value-free. If, however, Nazi Germany had developed and employed atomic weapons in 1942, the configuration of our world might be very different today and the effects on our values would probably be incalculable. Scientific descriptions of our universe—for example, the Copernican or Ptolemaic alternatives—are, in the abstract, value-free, but the new and successful Copernican variant had a very large impact on values when absorbed by European civilization at a time when the older variant was firmly interwoven with religion and culture. Societies have thrust upon them scientific theories and technological innovations which may either buttress or counter existing value preferences. The powerful groups in society are usually more successful than the weak ones in turning science and technology to their advantage, so there is a natural tendency for science and technology to buttress the values of ruling classes and political groups. In both Germany and Russia by the 1930s, these links between ruling value systems and contrasting theories of heredity had been successfully forged, after some initial uncertainties, and the links then became powerful weapons of propaganda.

The responsibility of scientists for their creations is not less because the effects upon values of their work usually derive from a changing relationship between science and society. It is greater. They must not only try to judge what *can* be done on the basis of their work but what, in all probability, *will* be done in view of the existing social forces. Suppose, for example, that a scientist in Nazi Germany had discovered Tay-Sachs disease, a disorder with genetic causes that is more common among Jews than other groups of the population. In this hypothetical situation, the decision whether to publish and to whom to send reprints (Hitler?) should become a moral act, not because of the values inherent in the scientist's scholarly work, but because of the possible impact in that particular political setting of this purely scientific (not technological) finding.

One last word on the relevance of these episodes to the present day. I have emphasized that the contributions of rival scientific theories to formation of values depend much more on external social forces and available technologies than they do on anything inherent in the theories. We now live in a time when the social circumstances and the available technology connected with the science of human genetics are very different from what they were in the twenties. We should not, therefore, assume that the value links of our time on the nature-nurture issue are identical to those of the 1920s. To ignore the genetic basis of human beings at the present time, when we understand that basis much better and are in some instances able to affect it in reliable ways or alleviate great suffering, would be inhumane. Eugenics is now a word in disrepute, but the use of genetic knowledge to benefit humankind is a far more viable possibility now than ever before. On the other hand, some of the dangers of the 1920s are still with us. We need to examine each issue relating to human genetics separately, considering the whole context of second-order value links, and decide the issues without a priori ideological commitments.

The Nazi biology and Stalinist biology that immediately followed the events described here were fraudulent interpretations and misuses of science that could have arisen only in their particular political and social milieux. However, the explosiveness of the nature-nurture debate that lay behind these episodes, and which, mediated through structures of political control, gave them such passion, has not dissipated even today. Indeed, the discussions in the sixties and seventies in Western Europe and North America about race and IQ (Jensen, Shockley, et al.) illustrate that the nurture-nature issue still has enormous potential destructiveness. Partisans of two extreme positions attempt to seize the commanding position: One group, represented by only a few geneticists of genuine merit, would capitalize on the convincing evidence of genetic diversity in human populations to draw unsubstantiated conclusions about native intelligence (however that would be measured) of ethnic groups or taxa. The other camp, traumatized by the crimes already committed in the name of genetics in this century, would question the right of geneticists to present evidence of human diversity and would discourage or even prohibit certain types of research.

chapter nine

BIOMEDICAL ETHICS

ISSUES IN biology with controversial political and ethical implications took on a new tenor in the decades after World War II. In the first half of the century, the potentialities for the biological modification of organisms, including human beings, were limited. By the last third of the century, however, these potentialities had increased so rapidly that Donald Fleming, historian of ideas at Harvard University, wrote of a "veritable Biological Revolution likely to be as decisive for the history of the next 150 years as the Industrial Revolution has been for the period since 1750."[1]

Ethical questions presented by the new developments were myriad and were often presented to the public by journalists in sensational forms: Will sexual reproduction by intercourse be replaced by artificial fertilization or even by cloning (vegetative reproduction)? Should experiments directed toward *in vitro* development of human embryos be permitted? If such experiments are successful and result in practical application, what will happen to the legal rights and ethical obligations involved in parenthood? Will transplantation of organs, or the development of artificial organs, lead to a "fabricated man" of indefinite life span? Who has the right to decide who will receive transplants or artificial organs if only a limited number are available or affordable? If human beings can be kept alive indefinitely with the use of heart-lung or other life-sustaining devices, who should decide if and when the machines should be unplugged? If human genetics soon permits the identification of "beneficial" and "harmful" genes, and provides the ways to engineer improvements in the genetic endowment of a child, what rules should govern attempts to achieve these improvements? Who takes responsibilities for failures? If neurosurgery and psychotherapy permit behavior modification, who is to weigh the risks involved and who is to establish the criteria upon which the decisions are to be made?

Each of these questions opens up such vast areas for discussion that there is no possibility of treating them individually in any detail. Behind them all, however, lies the one essential issue of the relation of science, mediated through technology, to ethical values. These developments have caused some people to maintain that the old seventeenth-century Cartesian separation of science and values has broken down as the result of the impossibility of continuing to regard values as an autonomous system of beliefs external to science. According to the older view, values could be applied to science, but they could not be found within science, nor could science itself create a new value. Now science seemed to be continually undermining traditional values, causing the emergence of new values, and, in sum, destroying the barrier between the two realms. Many examples of such processes could be given, but I will, for the moment, give only one: To continue to accept a religious prohibition of abortion becomes more difficult for a rational woman or her husband if it is learned after examination (amniocentesis) of the pregnant wife that her child, if born, will have Down's syndrome (Mongolism), and will experience mental retardation and an early death. Here is a case where a development in the new biology appears to have the power to force a change in values. There are others, and these forced changes are what the ethical revolution in science is all about.

In the 1960s and 1970s one heard over and over again observations that ran, "The advances of the biological and medical sciences have been so great that they have caused us to reevaluate our ethical systems, and to rethink our definitions of birth, life, and death, and to change our views on what is permissible and impermissible to do to human beings. Thus, science is revolutionizing our deepest values."

Such statements obviously make an enormous claim about the impact of science on our most deeply held moral values. If they are literally true we can justly say that we are in an age of scientific determinism that has no possible comparison with any other period of time, and which historians and philosophers have not even begun to probe or explain in an adequate way. If these statements are correct, science has emerged in history as an independent, autonomous force, pushing different societies before it in remarkably similar ways, changing religious and ethical values in fashions that are much more easily explained by the internal logic of science than by socio-economic forces or by the antecedent characteristics of cultural traditions.

But is science really this powerful and autonomous a force? And are our most fundamental values changing in this dramatic way? When we discuss the issues of biomedical ethics that have become popular in recent years, to what extent are we responding to the impact of science, properly speaking, and to what extent are we responding to other kinds of forces, including social, economic and political ones? It is quite striking that in all of the recent discussions of biomedical ethics, so little attention has been paid to these questions. Most of the literature on biomedical ethics is of two types: 1) sensationalist literature about the unsettling developments that will supposedly soon become possible; 2) policy literature discussing the wisest and most ethical course a physician or a private individual should take when confronted by a dilemma of biomedical ethics. The second category is a particularly worthwhile effort, but it is not primarily concerned with the origin of the dilemma, but with its consequences. In this book I have been examining the relation of science to values, and therefore the question of origin is a primary one. For my analysis, the question of whether science is changing our fundamental values must be addressed. And as a preliminary glimpse at my conclusion, I will postulate that only our derivative (not our fundamental) values are being changed by the current biomedical revolution, and that, furthermore, science is only one of many factors responsible for these changes. In sum, science's role in the determination of fundamental values is much less direct and important in bioethics than often described.

Exemplary Issues

AMNIOCENTESIS

Let us return to the example of a change of opinion on abortion as a result of amniocentesis. As I explore this question I will be examining two subordinate issues: 1) the significance of recent fundamental science in the procedure of amniocentesis; 2) the degree to which amniocentesis does, or may, lead to a change in basic values.

Prenatal diagnosis by means of amniocentesis is a biomedical issue with ethical dimensions that has attracted much attention in recent years. The amniotic fluid surrounds the human fetus and is contained within the amnion, the inner of two fetal membranes. For many years, the fluid was assumed to be an inert pool with little direct connection with the fetus itself. In the 1930s and 1940s, however, a series of researchers showed that the fluid is in a state of continuous circulation and that there is a rapid interchange between the amniotic sac and both the maternal and fetal blood systems.[2] This interchange raised the possibility that the amniotic fluid could be used as an indicator of the state of the fetus itself, if direct physical correlations between the fetus and the fluid could be found, and if a means of safely sampling the fluid could be devised.

It was soon found that the amniotic fluid does indeed yield information directly reflecting the state of the fetus. The fluid contains not only many chemical solutes but also a number of intact cells that developed with the fetus and are genetically identical with it. Since each cell in the human body contains the entire array of genetic information necessary for the development of the whole organism, the ability to examine fetal cells obviously extended an unusual opportunity for an early look at the fetus's genetic character, at least as far as can be ascertained with present knowledge.

Despite the fact that the technique of amniocentesis itself is not entirely without risk, in the last twenty years it has become a routine procedure, particularly in pregnancies where there are known Rh complications.[3] By the early seventies the maternal and fetal mortality and morbidity linked with amniocentesis was well under 1 percent,[4] a rate considered quite acceptable by many physicians and parents when balanced against the fact that there are known categories of pregnancies where the risks involved are much higher if prenatal diagnosis is not performed. A necessary part of the decision to perform amniocentesis is, of course, completely informed consent by the mother, and, where possible, by the father as well.

The technical aspects of amniocentesis are not nearly as difficult as the moral ones. Although the process is now routine, there are certain aspects of it which can still be questioned. It is, for example, much more common among parents of high socioeconomic status than those of lower status.[5] Adequate support and adequate counseling for women of lower educational and income levels are still not available in many areas of the United States. Amniocentesis also confronts problematic ethical and legal issues when the resulting abortion comes late enough to approach the threshold of viability of the fetus, an event that may become more common as medical practice pushes the boundary of viability back further and further. Amniocentesis can also determine the sex of the fetus, opening up the issue of whether that is a legitimate reason for performing it. Also, amniocentesis is being supplemented by other means of prenatal diagnosis, such as ultrasound, fiberoptic endoscopy, and measurement of α-fetoprotein levels in maternal serum. In short, prenatal diagnosis is growing rapidly, and so is the need for guidelines for the ethical, social and legal issues involved. At least one set of such guidelines, based on extensive research and examination of the relevant issues, has been published.[6]

Technically, amniocentesis involves the insertion through the abdominal and uterine walls of a hollow needle to a depth of 3 or 4 centimeters. A small sample (5 to 10 ml) of the amniotic fluid is withdrawn, the needle is removed, and the puncture is covered with a small dressing. The patient usually waits in an observation ward for an hour or two and then she leaves as she comes, walking through the door.

Examination of the amniotic fluid can not assure the physician and the prospective parents that the future child will be healthy and free from genetic flaws, but it can reveal a number of specific abnormalities. Best known is Down's syndrome, or Mongolism, a disorder familiar throughout history that was only in 1959 shown to be based upon a chromosome abnormality. Children with Down's syndrome possess three of a certain chromosome (designated no. 21, one of the smallest) instead of the normal two, for a total of 47 chromosomes (instead of the normal 46). By the procedure of "karyotyping" of fetal amniotic fluid cells the chromosomes of an individual fetus can be counted. This procedure was first used with cultured amniotic fluid cells in 1966.[7]

On the basis of amniocentesis let us now construct a hypothetical case for a strong link between the development of science and a shift in values. We can imagine devout Roman Catholic parents who are, as a result of their religious and ethical principles, opposed to abortion and even to birth control. The woman has had six or seven children, is now forty-five years old, and has assumed that her days of child bearing are over. Suddenly, she becomes pregnant once again, and her physician informs her that at her age the chances of having a child with Down's syndrome are much higher than they were when her other children were born. In fact, she is told that when she was twenty the probability of having a child with Down's syndrome was 1 in 1,500; but now the chances have risen to 1 in 25. The woman knows of a case of Mongolism in her family and has a personal fear of such an event. She assents to the physician's suggestion that she undergo amniocentesis in the sixteenth week of gestation.

The procedure is performed, and the diagnosis is given that the fetus does, indeed, suffer from Down's syndrome. This news causes great grief to the prospective parents and an intellectual and emotional crisis. Until that time they had accepted as an absolute moral principle that abortion is wrong, that it was the unethical destruction of human life. Suddenly they find themselves engaged in a debate with themselves in which the assumptions are almost inevitably those of utilitarian ethics: Which is the greater evil, to permit a child to be born who will never have a normal life, or to terminate the pregnancy before the fetus has become a sentient human? Although the answer to this question is not inevitable, the parents might decide in favor of an abortion despite the fact that on many earlier occasions they had opposed abortion. For the purpose of our example, we will assume that the parents decided on an abortion, after much soul-searching, and that it was successfully performed.

Can we not build a persuasive case here that a value change has occurred in the minds of these parents, and that the most important cause for the change was a new development in medical technology that is closely linked to new knowledge of fundamental biology? Is not this even an example of how a factual statement("The fetus which you are bearing suffers from Down's syndrome as revealed by the presence of three no. 21 chromosomes in the amniotic fluid cells") can lead to a shift in values ("Although I have previously opposed abortion in an absolute sense, I now believe that in this particular instance it would be unethical to allow this pregnancy to come to term")? In approaching this question we must try to unravel several implicit issues.

First of all, it is fairly clear that the above hypothetical case of the Catholic parents, while specific in its details, is typical enough to be of use in our effort to explore the impact of amniocentesis on parental decisions about abortions. Physicians and health care specialists who have worked with parents in hospitals where amniocentesis was available have agreed that the discussions about abortion are altered by sets of circumstances similar to those described above. As one researcher of the implications of amniocentesis wrote, "Even those who disapprove of abortion in general often favor it when a birth defect of some sort is involved."[8] Amniocentesis is one of our most effective ways for determining the existence of certain fetal genetic defects that will become birth defects. The link between medical knowledge, or diagnosis, and a change in value beliefs seem particularly intimate here.

Yet just how direct is this relationship in this instance? Is this case a refutation of the traditional philosophical argument that factual statements can not logically lead to value statements? After all, the crucial moment seems to have been the presentation to these parents of a strictly factual statement about the genetic state of the fetus being carried by the woman.

From a logical standpoint, one must see that factual knowledge has not, even in this case, led to the creation of a value position. Rather, it has led to the illumination of a conflict between two value positions, both of which were held before the new situation arose, but which were not previously seen as being in conflict. Out of that conflict a choice was made, but it was a choice among values, and not the transmutation of a factual statement into a value statement.

In the case of these parents the two ethical positions that came into conflict were probably (the parents may never have made them explicit) something like: 1. One should not take human life. 2. Parents bear a heavy responsibility for the welfare of their children, and should do all that is possible and practicable to spare them suffering.

Different people might resolve this conflict in different ways. If the parents decide to go ahead with an abortion, they can justify it on several different grounds. The easiest, perhaps, is to change one's definitions. The parents could decide that a four-month-old fetus is not "really alive" and that, therefore, the abortion which they decide to go ahead with is not "the taking of human life." If the parents take this position they have not infringed on either of the two initial fundamental values at all. They have solved the problem by eliminating it. We have assumed, however, that our hypothetical couple are devout Catholics, and they may accept the Church's position that such a fetus is a living human being. If they decide to go ahead with an abortion nonetheless, they have weighed what they consider to be one evil against another and decided which was the lesser. (If they decide not to have an abortion, they have made a similar weighing, with an opposite conclusion.)

In any case, it is difficult to maintain that the fundamental values of the couple have changed. They both come out of the experience believing in the same basic values with which they went into it. Since our couple went ahead with the abortion, they have bent the rules on "the taking of human life," and doubtlessly they feel guilty about this. But they still adamantly support the general principle that one should not take human lives.

Something has changed here, but what is it? It seems to me that the best way to describe the change is to say that there has been a shift in a secondary or derivative value, the couple's position on abortion. They earlier believed that abortion should never be permitted. They now believe that in some instances it is permissible. But their fundamental values are still unchanged.

Most of the value changes that occur in the realm of biomedical ethics are of this sort, i.e., they touch secondary, or derivative values much more closely than they do the fundamental ones. I see little evidence that the biomedical revolution has yet changed the innermost values of human beings. To be sure, as practices such as amniocentesis and abortion become commonplace a ripple effect is probably at work, resulting in some long-term changes in attitudes on other issues. The recent rise in concern about biomedical ethics in many societies is a wise response to the increasing powers of intervention provided by advances in medical technology. The revolution in biomedicine might actually have the effect of causing people to defend their closest values more strongly in order for them to avoid losing their bearings as they shift their secondary positions. Recent events in the United States, such as the setting up of ethical guidelines and institutional review boards (IRBs) in hospitals and research larboratories seem to point in this direction.

Turning now to the issue of the role of science in our hypothetical couple's decision on abortion, it is clear that factual knowledge played an important role in their decision. In order to decide what to do, they needed to know that the fetus had three no. 21 chromosomes, and they needed to be told what degree of deformity their child, if born, would likely suffer. Once they had been given this information, however, they still had a moral decision to make that was quite independent of the factual information. The strength of conviction with which the parents hold to the principles that taking a human life is wrong and that solicitude for children is right derives from sources independent from what they have been told by their physician. Respect for human life and care for children are certainly emotional in nature, and in the case of our example, are considered by the parents to be religious concerns as well; in addition, these commitments may even have a genetic base in all members of the human race. Such a foundation is outside the realm of conscious argumentation. Medical knowledge can so radically alter the context in which the decision is made that a person may find himself or herself shifting priorities but that knowledge has not created either of the two basic value positions of our couple, since they already existed. The strongest case that we can make is that, in this instance, medical knowledge has done damage to a secondary ethical stance, the belief that abortion is not permissible. If we restrict ourselves to this more modest claim, medical knowledge *has* facilitated a shift in values.

But how important and direct was the link between this shift and recent work in fundamental science? Many people, upon hearing amniocentesis described as an example of genetic diagnosis, and upon learning that it has become routine only in the last decade or two, assume that somehow it is connected with all the recent breakthroughs in molecular biology and the study of individual genes. To the contrary, the present techniques of amniocentesis leading to genetic counseling are not based on knowledge of the structure of DNA or individual genes, but rather on the study of whole chromosomes, each of which contains thousands of genes, and on the biochemical analysis of amniotic fluid and cultured amniotic fluid cells. The chemical and biological principles involved have been known for many years, although the efficacy of the methods is rapidly increasing. It would be possible for the above story of the anguished Catholic couple to have occurred even if biologists knew nothing of the most exciting event in biology of this half-century, the identification of the double-helix structure of DNA, an event that formed the paradigm research model for the following generation of biologists. In fact, the prenatal detection of genetic disorders was predicted by J. W. Ballantyne in 1902.[9]

By the late 1970s the process of amniocentesis permitted the detection of all gross chromosome abnormalities and approximately 80 metabolic disorders.[10] The examination was of three types:

1. Karyotyping of cultured amniotic fluid cells and the identification of chromosome abnormalities of number or structure.
2. Biochemical analysis of the constituents of the amniotic fluid.
3. Biochemical analysis of the cultured amniotic cells.

None of these methods rests on knowing how DNA or genes are actually structured. In methods two and three, the goal of the physician is to identify deficiencies in the organic constituents of the fluid and the cells, such as enzymes, hormones, and lipids. In that way the physician can often indirectly determine that the fetus has a flawed gene. Tay-Sachs disease results, for example, from a deficiency of the enzyme β-N-Acetyl hexosaminidase A.[11] But the physician can not determine the existence of the flawed gene on structural grounds. He or she knows it is there by its observable biochemical effects.

The distance between the cutting edge of biological research and the crises in personal ethics which ordinary citizens may have to resolve is considerably greater than the popular press often indicates. And the greater the distance is, the more important are the extrascientific mediating influences between the scientific breakthroughs and the clinical applications. These influences include the development of technology, the positive or negative influence of the social structure, and the institutional matrix. By the time a public bioethics issue emerges the original scientific advance is only one of many causal factors at work. Therefore, we should be cautious about maintaining that "science is forcing changes in our values."

In this particular instance of amniocentesis, the importance of such extrascientific factors becomes clearer if we ask ourselves what the necessary conditions were for this debate in the minds of the parents to occur. If it was necessary that physicians have the scientific knowledge and technical ability to perform amniocentesis, it was also necessary that society be ready to use this knowledge and this ability, and even to call it forth. What was the nature of some of these social preconditions?

This family crisis over what to do about the results of amniocentesis would have been unlikely in the United States without two earlier developments: the widespread use of the contraceptive pill and the legalization of abortion. Ironically, our devout couple had opposed both of these developments, but was fundamentally affected by them. To take the latter event first, one should see that performing amniocentesis does not make sense unless society has taken the prior moral and legal step of making abortions legal. It is true that one could possibly favor amniocentesis for purely diagnostic reasons ("one should be prepared psychologically and even materially for the birth of a child with Down's syndrome, even if one can not legally prevent its birth."), but such an imagined situation has an unreal air to it. Since the operation does involve some risks and expense, few parents would favor it for purely informational purposes. Abortion as an option is an assumption built into the clinical practice of amniocentesis. And the making of legal decisions and laws permitting abortion was a process that was overwhelmingly social and political in nature, with relatively small inputs of new scientific knowledge. Once again, the role of science in the chain of events leading to a shift in values is diminished.

The widespread use of the pill and other contraceptive devices was also a preparatory social condition for this particular debate over the ethics of abortion. Without the increasing belief that women should have greater control over their bodies than many earlier societies thought proper or necessary, the corpus of social attitudes that sanctions such techniques as amniocentesis could hardly have formed. The pill was an important causal factor here, although there were many others. Most important of all was a broad shift in attitudes about the place of women in society behind which lay a whole array of socioeconomic factors. What was cause and what was effect is extremely difficult to determine.

The pill *was* a development in technology with roots in scientific knowledge, research on steroid hormones. Steroid chemistry has a long history; the rate of work in the field greatly intensified in the 1930s and 1940s. We must include the development of this field of science as yet another level among those factors that prepared the way for our case of a debate in one family over amniocentesis. We should notice, however, that we are now talking about *different* roots than those behind amniocentesis itself, and our causal chains are becoming ever more remote and indirect. Furthermore, if the pill has roots in science, so also does it have economic and social ones, and our search, if continued, would rapidly fray out into ever more complex investigations of multiple factors.

Returning to our example of the couple who have decided in favor of abortion after amniocentesis, we see that there are limited but definite inputs of fundamental science to the situation in which they found themselves. First of all, it was not until the first two decades of this century that enough was known about chromosomes to see that aberrations in them lead directly to observable deformities. It was not until the 1930s and 1940s that researchers understood enough about human physiology to suspect that the amniotic fluid was a good carrier of fetal information. And it was not until 1959 that Down's syndrome was directly correlated to a observable chromosomal aberration. Each of these advances in knowledge, while not paradigm shifts in a Kuhnian sense, was crucial for the development of prenatal diagnosis of Down's syndrome through amniocentesis.

We are now able to discuss the role of science in the emergence of a particular debate over values in a clearer fashion than earlier. We see that no direct, logical connection can be made in this instance between a particular factual statement about genetics and a particular value statement about abortions. We see that the road between fundamental biological science and debates over abortions is somewhat longer than we may have suspected earlier, with a consequent greater opportunity for the introduction of extra-scientific factors. But when all this has been taken into consideration, we are still justified in asserting that there *is* a connection, indirect though it may be, between developments in science and the particular crisis which our Catholic couple suffered.

But how typical is our example of amniocentesis? Are there not other examples where a stronger case can be made that science is changing our fundamental values?

THE LIFE-SUPPORT MACHINES

When I once asked a physician who had long been in a leading university hospital what he thought had initiated the intense discussions about biomedical ethics among physicians—rather than the general public—he thought for a moment and then replied, "The advent of the machines." He continued that the presence of respiratory, dialysis, and other life-sustaining machines in the hospitals had forced physicians to ask questions about which lives to save and how long to try to save them that they had earlier not faced in nearly as pressing a fashion. The physician confessed that he found some of the decisions so difficult that for the first time in his life he felt the need to look outside his profession for advice, turning to philosophers, theologians and lay people.

This issue found its most famous expression in the case of Karen Ann Quinlan, the hopelessly brain-damaged young woman who seemed (erroneously, it later turned out) to be totally dependent on a life-support machine. The act of disconnecting the machine came after an agonizing consideration of the ethical significance of the available alternatives.

As was the case with our couple deciding what to do after

amniocentesis, the essential issue with life-support machines is the reconciling of conflicting prior value commitments, not the determination of new values. As a specialist on medical ethics has noted, the Hippocratic oath, to which most Western physicians are expected to adhere, "includes both a promise to relieve suffering and a promise to prolong and protect life."[12] In some circumstances these commitments are in opposition.

The typical modes of coping with the machines have been 1) redefining the legal definition of death, such as accepting the concept of "brain death," so that turning off the machines may not be, technically, the taking of a human life; 2) adopting guidelines, accepted by patients and relatives, on the issue of whether "extraordinary measures" should be taken to prolong a life. The decision about extraordinary measures varies with the patient's desires and the prognosis for recovery.

These policies have been difficult to formulate and still present problems. But the basic commitments to preserving life and avoiding suffering have remained strong. There is reason to believe that secondary or derivative values can be altered without undermining fundamental commitments.

How important has science been in the origin of the life-support machines? A specialist on the machines wrote in the late 1960s:

> We can replace only three bodily processes by machines: pumping blood, exchanging its gases, and removing its wastes. We can do so without having to imitate a single metabolic activity, and the principles involved were known to nineteenth century physics, chemistry and engineering.[13]

Very little basic science is involved in the life-support machines, although there is a great deal of exquisite engineering. Dialysis, upon which the kidney machines are based, was named and investigated by Thomas Graham (1805–1869). The first workable artificial kidney was developed by Willem Kolff in World War II when he was working with wounded Dutch partisans. The mechanical and chemical principles involved in heart-lung machines are also quite old. The first attempts to develop an artificial lung (oxygenator) and a substitute heart (pump) were made in the 1860s and 1870s, but only after decades of work with animals was the first heart-lung machine successfully used on man in 1953.[14]

One should not push the point too hard that these efforts had little to do with recent science, since the development of materials for the machines involved high-order research in polymer chemistry and metallurgy. Primary problems were finding suitable materials for connecting the machines to the patients and the materials in the machines themselves. Human blood cells are easily damaged by pumps and even by the interior surfaces of the tubes; it turns out that the important characteristics of human blood vessels are not easily duplicated. Until such plastics as methyl methacrylate and polyvinyl chloride became available it was impossible to go ahead with machines whose basic principles were well known and tested. Most of the materials were developed elsewhere, and only later adapted for use in artificial organs. In the case of the artificial heart, for example, a specialist wrote:

> All materials presently in use for fabricating artificial hearts were developed, not for medical use, but for industrial or commercial purposes. They have simply been adopted for medical uses. . . . The main materials which have been actually used are Nylon (polyamide), Dacron (polyester), Teflon (fluorocarbon), Silastic (silicone rubber), polyvinyl chloride, polyester-type and polyether-type polyurethane, polypropylene, polystyrene, Lexan (polycarbonate), Plexiglass (methyl methacrylate), natural rubber, epoxy resin, fiberglass, stainless steel, titanium, and stellite 21 (a cobalt chromium alloy).[15]

High-order chemical and metallurgical engineering, then, was an important part of the development of these artificial organs. Nonetheless, basic science played only an indirect and fairly distant role.

The recent public and academic discussions over death and dying are examples of issues rather far from fundamental science. They were initiated, in some instances, by the advent of the life-support machines, a form of technology, but the discussions have now gone far beyond this stage to analyses of "personhood," of what makes a life worth living, and to consideration of the social and financial burden that society is willing to shoulder in order to provide full support to all who might possibly use it. These important discussions make up a part of societal reevaluation of life and death in which science is only one of many powerful factors at work.

Nevertheless, there are several categories of biomedical issues in which the links between value dimensions and fundamental science seem to be much closer than those discussed so far. Two examples are genetic intervention and the ethics of biological research. Abortion for genetic reasons following amniocentesis is an illustration of genetic intervention, but not one involving the most recent advances of genetics. But the possibility for applying more advanced genetic research is growing. If one can examine chromosomes today the possibility of examining genes for flaws tomorrow is present. True genetic engineering of human beings (alteration of their genes) is still largely a myth, but we are justified in saying that in the area of human genetics the length of the link between scientific advance and ethical impact is growing progressively shorter.

The second category where the link is getting shorter, the ethics of biological research itself, is a very special case. It is the only area where the impact of science on an ethical position is direct. In this area we can, indeed, say that science is the single most important causal factor for value change. The reason for this direct impact is that the ethical principles involved are internal to science itself, rather than being a part of the larger society.

By the internal ethic of science we mean that body of value statements which a scientist must support in order to obtain the best quality science. They include support of freedom of inquiry, objectivity, honesty in reporting results, and free communication of research among colleagues. A scientist supports these value judgments *qua* scientist, since he knows that their denial will do harm to science. Other value positions which he may hold—political, moral, religious—do not impinge on his science in an obvious and necessary fashion, although there may be indirect links.

The best example of the way in which science can impinge directly on its own internal value code is that of the issues arising from recombinant DNA research.

THE RECOMBINANT DNA DEBATE

It is important to see that although this issue involves ethics, the relationship between ethics and research is different from the majority of discussions of the "ethical responsibilities of scientists." In the majority of such discussions the main problem involved is whether scientists should assume responsibility for the results of the future applications of their work. One of the clearest instances of this broader class is the case of a scientist engaged in military research, or in research with known military potentialities. The scientist here has difficult ethical problems, but they revolve around whether he or she has responsibility for what may be done with the research after it leaves the laboratory. In the case of recombinant DNA the original ethical discussion was not primarily about subsequent applications of the work but about the possible accidental results of the fundamental research *itself*. It is for this reason that the link between fundamental research and values appears so direct and clear in this instance.

Developments in biological theory responsible for the controversy of the 1970s over recombinant DNA stretch back almost a generation, certainly to the pioneering work on the structure of DNA of the 1950s. Only in the late 1960s and early 1970s, however, did scientists learn how, in some instances, to splice DNA in a controlled way and to introduce DNA from one species into another. This spectacular accomplishment not only represented a breaching of the barriers in nature that normally prevent the exchange of DNA between different species, but also gave molecular biologists a powerful tool for determining the genetic structures of different organisms. This procedure soon earned the name of "genetic engineering" because of its potential for creating novel genetic combinations. As we will see below, however, the term was grossly misleading when some writers used it to imply that genetic engineering of human beings would soon be common.

The ability to introduce a known segment of foreign DNA into a bacterial cell depends on a four-step procedure: 1) breaking and joining strands of DNA taken from different sources; 2) finding a suitable gene carrier that will replicate both itself and foreign DNA attached to it; 3) introducing this composite DNA molecule into a bacterial cell; 4) identifying later generations of cells containing the foreign DNA among a population of normal cells.[16] Each of these steps required research of a very high order and an exquisite knowledge of certain organisms. It is no surprise, then, that the organisms that were used in the first successful experiments were those which had already been studied in the most thorough ways: the famous colon bacillus *Escherichia coli*, the bacterium *Staphylococcus aureus*, the fruit fly *Drosophila melanogaster*, and the toad *Xenopus laevis*. The familiarity of these organisms is already an indication of the great gaps between using them and thinking of doing the same thing to humans, not to speak of the sobering ethical issues involved in experimenting with human subjects.

The first step primarily involved work with certain enzymes that have the properties of severing or annealing strands of DNA. In 1967 a group of enzymes called DNA ligases were discovered that can repair breaks in DNA and even join different strands together. Different types of enzymes worked in different ways in making genetic manipulation experiments possible. Another group called restriction endonucleases severed DNA at certain sites in such a way that the ends had nucleotide sequences with complementary overlapping projections; these "sticky ends" could join with any other fragment of DNA produced by the same enzyme.

Since strands of DNA will not automatically replicate in any host, a carrier that *will* replicate is necessary in order for functioning foreign DNA to be introduced into another organism. To find such a carrier, biologists relied on their previous research into the actions of certain plasmids (extrachromosomal DNA) on the bacillus *E. coli*. It was known that a group of plasmids could confer resistance to antibiotics (*R* factor) to *E. coli*, indicating that these plasmids entered into the genetic constitution of the bacillus. The same plasmids were likely candidates for foreign-DNA carriers.

In 1970 groups of researchers in Hawaii and California found a way of making the cell membranes of *E. coli* permeable to viral DNA and developed a reliable method for introducing *R*-factor plasmid DNA into *E. coli*. The resistance to antibiotics bestowed by the *R* factor also gave a method of identifying the affected cells. Upon treatment by antibiotics those cells that had not taken up the foreign DNA died.

In 1972 and 1973 a group of researchers at Stanford University and the University of California School of Medicine at San

Francisco used these methods to combine DNA information from two different plasmids in *E. coli* and then inserted the composite DNA into other *E. coli* cells, where it replicated in a normal manner. This work led to further experiments in which segments of DNA taken from a rather large variety of organisms—viruses, fruit flies, toads—could be added to a carrier plasmid and then inserted into bacteria cultured in a laboratory. In principle, the combination of DNA segments from the most diverse sources—animals, plants, microorganisms—seemed possible.

The potentiality of such research to lead to useful results is very great indeed. First of all, the controlled insertion of different DNA segments into bacteria should produce an enormous increase in knowledge about genetics itself, including identification and isolation of particular genes. As Joshua Lederburg observed in late 1975, "We can now fragment animal or human DNA into perhaps a million segments and transfer a single segment to a bacterial host for study in a microcosm or for production of large quantities of a specific DNA segment. This allows more elaborate analysis than has ever been possible with the enormously complex, original, unfragmented source material."[17] The production of new antibiotics, insulin, and human antibody globulins is all possible as a result of such research. Further possibilities are improved agricultural products (food protein sources) and means for controlling certain types of pollution. Even progress in the control of cancer might be made in these studies. In view of such beneficial potentialities, many biologists would maintain that there is an ethical imperative to proceed with such research, completely aside from the advances in fundamental knowledge it will bring.

A major problem was that one could not predict very well what the results of such research would be, and there was at least some ground for believing that one might accidentally produce destructive, even life-threatening organisms. This fear was increased by the fact that by far the most convenient bacterium for use in the experiments, *E. coli*, is a resident of the human intestine. A new strain of *E. coli* might possibly develop that could harm human beings and yet still live comfortably within them. It was feared that a researcher might accidentally carry the new strain from the laboratory to the larger community, where it might spread.

This sort of alarming scenario led to controversies both within and without the scientific community. In 1973 a number of specialists in recombinant DNA wrote letters to the two leading scientific periodicals, *Science* and *Nature*, warning of the danger. The next year a group of researchers called for a voluntary moratorium on recombinant DNA research that might lead to novel strains of bacteria resistant to antibiotics or to replication of tumor viruses. In February, 1975, an international conference on recombinant DNA was held at Asilomar, California, where the participants issued guidelines for research in the area. After much subsequent analysis and revision, the guidelines were adopted in 1976 by the National Institutes of Health. In the meantime, the issue of recombinant DNA spilled over into the news media and led to controversies in such university communities as Cambridge, Massachusetts, and Ann Arbor, Michigan.[18]

In 1978 and 1979 the controversy decreased dramatically. Safe means of conducting the research seem to be quite feasible after all. A particularly important safeguard is the use of organisms that cannot survive outside laboratory conditions. Most scientists now consider the risks much less than earlier feared.

From the standpoint of our analysis of the relationship of science and values, we have an example here of a link between fundamental research and what is normally considered to be a value-position: the freedom of inquiry. Newspapers and journals announced the "moratorium" as "the first time in history, so far as is known, that the scientific community called upon its own members to abstain from a line of research not involving human subjects."[19]

Among issues involving science and values, the striking feature of this one is its specificity. It may help us to understand relations between science and values, but primarily in exposing the extreme end of the spectrum: that point where the main value issues are inherent to the scientific enterprise. Indirect relationships characterize most other examples in bioethics.

And even on the issue of recombinant DNA research the presence of extrascientific elements was clear. The fact that the problem first came to prominence in the United States illustrates not only that much of the research enabling these particular experiments to be carried out was performed there, but also that scientific research was considered to be a legitimate topic for public criticism to a greater degree there than in other countries. Recombinant DNA research was being performed in 1975 and 1976 in New Zealand, Poland, Rumania and the Soviet Union under no guidelines at all.[20] A search of Soviet literature revealed little public discussion until after the issue was highlighted in the United States.[21]

HUMAN GENETIC ENGINEERING

The issue just discussed—the ethics of conducting recombinant DNA research—was specific by reason of the fact that the original problem being raised was not about the application of the research, not about technology, but about the possible accidental results of the research itself. When we turn to questions about the impact on ethics of proposals for human genetic engineering we will be dealing, on the scientific side, with many of the same techniques for manipulating DNA. The methods described in the preceding section for DNA splicing could be used on human beings. Despite this common background, however, it is important to see that we have now transferred our attention to a different type of issue, one dealing with applications. *If* scientists develop the knowledge necessary for full-scale genetic engineering of animals and plants, would it be ethically justifiable to apply that knowledge to human beings? What would be the impact of such genetic engineering on our ethical systems if it should become an accepted practice?

In this field, the link between the new molecular biology and ethical crises appears to be obvious. Was not the very idea of human genetic engineering based upon exact knowledge of the structure of DNA and its gene sites, the capability of altering that structure and the genes, and the pursuit of the goal of healthier and happier human beings as a result of this knowledge and its associated technology?

Once again, the distance between the theoretical knowledge and the practical realization of its application is important to notice. If "human genetic engineering" is defined as controlled intervention in human beings at the level of individual genes, the field hardly yet exists. The first case of attempted gene therapy on humans on record was in 1975, and it was apparently an unsuccessful one. (It will be described below.) To be sure, manipulation of genes in plants and nonhuman animals will undoubtedly become ever more common, and progress in human genetic therapy is also likely on a longer time scale.

Human genetic engineering possibilities should be divided into two categories which differ sharply in feasibility and ethical impact. The first, or "therapeutic" category consists of efforts to eliminate or control genetic flaws in individual humans. Often these flaws are due to a single dominant gene or a single pair of recessive genes, as in Huntington's chorea, phenylketonuria, alkaptonuria, galactosemia, thalassemia. The second category, much more questionable, can be called "eugenic" and consists of efforts to change gametes or sex cells in order to influence the resulting progeny. Genetic engineering attempts within the last category would be much more difficult than the first and raise a whole series of basic ethical problems as well.

If scientists and physicians should obtain reliable ways of correcting single-gene or single-gene-pair defects in individuals without causing undesirable side effects some of our present ethical problems might actually be eased, rather than deepened. This therapy would cause no change in our basic values. The number of preventive abortions because of genetic flaws in fetuses would decrease and the problem of genetically flawed neonates would probably be less serious than it presently is. Paul Ramsey, a professor of religious ethics who is a strong critic of the second category of genetic engineering, has written of the treatment of individuals in the first category in quite a different vein: ". . . treating the genes [in these instances] will raise no unusual moral questions, no issues not already present in investigational therapeutic trials, or hazardous or last-ditch surgery of other sorts."[22]

Ramsey was analyzing here a hypothetical situation in which a reliable and safe means of treating serious individual gene defects had already been worked out. Theodore Friedmann and Richard Roblin of the Salk Institute for Biological Studies warned in 1972 in *Science* that this situation does not yet exist:

> For an acceptable genetic treatment of a human genetic defect, we would require that the gene therapy replace the functions of the defective gene

segment without causing deleterious side effects either in the treated individual or in his future offspring. Years of work with tissue cultures and in experimental animals with genetic defects will be required to evaluate the potential side effects of gene therapy techniques. In our view, solutions to all these problems are needed before any attempt to use gene therapy in human patients could be considered ethically acceptable.[23]

These same authors called for researchers not to attempt gene therapy in human patients unless the following criteria are met: 1. "There should be an adequate biochemical characterization of the prospective patient's genetic disorder." For example, if a person seems to be suffering from lack of a certain normal protein, one should determine in advance of genetic therapy whether the patient is not producing enough of the protein due to genetic defect, or has an inability to utilize the protein even though it is being produced. 2. "There should be prior experience with untreated cases of what appears to be the same genetic defect so that the natural history of the disease and the efficacy of alternative therapies can be assessed." It is known, for example, that the effects of galactosemia can be avoided rather well through dietary restriction, obviating the need for radical genetic therapy. 3. "There must be an adequate characterization of the quality of the exogenous DNA vector." The authors suggest that the Food and Drug Administration in the United States control standards for the preparation of DNA. 4. "There should be extensive studies in experimental animals to evaluate the therapeutic benefits and adverse side effects of the prospective techniques." 5. When possible, *in vitro* tests of the prospective gene therapy on excised skin cells of the prospective patient should be carried out before treating the whole patient.[24]

This rather long list of precautions is cited here to show that even in that one category of human genetic engineering which is considered to be *least* controversial from an ethical standpoint—the treatment of individuals for specific genetic flaws—the need to meet complicated criteria requiring both further research and official approval is so great that there is little likelihood that the next few years will bring a rush of genetic engineering applications in humans (as opposed to plants and animals). The time lag between theoretical possibility and practical application will probably be particularly marked in this instance, since the risks are large. (It should be noted that *in vitro* fertilization, as in the case of Louise Brown, in no way involves genetic engineering, since no alteration of genes occurs.) The long period of time that will likely elapse between the moment that human genetic engineering becomes possible and the moment when it finds wide application illustrates that science's impact on our values is not direct and automatic; it is also a sign of freedom from scientific and technological determinism.

The requirement for federal approval, which seems highly likely, would so slow the process in the United States that one knowledgeable American scientist has written, "I believe that the tremendous barrier of externally imposed restraints by government regulatory agencies . . . will probably be the most secure insurance that little, if any, of this research will actually be performed on humans."[25]

If this range of obstacles faces the rather narrow field of single-gene or single-gene-pair therapy, how much greater the obstacles are for any kind of eugenic engineering of the parents' gametes or germ cells! The range of difficulties presented by such attempts is of an entirely different order than therapy of one or two genes. Multiple-gene modification of human gametes without enormous risks seems difficult to visualize. In the face of these considerations, one has some justification for saying that marvelous as the achievements of modern genetics have been, we are a very long way from the biologically engineered futures portrayed by popular magazines.

The main point here is not that human genetic engineering, even of gametes, is impossible. Nor is it that such measures would inevitably be ethically impermissible. The main point is, instead, that we have time, and we should insist on using it, to discuss each of the new possibilities as it presents itself in concrete and realistic terms. Science is not coercing us. We can choose which of its possibilities we wish to try to fulfill, and we can take the time necessary to ensure that the applications are safe and ethical.

To be sure, there is a real danger that eager researchers will attempt human genetic engineering prematurely. At least two examples of such haste (to be discussed) already exist. The physicians in both cases jumped ahead with treatment of humans at a time, and under circumstances, when many of their colleagues believed that inadequate preparatory research had been done. And yet, in defense of the physicians involved, one should notice that neither case was a horror story involving the abuse of patients, but instead each was an attempt to help a terminally ill patient with procedures that everyone recognized were unlikely to succeed, yet offered a slim hope.

TWO PREMATURE CASES OF HUMAN GENETIC ENGINEERING

In 1969 a German physician, Dr. Heinz Terheggen of the University Pediatric Clinic in Cologne, reported in the medical journal *Lancet* a case involving two young sisters with what appeared to be a hereditary deficiency of the enzyme arginase, which is normally produced by the liver. Arginase degrades a toxic substance in the urea cycle, arginine. An accumulation of arginine because of arginase deficiency (or for any other reason) is associated with severe mental retardation. The two sisters suffered from spastic diplegia, epileptic seizures, and serious mental retardation. Their condition was deteriorating and, in the opinion of their physician, "the argininemic sisters were certainly doomed in the absence of any known alternative treatment."[26]

The case of the two sisters was novel in medical history, and no treatment was thought possible before Dr. Joshua Lederberg pointed out that extensive work done on the Shope papilloma virus in past decades had shown that this virus induces the production of arginase in rabbits. This virus, as all viruses, is almost pure DNA and the circumstantial evidence therefore seemed strong for believing that Shope papilloma viral DNA carried a gene for the production of arginase and could be used for this purpose. It was further known that researchers who had worked with Shope papilloma virus had abnormally low serum arginine, a result, evidently, of self-infection with the virus. While investigating the virus Dr. Richard Shope had innoculated himself with it, with no ill effects. All this work had been done long before the discovery of the two sisters. One of the early researchers of the virus, Dr. Stanfield Rogers, had concluded that the "Shope virus in man is acting as a harmless passenger," but that it apparently gave science an "ability to modify arginine blood levels in animals and man—a form of treatment with no known disease."[27]

The appearance of the two German sisters was seen as the disease that fitted the treatment. Dr. Terheggen sent skin biopsies of the girls to Oak Ridge, where *in vitro* experiments (using control cultures as well) indicated that the papilloma virus could, indeed, induce production of arginase. The girls were subsequently treated with the virus twice, and the virus-borne gene therapy caused a 20 percent drop in the level of arginine in their blood. However, the condition of the girls did not improve, evidently because the treatment was administered too late, or because the source of the hyperargininemia was not correctly identified.[28]

Several medical researchers criticized what they called this "uncharted stab at genetic engineering." It is clear that Dr. Terheggen did not meet the criteria for gene therapy set up by Friedmann and Roblin, but of course Terheggen's work preceded their formulation of precautions. Terheggen had nothing to go on but his own judgment. The path he chose deviated from Friedmann and Roblin's later recommendations most clearly with respect to the second guideline, the requirement that "there should be prior experience with untreated cases of what appears to be the same genetic defect so that the natural history of the disease and the efficacy of alternative therapies can be assessed." The case of hyperargininemia in the German girls was the first ever reported in the medical literature. Furthermore, the genetic disorder was understood only in the roughest outlines. No one knew whether the viral DNA produced arginase directly or through cellular genes. Little was known about whether the viral DNA injections would cause deleterious side effects, including possibly hereditable ones (although there were the hopeful examples of Dr. Shope and other researchers who either had papilloma virus injected or picked it up accidentally). In rabbits, Shope papilloma has produced skin papillomas, some of which developed into cancerous lesions. Whether cancer would result in humans was not known. Furthermore, it is somewhat questionable whether the experiment was being carried out for the sake of the children or for science, since brain damage had so progressed in the sisters that the chance for recovery was evidently very small. And, lastly, Dr. Terheggen administered the viral DNA on his own authority, after consulting with the parents. The opinion is growing among geneticists and physicians that gene therapy should never be attempted without the approval of larger public bodies taking into consideration concerns "beyond the principal physician-investigator."[29]

In all these regards, the case of the German sisters has been very instructive. It has resulted in greater awareness among the scientific community and the public of the need for guidelines of the type suggested by Friedmann and Roblin. Furthermore, the federal government in the United States (and many other governments as well) have in recent years established guidelines for government-financed research and treatment of human subjects and/or recombinant DNA. Universities and hospitals have established "human subjects committees," "biomedicine advisory committees," and "institutional review boards" aimed toward preventing unethical experiments on humans, including those involving genetic engineering.

Yet borderline cases continue to exist, since great disagreement continues among researchers on where the line between the permissible and the impermissible should be drawn. The team of UCLA researchers led by Drs. Martin J. Cline and Winston Salser encountered heavy criticism in October 1980, when it became known that they had used recombinant DNA techniques on humans (evidently for the first time in history) in an attempt to treate the genetic defect that causes beta-zero thalassemia. The facts that the gene-therapy attempts were carried out in hospitals in Italy and Israel, where the guidelines were not as strict as in their own institution in California, and that published animal experiments had not yet supported the human gene-therapy attempt, caused many critics to condemn Cline and Salser.[30]

Sufferers of beta-zero thalassemia are incapable of producing the protein molecule that forms the beta-chain of hemoglobin. Consequently, they do not have normal red blood cells, a deficiency that limits the ability of the blood to carry oxygen to the rest of the body. Frequent blood transfusions help such patients, but most of them die in their late teens or early twenties as a result of the accumulation of iron in the body, a condition that arises from the repeated transfusions. Beta-zero thalassemia is classified by physicians as a fatal blood disorder.

The UCLA team removed bone marrow cells from two patients (a 16-year-old girl at the University Polyclinic in Naples and a 21-year-old woman at Hadassah Hospital in Jerusalem) and treated these cells by adding a normal version of the gene for hemoglobin production. The clinicians also added a second gene to increase the proliferation rate and viability of the treated bone marrow cells. Both genes had been created by recombinant DNA techniques. The treated bone marrow cells were then returned to the patients. To create space in the bone marrow for the new cells, the patients were irradiated in the thigh bone at a level considered by Cline to be "not harmful."

The patients gave informed consent for the experiments, and they were told that beneficial results for them were not likely. Dr. Cline told the patients that this experiment was the first of several planned, and it was hoped that later ones might have greater chance of success, either with them or subsequent patients. Dr. Cline acknowledged that much about the experiment was unknown. The purpose of the experiment, he said, was to test the delivery system and watch for any possible toxicity.[31]

It is quite clear that the UCLA team did not fully observe guidelines as strict as those recommended by Friedmann and Roblin (nor were they legally obligated to do so). In particular, no published animal studies existed that supported the view that such an experiment on humans is likely to succeed. Thus, many critics agreed with Richard Axel of the Columbia College of Physicians and Surgeons who observed, "There is simply no scientific basis for expecting the experiment to work in people. A lot more has to be experimented with in animal systems."[32]

Nicholas Wade later wrote that "Cline and Salser took something of a gamble, skating close to the edge of what was scientifically reasonable and publicly acceptable."[33] Yet it should be noted that Drs. Cline and Salser—contrary to Dr. Terheggen in his earlier work with the German sisters—applied for permission to all responsible authorities before carrying out the experiments, and they actually performed the experiment only where these authorities had given their assent. The Hadassah hospital and the Naples

Polyclinic gave permission and the experiment was carried out *before* the UCLA Human Subjects Use Committee denied permission for the experiment to be done at UCLA. Cline maintained that he went to Italy and Israel not to evade the rules, but simply because beta-zero thalassemia is a disease much more common in the Mediterranean area than in the United States; therefore, the task of finding patients who were in a condition to give informed consent was easier there.

The story of the UCLA team leaves the impression that poor judgment was exercised by the principal investigators, even though they may not have violated any specific guidelines or regulations. Cline and Salser knew how controversial their experiment was, and they should have realized that any actions by researchers or clinicians that increased the distrust of the public could hinder the later development of even those gene-therapy treatments that were fully justified in terms of ethical and legal criteria. Cline and Salser, however, were in a hurry.

But even this case of haste is a far cry from the terrifying genetic engineering scenarios portrayed by some critics of the potential uses of human genetics. The specter of 1984 is a real one that we must guard against, but gene therapy directed toward correcting single flaws will probably be much more similar to traditional extreme surgery, with all the risks and problems involved in such heroic measures, than it will be to true genetic engineering, as the popular press understands that term. In cases of heroic surgery of terminally ill patients, there are some patients who want to say to the attending physician, "Go ahead and try if you think there is any chance of success." One hopes that regulations never close that option entirely.

EUGENIC GENETIC ENGINEERING

If one shifts one's attention from gene therapy, as discussed above, to attempts to engineer sex cells or gametes themselves, and in that way improve human progeny, both the ethical and the technical issues radically shift. Many specialists in biomedical ethics, such as Paul Ramsey, draw a sharp line between genetic therapy on individuals and genetic engineering of future generations.[34] Informed consent, usually required in biomedical experiments, is obviously not possible. Neither is a risk-benefit calculation based on avoiding pain or suffering to the child, since there is no assumption of the necessity of the child's birth. Furthermore, the whole question of engineering sperm or ova loses much of its importance when one sees that the same goals can be achieved, probably more simply, by starting with zygotes from different biological parents, by selections among zygotes of legal parents, or by prenatal diagnosis and preemptive abortions. In the present state of biomedical science there is no ethical defense for the wholesale engineering of human sperm and ova since so many other approaches of less medical risk and greater justification are either already available or appear to be feasible. It is entirely possible that this situation will change at some time in the future, but it would be pointless to try to predict the concrete circumstances of that change at this moment in time.

One of the most radical genetic engineering proposals is human cloning, the asexual reproduction of an exact genetic copy of a person. Ever since a mature frog was cloned from a single cell in 1965, discussions of the prospect of human cloning have raged in the press. The technical and ethical problems involved in any putative attempt to do such a thing are huge. The argument that a qualitatively superior type of human being would result meets with all the normal objections to positive eugenics (e.g., "Who will decide what are the desirable characteristics?") plus a whole range of new problems ("Who will take responsibility for failures?" "What would be the psychological effects on a young person of knowing he or she is a genetic copy of a genius and is expected to perform at a similar level?"). Furthermore, the risks incurred by narrowing the genetic base of the species, and the consequent elimination of adaptive reserves, are worth noticing. One is tempted to say that nature hit upon sexual reproduction, the combination of DNA from two parents in ever novel ways, for very good reasons.

Summary and Conclusions

Upon examination we see that the oft-repeated claim that biomedical science is changing our fundamental values is misleading and inaccurate. Biomedical technology often leads to serious ethical dilemmas, but it does not alter our fundamental values nor does it create new fundamental values. Instead, it illuminates conflicts between our earlier value positions, often causing us to decide what is "the lesser evil."

Biomedical technology does, however, play a role in shifts in our secondary or derivative values. As a result of an encounter with new medical technology, a person may modify his or her attitudes toward abortion or terminal care, justifying that new position by reference to more fundamental values that are still retained, such as responsibility and solicitude for one's children, relatives, or intimate friends. These latter bonds rest on our deepest values and it is difficult to imagine a development in biomedicine that would cause us to abandon them. In fact, the advances of biomedicine are probably causing us to support them in more intelligent ways than ever before.

Often a shift in a secondary value that appears at first glance to have been caused primarily by an advance in medical technology is actually a result of a whole series of social and cultural changes that either called the medical technology forth or permitted it to be used. Technology is often less important here than the social matrix. Oral contraceptives were widely used because they filled an important social need when they appeared; artificial insemination is only rarely used, even though on strictly scientific grounds its wider use could be justified. Social arguments, however, are (and should be) more powerful than scientific ones.

The ethical dilemmas arising from biomedical applications are of great importance, but they are not as interesting from an intellectual standpoint as the efforts earlier described in this book to bring human values themselves within the area of scientific research and explanation. Biomedical technology raises questions of ethical consequences; the scientific study of values, on the other hand, raises questions of the origins of ethics. Thus, when scholars like B. F. Skinner and E. O. Wilson, in their distinctly different ways, attempt to describe the sources of our most fundamental values, they are engaged in true expansionism. When bioethicists discuss the ethical consequences of certain medical technologies they usually still assume a division of science from ethics. They worry that ethical standards will be eroded by the adoption and widespread use of the new medical technologies, a very proper worry. But they do not base themselves on the assumption of a naturalistic explanation of ethics, as true expansionists do.

Some of the sociobiologists maintain that the deepest human values are under genetic control. I admit my skepticism about the fullness of this control. But if they should turn out to be correct, and if it also happens that the genes influencing these values can be modified through genetic engineering, then the concerns of the biomedical technologists and the sociobiological theoreticians of human values will be united; a technology of human values would be conceivable. An ultimate expansionism might be upon us.

Disregarding for the moment the inherent unlikeliness of this event, we can hypothesize its occurrence while benefiting from our discussion of past medical technologies. The period of time between the theoretical breakthroughs and clinical applications is usually rather long, and we have reason to believe that it would be in this case also. During this period of time society would have to discuss how much of the technology it wishes to use. It might decide it wishes not to use it at all, just as we do not apply many of the known principles of animal breeding to humans, even though it is theoretically possible to do so. Or we may wish to use it only rarely, as is the case with artifical insemination. Retaining the degree of autonomy necessary for making such decisions is the most important goal of all.

PART THREE

REACTIONS TO THE DILEMMA OF SCIENCE AND VALUES

chapter ten

PUBLIC CONCERNS ABOUT SCIENCE AND TECHNOLOGY: THE QUESTION OF LIMITS OF INQUIRY

CONCERNS AND anxieties about science and technology are not novel in the history of science, but a persuasive case can be made that a new stage was reached in the late sixties and seventies of this century. This was a stage marked not only by a growing suspicion of science among segments of the lay public, but also by an expanding belief among scientists themselves that questions about the social values and responsibilities of researchers must be faced more directly than had previously been the case. The issue of the relationship of science to social values was often posed in new ways in these debates, but underneath the controversies the old issue of expansionism versus restrictionism was often present.

The best publicized example of the new concerns about science in the early 1970s was the debate over recombinant DNA, but other challenges appeared in a great variety of areas, including those of biomedical research, military research, and nuclear energy. As a result of all these controversies, some scientists voiced their worry that "scientists and scholars have long had a bargain with society by which they have produced ideas and devices with few constraints, but that now this bargain is in danger of breaking down or in need of revision."[1] At a discussion at the Massachusetts Institute of Technology of this new resistance to science and technology, one participant spoke favorably of proposals to "regulate science as one regulates the railroads," while others met this concept with sharp criticism.

Discussions of proposals to regulate or restrict science do not have general value until a number of basic distinctions have been made among types of concern about science and types of proposal for its restriction. What we need as a first step toward a discussion of possible limits to inquiry is a typology or taxonomy of concerns about science. We can then assess the validity of each concern and address the problem of restriction or regulation in a more specific and informed fashion.

The main goal in classifying the concerns about science is not merely to introduce order, or to engage in a possibly interesting academic exercise. The goal is much more pragmatic and is, indeed, clearly policy-oriented: to sort out those categories of concern about science and technology where we are coming to recognize the necessity or inevitability of controls of one sort or another, while retaining other categories where freedom from control is still defensible and necessary. There is still a large section of science where regulation should not be permitted. This autonomy of science should be defended not as a privilege for an elite, nor as an absolute right, but as a need of society itself. A conclusion will be, then, that some aspects of science and technology should be controlled for the same reason that other aspects should not be: namely, it is better for society that way, society taken in the broadest terms of all of its interests.

A limitation of this analysis is that many concerns about science do not easily fit into a defined framework, because they are expressed in poorly defined or even irrational terms. Some people, for example, oppose expansionism in any form whatsoever—oppose any scientific endeavor that may explain or alter human values. And the most basic fear of science is the very fear of rationality itself. Because such anxieties are difficult to classify except in the simplest way, I will discuss in this chapter only the "rational variable" in the complex cluster of recent concerns about science. There are at least three reasons for such an approach: 1) by isolating the anxieties about science that can be phrased in rational terms, we may decrease the apparent dimensions of the irrational residue; 2) the rational criticisms of science need to be evaluated on their own ground; 3) many of the concerns about science and technology are legitimate, in the sense that they represent genuine fears for the safety, morality, and the sense of human purpose and worth of society. If we dismiss all anxieties about science as "irrational," we will not be listening to some important debates.

The following discussion will be based on a number of broad categories of concerns about science and technology, some of which have subcategories;

I. Concerns about Technology	
A. Concerns about the physical results of technology	Destructive Technology
B. Concerns about the ethical results of technology	Slippery Slope Technology
1. Biomedical ethics	
C. Concerns about the economic results of technology	Economically Exploitative Technology
II. Concerns about Science	
A. Concerns about research on human subjects	Human Subjects Research
B. Concerns about distortions in allocations of resources for science	Expensive Research
C. Concerns about certain kinds of fundamental knowledge	
1. Knowledge itself	Subversive Knowledge
2. Knowledge "inevitably" leading to technology	Inevitable Technology
D. Concerns about accidents in the research itself	Accidents in Science
E. Concerns about the use of science to excite racial, sexual, or class prejudices	Prejudicial Science
F. Concerns about certain modes of knowing	Ways of Knowing

The list given above sorts the concerns into the two broad divisions, technology and science. Some critics will respond that this division is no longer tenable. In an eloquent article Hans Jonas maintained, "Not only have the boundaries between theory and practice become blurred, but . . . the two are now fused in the very heart of science itself, so that the ancient alibi of pure theory and with it the moral immunity it provided no longer holds."[2]

Although I appreciate the argument Professor Jonas has advanced, I do not consider it a serious challenge to the classification. First, I do not maintain, as Jonas' comment seems to imply, that problems of "science" enjoy a "moral immunity." I will consider ethical and moral issues under the headings both of technology and of science. (Note, for example, "human subjects research.") Second, although I agree that there are some problems which cannot easily be classified as either science or technology, it is my opinion that the great majority of recent issues can be so classified without much difficulty. The division between fundamental studies of physical and biological nature, on the one hand, and technological studies directed toward social or economic goals, on the other, is still a useful one. Third, I will consider issues under "Inevitable Technology" where the boundary between theory and practice has become blurred in the way in which Jonas emphasizes.

Concerns about Technology

DESTRUCTIVE TECHNOLOGY

Concern about the physical results of technology is one of the most familiar and easily identified categories of concern. A prime example is the damage to the environment caused by industrial civilization. A specific case would be damage to the ozone layer alleged to result from the escape to the upper atmosphere of fluorocarbons used in aerosols. Others would be the polluting effects of DDT, supersonic transports, and various energy sources. The large array of issues under this category also includes many that are not best described as environmental, such as regulation of the distribution and use of pharmaceutical drugs, food additives, and chemicals (like saccharin), explosives, and radioactive materials. All are examples of society's need to prevent the products of technology from spreading in an unmonitored or unregulated way.

This category is the one where the observation, "If we can

regulate the railroads, so can we regulate science," is most apt, although it ought to be made clear that we are speaking here of technology, not science. The imposition of controls on the use of destructive technology is clearly within the rights and traditional practices of government regulatory agencies. Important questions about the wisdom of the regulations, economic and otherwise, can be raised, but the principle of government regulation is well established. Whether the sale or use of aerosols with fluorocarbons should be controlled is a problem of technical facts and assessments, not one of principle about limitation on freedom of inquiry.

Some concerns about military technology also fall within this category. Individual scientists differ widely on whether they should engage in work on destructive weapons, but few people question the principle that governments have the right—indeed, the obligation—to regulate and control this activity. In the United States, the original McMahon Act of 1946 placed strict controls on the ownership and production of fissionable materials, although it was much less restrictive (and more intelligent) than the May-Johnson Bill, which favored military rather than civilian controls.[3]

Controls in this area can extend down to the level of the individual, as the McMahon Act did. If it is possible to discuss regulating hand guns, it is certainly possible to continue to control possession by private individuals of materials necessary for manufacturing nuclear weapons and even to prohibit unauthorized research on methods necessary for such manufacture, such as laser separation of isotopes.

SLIPPERY SLOPE TECHNOLOGY

Whereas the previous category dealt with the physical effects of certain technologies, the present one centers upon the ethical effects. Obviously, every act of physical damage contains an ethical dimension, and no clean separation of the two can ever be made. Nevertheless, there exists a significant difference between, on the one hand, a concern whether the physically damaging effects of a certain technology can be excused within existing ethical systems, and, on the other, whether a certain technology may be destroying the ethical system itself. The present category deals with the latter concern. Issues in this category can be described as "expansionist," since they are instances in which technology is affecting human values.

We have already discussed in the previous chapter how the use of new technology in the biomedical area has raised many difficult ethical issues. New possibilities for prenatal diagnosis by amniocentesis, prolongation of life through the use of dialysis and respiratory machines, psychosurgery, *in vitro* fertilization, and DNA therapy, or genetic engineering, are only a few examples of issues that have been widely discussed in the press. A whole new field of discussion, biomedical ethics, developed as a result of the advent of these new technologies. The Institute for the Study of Society, Ethics and the Life Sciences in Hastings-on-Hudson in New York state, and the Kennedy Institute Center for Bioethics in Washington, D.C., are two of the more active centers among the many where the issues of medical ethics are now under close study.

One of the major concerns expressed by observers of technological developments in the biomedical field has been that by blurring or erasing ethical boundaries that were earlier considered absolute, we will go onto a "slippery slope" of relativistic ethics on which we may lose our balance and tumble to the bottom. Critics often raise the specter of eugenics and medical experimentation under German National Socialism as the ultimate point of descent. They ask, if we sanction the disconnection of life-support machines in terminal cases, or perform abortions on genetically defective fetuses in the last months of pregnancy, what ethical limitations exist preventing us from taking a more active role in ending the lives of terminal patients suffering pain, killing a deformed child immediately after birth, or even several months after birth? The possibility of becoming callous to moral values through adoption of the radical proposals or repetition of the less radical ones is worrisome, and properly demands our attention. On the other hand, if we simply prohibit abortions or termination of extraordinary care, we will encourage worse ethical abuses (illegal abortions under frightful conditions, economic injustice, or even infanticide by parents). The slippery slope slants in more than one direction.

Controversial as these issues are, they do not, in themselves, directly involve the question of limits to inquiry. They are immensely difficult questions of the wise use of medical technology. Guidelines are developing which help to prevent abuses while still taking advantage of the benefits of the new technologies, benefits which in many instances are significant and deeply humane.

Whether or not regulation in this area is justified is not in doubt (it *is* justified), but what the regulations should be and what person or institution should devise or enforce them raise fundamental problems. Current practices in regulation differ widely. For example, where chemicals might be used, as in proposals for DNA therapy (DNA is, after all, a chemical, and therefore subject to more rigorous controls than pure surgery, which is up to the doctor and the patient) the Food and Drug Administration in the United States may play a large role.[4] The rulings of the courts are already important in abortion and disconnection of life-support mechanisms. Regulation of technology falling under this category is already an accepted and necessary practice, albeit one requiring enormous care.

ECONOMICALLY EXPLOITATIVE TECHNOLOGY

Enormous costs are involved in research and development in certain high-technology areas. Should large public sums be spent for the development of a supersonic transport when only a small portion of the population would ever utilize such transportation? Should large amounts be invested in developing exotic and cosmetic medical treatments if these sums detract from public health programs? How does one decide how much money should be given research on disease such as cancer and heart diseases which afflict highly industrialized societies to a greater degree than underdeveloped societies (where, for example, kwashiorkor and schistosomiasis may be more serious threats)? From a health standpoint, how does one decide the relative importance of research on curative medical technologies, as opposed to improvement of environmental conditions, which are increasingly seen as the source of many illnesses?

All the above examples refer to control of research and development for future technologies, not the deployment of existing technologies, and in that sense relate rather directly to the question of limits to inquiry. Yet the general public usually does not make this distinction. When pollsters for the National Science Foundation asked American citizens what, if anything, they feared about science and technology, they sometimes voiced an economic concern quite different from the "priorities problems" discussed in the previous paragraph. They said technology "puts people out of work by replacing them with machines" and "it has forced a lot of small businesses out of work." This concern is, of course, one of the oldest in the history of modern technology, and can easily be traced to the very beginnings of the Industrial Revolution. The fact that the overall result of industrial expansion has been greater employment with greater variety of tasks is little consolation to the worker made surplus by the disappearance of his or her craft or task.

A final example of economic results of technology is the distortion in the use of natural resources which certain technologies entail. The recent effort to slow the exhaustion of fossil fuels by automobiles and home-heating is an example of control of expensive or wasteful technology.

In the examples discussed above, the principle of control for economic reasons is already established in the United States, although great controversies about the best means of control remain. In contrast to the centrally planned economies, where direct limitation on outputs of certain technologies is often imposed, in the United States the emphasis has frequently been on economic penalties or rewards (taxes, deductions, fines) on the technology, and on controlling research and development through distribution of federal research grants. This approach works well in some instances, but the need for controls of a variety of types, particularly in response to crisis situations, is apparent. If these controls are carefully executed within a democratic framework, they are conceivable without damage to political or academic freedom.

As we move now from the subject of concerns about technology to that of concerns about science itself, we come much closer to questions of principle that are essential to the concept of free inquiry.

Concerns about Science

HUMAN SUBJECTS RESEARCH

The methods of procedure in certain types of research may do damage to human subjects of the research, or be suspected of doing such damage. The adoption of guidelines in research on human subjects in the United States by the National Institutes of Health is an example of efforts to avoid damaging effects of this type.

Surely we all admit that limits to inquiry *do* exist within this category. One cannot, for example, morally justify injecting human subjects with pathogenic organisms or toxic chemicals to determine lethality. Nor can we determine the outer limits of human toleration of physical or psychological deprivation, or of physical pain, by direct experimentation, even if subjects would volunteer for such experiments. The history of medical experiments in Nazi Germany, such as the ones designed to help the Luftwaffe decide how long a downed pilot could survive in icy waters, is still a recent memory warning us against such iniquitous and crude experiments. Even more recent is the CIA's giving of behavior modification drugs to unsuspecting subjects in order to learn more about the characteristics of the drugs.

This category of concern about science is occasionally lumped together with "slippery slope technology" issues in the biomedical field, but a distinction between the two is useful in discussions of possible guidelines or regulations. The essential question already discussed in the category of slippery slope technology is the direct result to the individual patient and the indirect ethical results to society at large of the use of certain types of medical technology; the goal there is therapy, not acquisition of fundamental knowledge. The category presently being discussed is research involving human subjects in which the goal of therapy is either absent or secondary to the acquisition of information.

Since the main question here is of research design, the present category touches more closely than the earlier ones the problem of freedom of research. Some of the debates in this category have been quite heated (for example, the one on XYY chromosome research), with some scientists fearing the intrusion of public controls into fundamental research itself.

Separating the category of "human subjects" research from that of "slippery slope technology" reveals several interesting trends which have significance for policy recommendations. Usually we say that fundamental research is distant from moral and legal considerations while applied science is closely bound to such restraints. Yet a comparison of this type of fundamental research with medical practice in recent years reveals some interesting departures from this usual situation. The advent of life-support technologies has de facto given physicians greater and greater leeway in making life-and-death decisions, while in fundamental research on human subjects the ethical leeway granted to researchers has been progressively diminishing. Popular opinion has often supported this divergence. The physician who, either through action or inaction, may have facilitated the early death of a horribly suffering terminal cancer patient is usually not questioned closely about his or her practices. Often the close family members do not know exactly what the physician has done, and they do not wish to know. The situation may be one of unconscious complicity, with both physician and family desiring the early death. The law may have been broken in some strict sense, but a "higher morality" was supposedly served.

In research directed toward acquiring information rather than treating a particular patient, however, the ethical restraints are growing, not relaxing. Finding a cure for yellow fever in the way in which Walter Reed and his colleagues actually did it would now be considered ethically questionable, since his controlled experiments exposed human volunteers to the disease.

Despite the new restrictions, some of which have troublesome implications, most of us would agree that greater ethical awareness about research is a laudable development. Research scientists have not only a moral responsibility to avoid damage through individual and corporate examination of the possible effects of their research procedures on human subjects, but also an interest in avoiding such controversies by self-regulation, since the alternative is increasing regulation by bodies outside the scientific community. The fact that the law already provides some avenues for redress to individuals actually harmed is not a sufficient answer to the problem. The responsibility of scientists themselves for the alleviation of public concerns is heavier in this category of research than any other so far discussed.

EXPENSIVE RESEARCH

This is a category of concern about distortions in the allocations of resources for science. One of the most frequent observations made about contemporary science is that research in many areas is no longer possible on individual budgets, or even small institutional budgets, and that large external support is now necessary. Although external support of scientific research is centuries old in several countries, the degree to which such support is obligatory if high-level research is to be continued has changed markedly in this century. Robert Boyle, Antoine Lavoisier, Charles Lyell, and Charles Darwin were all outstanding scientists who were able to support themselves entirely or partly on the basis of their own funds. Now even a member of the Rockefeller family who was a radio astronomer would probably not be eager to undertake the building of a very large array radiotelescope.

As external support of science has grown, the degree of involvement of governmental organizations and large private foundations in making decisions about allocations of resources has increased in step. The United States Congress has a proper and natural interest in the size of the budget of the National Science Foundation. The question of freedom of inquiry does not arise here in a pointed way so long as the external bodies are generally favorably disposed to science and allow qualified specialists to evaluate individual proposals. Possibilities for conflict involving academic freedom do exist, however, if the budget-granting bodies try to evaluate the quality of individual applicants. The attempted Bauman Amendment to the NSF authorization in 1975 would have permitted Congress to review individual NSF research grants on a systematic basis and would have been a serious threat to research autonomy.[5]

The concerns of fund administrators about scientific research impinge on freedom of inquiry in several indirect ways not yet mentioned. If a line of research has no visible practical value, the scientist in the field will probably have fewer possible sources for funds than if he or she can point to a potential valuable technology issuing from that research. If the research is highly unorthodox, it may not receive support even if potential applications are visualizable. Fund administrators traditionally fear controversy, and they fear even more the possibility that the work which they support will bring discredit to the funder (as did the Carnegie Foundation's support in the early decades of this century of the eugenics movement). When an administrator of another large private foundation recently offered to show its early archives to an historian, he acknowledged that he had records only on those who received grants, not on unsuccessful applicants. The discarded records might have been more revealing.

At the level of decision making about the construction of the most expensive research installations, such as mammoth high-energy accelerators, a worry about distortions in the allocation of resources is a serious one. The important discussions of social priorities necessary for making such decisions do not present a fundamental limitation on individual freedom, even though in such areas work cannot be done without outside funds. Society obviously does not have an obligation to build a facility costing millions of dollars for every good scientist who wants one. On the other hand, a prosperous society that decided to stop building such facilities would be doing damage of a double nature: it would be inflicting harm on its scientific community and it would be limiting its own curiosity and creativity.

SUBVERSIVE KNOWLEDGE

All the above categories of concern about science and technology involved side effects based on the physical, ethical, or economic results of individual technologies, the effects of the design of fundamental research on projects involving human beings as subjects, and the economic costs of large fundamental research projects. The concern to be discussed now rests on fundamental knowledge itself.

Some of the best-known cases of interference with science in past centuries have been the results of resistance to fundamental knowledge. In the classic instances the anxiety has had two sources:

the new fundamental knowledge was seen as conflicting with the theories of ruling authorities, and it appeared to demote the place of man in nature. The affair of Galileo's censure by the Church is one which can be interpreted in terms of these concerns. The Catholic Church had, undoubtedly without logical necessity, incorporated the basic features of the Ptolemaic system into its body of theological teachings. Furthermore, the rival Copernican theory seemed to demote the place of man by situating his abode outside the center of the universe.

Historians of science will be quick to point out that the situation was far from being as simple as the foregoing description might seem to indicate. Copernicus dedicated his treatise to a Pope, and his work was not criticized by Rome until many years after his death; Protestant leaders such as Melanchthon and Calvin protested at first more vociferously than the Catholic leaders;[6] the Copernican system was actually not heliocentric but heliostatic; the Catholic Church might not have taken the position it did were it not under the pressures of the Reformation and Counterreformation.

All these qualifications are valuable, but the fact still remains that the Copernican system as espoused by Galileo was perceived as both contrary to theological teachings and demeaning to the place of man. It was a form of pure knowledge that was seen as a threat in itself. It was an expansionist incursion into the realm of religious and social values. Debates over Darwinism in the last part of the nineteenth century also contained these two elements: conflict with at least some interpretations of religions, and an apparent denigration of the uniqueness of man. The controversies over Darwinism were unsuccessful efforts by restrictionists to hold back the incursions of expansionists.

We often assume that such conflicts are now elements of the past, that a mature world that attributes such an important place to science has surely outgrown adolescent anxieties of this type about science. And, indeed, we have improved very much in this regard. The concerns are still there, however, and they have appeared in several new forms. Furthermore, some of these concerns are not trivial.

Some of the resistance to studies in the fields of ethology and primatology discussed in chapters 6 and 7 can be related to concerns about the diminution of the uniqueness of man, the narrowing of the distance between humans and the rest of the animal world. The possibility of increasingly successful explanations of human behavior in terms of animal behavior often evokes resistance among intellectuals who would usually consider themselves far from being antiscience. One does not have to accept the vulgarized interpretations of ethology of the type of Robert Ardrey or Desmond Morris to agree that at least part of the resistance here is the ancient restrictionist one of concern for man's place in nature.

Another example of current concern about the uniqueness of humans is that expressed by the biophysicist Robert Sinsheimer at an Airlie House conference, where he commented,

> Should we attempt to contact presumed "extraterrestrial intelligences"? I wonder if the authors of such experiments have ever considered the impact upon the human spirit if it should develop that there are other forms of life, to whom we are, for instance, as the chimpanzee is to us. Once it was realized someone already knew the answers to our questions, it seems to me, the impact upon science itself would be especially devastating. We know from our own history the shattering impact more advanced civilizations have upon the less advanced. In my view the human race has to make it on its own, for our own self-respect.[7]

This particular objection to possible research cannot be described, it seems to me, as anything else than another variation of restrictionist anxiety about man's place in nature of the type often seen in past centuries, though not in this instance in a religious context.

On this topic of resistance to fundamental knowledge in itself my position is a simple one, and I would maintain, of great importance to intellectual freedom. That position is: "Critical discussion, yes! Regulation, no!"

In fairness to recent commentators such as Sinsheimer who are responsible and helpful critics, I should add that many of them are not suggesting censorship or a ban on research in these areas, but a deep questioning of the amount of national resources that should be invested in certain research. But to ask whether millions of dollars should be directed toward possible contact with another civilization is to transform the question from one of limits to fundamental inquiry to one of allocation of resources, which is the separate category "expensive science."

INEVITABLE TECHNOLOGY

Another form of concern about fundamental science is the one that states, "Anything that *can* be done, *will* be done," and therefore, the critics say, it is not justifiable to draw a line between science and technology when one is attempting to discuss the question of limits to inquiry. These commentators continue that since some sorts of knowledge will "inevitably" lead to technology which will "inevitably" be used (even though we all might agree that the specified use is not desirable) we ought to impose limits on the original inquiry. Holders of this viewpoint might even classify certain types of knowledge as "forbidden" in principle.

Although I believe that this form of argument is incorrect as a general position, unnecessarily condemning our civilization to technological determinism, I would like to present it in the strongest form before criticizing it. We will all agree, I think, that fundamental physical research in the first four decades of this century provided the knowledge necessary for the construction of nuclear weapons; most of us would agree that the major political powers in the world now possess these weapons in sufficient numbers that if all were exploded in the atmosphere, either civilization would be destroyed or it would be so altered that many of us would not be eager to survive in whatever remained. If that holocaust should occur the following argument would ring hollow in the ears of anyone able to hear it: "Science itself has no moral responsibility for what happened, since it was not necessary that nuclear weapons be constructed or used, even though the knowledge necessary for these events was produced by science."

Logically correct though the argument may be, we are all familiar with the failure of logic to handle adequately the most extreme human dilemmas. The simple fact would be that without the development of nuclear physics this ultimate disaster would not have occurred.

But on the basis of this realization should we try to regulate fundamental research in the future in order to avoid such events? I believe that we would not know how to do so even if we wished, and that in the effort we would cause immense damage to our existing values as well as deprive ourselves of many benefits, both material and intellectual. Frightening as our present situation is, it may just force us to be more responsible than we have ever been before.

The example of nuclear science may be misleading for investigations of "inevitable technology" in those less apocalyptic areas where we are trying to make decisions about the relationship of science to technology. Usually we have a little more freedom of maneuver, a little more space in which to make a few mistakes, than we do in the area of nuclear weapons. To a much greater degree than ever before we are adjusting to the need for controlling our technology instead of letting it control us, and therefore decreasing the strength of the link of necessity between science and technology. In areas of transportation, the environment, energy use, and city and town planning, we are trying to bend technology to our social needs. If we can succeed in this effort, the argument that "whatever can be done, will be done" diminishes in intensity. Society obviously needs some things to be done much more than others.

Taken literally, the "inevitable technology" argument is clearly false. Examples can be given of available technologies which have never been employed on a wide scale. As Harvey Brooks has reminded us, "Artificial insemination among humans is a good example of a technology which, though still in limited use, has never 'taken off.' People prefer the conventional method, and probably always will, when it works."[8] In vitro fertilization is a newer technology with a probable similar outcome. It is possible that human cloning—if it becomes possible—will lose some of its dramatic overtones if it becomes clear that not many people are interested.

The alternative of controlling fundamental research instead of technology is illusory, because it assumes the impossible: the foreseeing of the results of fundamental inquiry. Controlling technology is extremely difficult, but it is not impossible.

ACCIDENTS IN SCIENCE

The controversy over recombinant DNA is one that needs to be analyzed on several different levels. The first level is the one that addresses the explicit and most common concern expressed by people who object to this research, namely, that as a result of an accident in fundamental research the public safety would be threatened. This concern is the one I have classified as a separate category, "accidents in science." The second level has to do with those largely unspoken, but nonetheless real concerns that relate to other categories. I will treat them separately.

On the first level, it is important to notice that the opponents of recombinant DNA research were not (at least originally) objecting principally to what could be intentionally produced by a successful recombinant DNA technology, but to the possibility during fundamental research of the accidental production of a pathogenic organism immune to normal antibiotics. It is true that one *could* express concerns of the type of Category I ("technology") about recombinant DNA work (e.g., its possible use for biological warfare, or an objection to what might intentionally be done to alter the life of plants, animals, or people), but this type of resistance was not the major one. In Cambridge, Massachusetts, for example, where one of the major controversies occurred, the City Council explicitly excluded these latter concerns from its considerations.

The topic of accidents in research is not a new one; indeed, accidents of various types, ranging enormously in the scale of potential damage, have long been evident. Explosions occur even in undergraduate chemistry laboratories, and electrocution is often a possibility. Even some common chemical elements, such as mercury or sodium, must be handled with care, and the dangers increase with volatile and explosive chemical compounds, radioactive materials and high-energy experiments. The transportation, distribution, and use of the most hazardous of these materials are already under controls of various types. In biology, unwanted organisms have escaped to cause subsequent damage; the accidental release from a scientist's collection of the gypsy moth in New England many decades ago continues to disturb us today.

One can include under this same category the possible inadvertent results of an introduction, for scientific reasons, of foreign materials into a previously undisturbed milieu. Pumping water under high pressure into the strata of the earth in an earthquake zone for purposes of seismological research might, for example, have undesired effects. Incurring responsibility for "causing" an earthquake may not be just a speculative possibility. Satellites launched around the earth to gather scientific information may fall back to earth at a later date and cause damage.

The second level of analysis that needs to be applied to recombinant DNA research goes far beyond the issue of public safety. As the controversy progressed, the objections to this type of research broadened beyond the original issue to a whole series of concerns that belong outside the category "accidents in science." Jeremy Rifkin remarked, "The central issue is the mystery of life itself. It is now only a matter of time until scientists will be able to create new strains of plants, even alter human life." Rifkin's concern was a combination of several types: "destructive technology," "slippery slope technology," and probably a deep fear of science itself, one with nonrational roots. Ethan Singer commented that DNA research "will eventually tinker with the gene pool of humanity. So the public, like the subject of any experiment, must give its informed consent—but willingly, not by coercion." This statement was a mixture of concerns about "inevitable technology" and "human subjects research." Roger Noll added his view that since the federal government sponsors much of the research, "I suppose the question, 'Is this worth buying?' is going to be the real issue." This opinion represents a concern for "expensive research," but it is not clear whether Mr. Noll considered it science or technology, and it is furthermore not clear whether the financial concern was a cover for other worries.[9]

It is this mixture of concerns of various categories that made the recombinant DNA case so inflammatory. Any one of the concerns, by itself, could probably be resolved. Some of the anxieties were legitimate, although still not defined (e.g., public safety, which falls under my category "accidents in science"), while others were highly dubious or unacceptable to the scientific community (e.g., concern for the "mystery of life," which falls under my category "subversive knowledge").

An important step in trying to deal with the concerns expressed about recombinant DNA was the attempt to sort out and evaluate them one by one. This step could not be accomplished without first dividing "concerns about science" from "concerns about technology," the two most general categories. Part of the difficulty of discussing recombinant DNA research was the interweaving of these concerns, which in large part resulted from the fact that fundamental research in recombinant DNA is based on a novel "technique"—the introduction into a bacterial cell of a segment of foreign DNA by means of a now familiar multistep procedure—and therefore seems to be technology. Yet the development and use of novel techniques for conducting fundamental research is an old feature in the history of science (e.g., mass spectrometry in chemistry), and in no way defines the work as technology. It is true that the distinction was more complex with recombinant DNA than is usually the case, because the technique that is important to the fundamental research is also, apparently, the main method by which technology will probably proceed; nonetheless, the difference between exploring the nature of living organisms and intentionally trying to produce ones with specific socially useful characteristics is still important.

The separation of recombinant DNA fundamental research from recombinant DNA technology may be somewhat difficult to perform, but it is my opinion that the distinction should be made wherever possible. The policy implications are considerable. In the case of science, the emphasis should be upon keeping the work as free from controls as is consonant with public safety; in the case of technology, the emphasis should be upon the question of whether the application of the science serves a social need. The greater the risk of possible damage, the greater the social need should be before the application is developed and employed in society.

If one concentrates on fundamental research itself, the only concern about recombinant DNA that commands our attention persuasively is that of accidents that might endanger the public. Adequate protection against such accidents now seems feasible, but whatever the outcome of future decisions in this developing area, it seems rather clear that the recombinant DNA case is not typical of many other areas of fundamental research. It would be a great mistake if recombinant DNA were taken as a paradigm case for regulating fundamental research. There have been relatively few actual cases in the recent history of science in which the primary concern expressed by knowledgeable critics was not about an application of science, nor financial, nor an ethical issue in biomedical research, but instead a worry that during the process of fundamental research an accident would result in damage to the public. The closest parallel to the recombinant DNA case is probably concern about accidents in nuclear research (e.g., the controversy over the Triga reactor at Columbia University), not to be confused with accidents in nuclear power installations (a "destructive technology" fear). The category "accidents in science" will probably broaden in coming years, but it is difficult to conceive that its purview would include more than a small portion of all fundamental research.

The early recombinant DNA controversy was significant in two other senses: 1) the authentic issues it raised about who should bear the responsibility of devising and enforcing regulations on research, and 2) the possibility that underneath the specific concerns expressed by the public on recombinant DNA lurk much more significant and deeper irrational fears of science. These two issues are extremely difficult ones to resolve (and more will be said below on the first), but little progress can be made in handling them unless distinctions have been made about various types of concern about research, as I have been attempting to do here.

PREJUDICIAL SCIENCE

A type of concern which is separable from those discussed so far is that scientists will present evidence or arguments that exacerbate racial, ethnic, sexual, or class prejudices and which might be used in the service of a particular ideology. The old controversies over eugenics and the newer ones over intelligence tests and race can be related to this category, as can, to a certain extent, the recent one over sociobiology. Several of these episodes were discussed in chapters 6 and 7. Here we see resistance to expansionism of a particular type—a fear that science will be used to support political programs serving one group of society more than another.

So far as formal regulation is concerned, my position on this category of concern is the same as on "knowledge itself:" critical discussion, yes! Regulation, no! I would not be candid, however, if I did not admit that I consider this category of concern to be one in which a pure separation of facts and values will never be possible, and may not even be desirable. If I were at present an executive of a research foundation and I were asked to vote on funding a program for projects on race and intelligence, I would vote negatively. I would refer to the fact that current definitions of "intelligence" and "race" are inadequate for the task, that the problem is not one that seems currently solvable, and that money could be used to greater advantage in other areas. And yet I know, and would acknowledge, that my negative opinion on the matter was influenced by my own values, my belief that the more potentially dangerous to society the results of research might be, the more rigorous one should insist that the methodology for that research must be.

If other people have the same opinion, and if the personal views of fund administrators at foundations and elsewhere are influential on this topic ("We are not going to do any work in this area"), is this not, in effect, the denial of freedom of inquiry in an area of fundamental research? In my opinion, freedom of inquiry is not foreclosed so long as: 1) no regulations are made prohibiting the research or the publication of the results; 2) fund administrators do not prohibit giving money for such research within "general category fellowships and grants," even if they refuse to support special projects of this type; 3) administrators do not interfere with the conduct of such research in institutions.

In a society where class, race and sex prejudices are already widespread, it is very nearly inevitable that research on topics such as intelligence, racial differences, and homosexuality will be used against the underprivileged, racial minorities, women, homosexuals, and lesbians. In a previous chapter on eugenics I pointed out that political loyalties and traditional belief patterns are often far more important in determining the results of scientific discussions than logical rigor or scientific validity. Every society takes a risk of increasing discrimination, then, when it permits such research to be done. Greater damage would surely be done, however, if such research were prohibited, since the limits to inquiry would have been increased in a way that presents almost no safeguards against greater and greater limitations in the future. In a free society the prudent response to the prejudicial misuse of science is vigorous criticism, not administrative control.

The problems presented by the category "prejudicial science" are immense, and I do not pretend to have solved them here. In the final analysis, the tasks of preserving both freedom of inquiry and social justice in this area depends on the existence of a nonpolarized society. A social environment so hostile to research of this type that no work could be done would pose, in fact, a true limit to inquiry of a sort that would be a dangerous precedent; on the other hand, a social environment in which certain political groups eagerly seized and successfully exploited arguments linking, for example, intelligence and race would present an extreme threat of another sort to society.

WAYS OF KNOWING

A final type of concern about science stems not so much from any piece of information that science might produce nor from any technology that might result from it, but from a critique of its mode of cognition. There are many people who maintain that science is only one of several avenues to knowledge, and these people often resist the tendency of science to expand its claims. Their resistance to science is the ultimate form of restrictionism, what might be called "cognitive restrictionism." The critique of science which these people advance attacks the epistemological bases of science on the ground that it is, at best, so specialized that it misses the most significant modes of reality, and, at worst, fundamentally alienating to the human spirit. The supporters of this view, who vary greatly in their sophistication, often call for supplementing the scientific approach with "other ways of knowing."

A recent example of such a critic was Theodore Roszak, who observed in his popular *Where the Wasteland Ends* that "science is not, in my view, merely *another* subject for discussion, it is *the* subject." He spoke of the "intimate link between the search for epistemological objectivity and the psychology of alienation: that is, to idolatrous consciousness." "It is no mere coincidence," he continued, "that this devouring sense of alienation from nature and one's fellow man—and from one's own essential self—becomes the endemic anguish of advanced industrial societies."[10]

One should not assume that all proponents of "other ways of knowing" come from outside the scientific community. We have seen in Chapter 2 that the distinguished astrophysicist and mathematician Arthur Eddington was a cognitive restrictionist; he surprised and irritated some of his scientific colleagues by writing popular books about science (which were adopted in many universities) in which he spoke of a world of "intimate knowledge" not accessible to scientific methods. Eddington wrote that when we make use of the "eye of the soul" as an avenue of cognition we should not think that we are betraying the responsibility of scientists. On the contrary, he believed that the vision of the scientist and the mystic are equally necessary for an understanding of the world.[11]

Such viewpoints have always existed in our history and undoubtedly always will. Whether one finds them valuable or not is a deeply personal characteristic. The views, fears, and premonitions of Goethe and Blake are only prominent historical examples of the many that could be associated with this category.

There is at least one area where this category of concern about science has current practical significance and where the term "regulation" may arise in at least some peoples' minds. In California and elsewhere, proponents of "creationism" criticizing Darwinian evolution have had some political success in their efforts to revise school textbooks by making the restrictionist argument that since religious interpretations of the universe and scientific interpretations are based on strikingly different or even incommensurable ways of knowing, the contrasting interpretations should be given equal time. (The very phrase "equal time," taken from television, illustrates the impact of technology on debates over what are essentially values.) The creationists say that since neither they nor the evolutionists can disprove the views of the other side, both sides should be represented in high school and elementary science textbooks. This specious argument has an appeal to some listeners because of its superficial resemblance to the principle of giving all sides in a true debate opportunity for expressing their opinions.[12]

It is conceivable that this threat to science could become a serious one; the controversies in the United States over the MACOS (Man: A Course of Study) project and the resulting attempts to restrict the National Science Foundation's autonomy are recent events in which this type of concern played an important role.[13] At the present moment in the United States, however, the strength of "alternative modes of cognition" may be receding with the passing of the peak of interest in mysticism and the occult in the late 1960s or early 1970s. At any rate, a form of special regulation that goes beyond the well-established separation of church and state does not seem appropriate. On the whole, school boards are more sophisticated than they were in the days of the Scopes trial, and one can only hope that the efforts to oppose religious, romantic, or mystical viewpoints to those of science will not receive official support.

Conclusions

The first major conclusion to be drawn from the analysis presented in this chapter is that there exists a core of categories of concern about research where regulation should still not be permitted at all. All of them fall under the heading of science, and not technology, but not all categories under science are in the core. This core consists of concerns I have labeled as "subversive knowledge," "inevitable technology," "prejudicial science," and "ways of knowing." The most inviolable category of all is "subversive knowledge," for this is the area where the fundamental restrictionist threats to science have emerged in past centuries, and where there is always the possibility of new dangers.

In the other core categories a democratic and healthy society should be able to avoid controls, but I will admit that I can imagine extreme and unlikely situations in which a temporary control of "inevitable technology" might be conceivable (a bit of fundamental knowledge which could lead easily to a serious and uncontrollable destructive application), and perhaps even "prejudicial science" (a bit of fundamental knowledge that would exacerbate social conflict in a temporary and extremely destructive fashion), but our society is presently far from being in such situations. (The circumstances would have to be roughly equivalent to the crying of "Fire!" in a crowded theater; for a hypothetical example earlier mentioned, see note.)[14] And in the category "ways of knowing," we might possibly need additional regulation *in extremis* to *protect* science, not limit it (which is what the principle of separation of church and state already does).

I am not willing, however, to admit that controls of any kind are justified on "subversive knowledge" under circumstances not covered by one of the other categories. It is my opinion, furthermore, that in the three other categories just mentioned (inevitable technology, prejudicial science, and ways of knowing) there is no justification at the present time for regulation of research, and also that the likelihood of the emergence of circumstances which would make such regulation advisable is remote. These four categories (together with subversive knowledge) form the inner core where regulation should be stoutly resisted.

The second major conclusion proceeding from this classification and analysis is that outside the inner core we have a group of categories (destructive technology, slippery slope technology, economically exploitative technology, human subjects research, expensive research, accidents in science) in which controls are not only conceivable and justified, but where regulations of one sort or another are already in effect. The most important debate within these categories of concern is not about academic freedom in principle, but about a compromise between the effort to avoid destructive social effects, on the one hand, and the effort to promote scientific and technical creativity and new advances in human welfare, on the other.

The third conclusion I would draw from this analysis is that even within those categories of concern where controls are clearly justified, there exists a danger that creativity will be regulated to death by bureaucracies with momentums of their own. Just as we are wary of the "slippery slope" in biomedical ethics, so also should we resist slipping inadvertently into increasing controls over fundamental science, since such controls can easily lead to abuses. Tremendous damage can be caused by the regulators, while the lack of regulation over some types of research would be quite dangerous to society.

To achieve a balance between these interests, we should look at the categories of concern where controls of some kind seem to be, in principle, both permissible and justified, and then ask the following questions: At what level should controls be imposed? Who should establish and administer the controls? The answers to these questions should be considered separately for each category of research and for each particular problem of regulation which has arisen.

Possible answers to the question, "At what level should controls be imposed?" include the following levels:

1. The individual researcher
2. The laboratory, university, or industry
3. Funding
   a. Government support
   b. Private support
4. Construction of research facilities
5. Actual performance of research
6. Publication of results
7. Application
8. Production
9. Sales
10. Use

Possible answers to the question, "Who should establish and administer the controls?" include:

1. The individual researcher
2. The research laboratory, university, or industry
3. A group of researchers in the field
4. An established professional organization
5. Mixed boards of specialists and lay people
6. Established governmental bodies
   a. Local, state, or federal
   b. By means of guidelines or by normal legislation

The main purpose in listing such a large number of levels where regulations could be imposed and on which regulatory agencies might operate is simply to point to the great diversity of possible approaches to individual problems of regulating science and technology; the lists do not represent any sort of "ascending scale" in terms of desirability or undesirability. The levels at which recombinant DNA research has been limited or regulated skipped from "the individual researcher" (before the Asilomar conference) to "a group of researchers in the field" (the Asilomar conference) to "guidelines by federal bodies" (the NIH guidelines). Professional researchers in the field seemed to fear most of all regulation by "local governmental bodies" (such as the Cambridge City Council). If legislation were inevitable, the professional scientists, by and large, preferred a federal law to a local one. Eventually, however, they won increasing support for the view that such legislation is not necessary even on the federal level and that existing guidelines for research are adequate. This issue is still not settled, however, particularly for industrial research laboratories, which are not necessarily affected by the guidelines.

It would be misleading to assume that the professional scientists always prefer regulations or guidelines administered on the federal level rather than local levels, or that a federal approach is always the best one in terms of general societal interest. In the categories destructive technology and accidents in science local and state regulations play important roles in protecting against fire and in controlling the use and transportation of certain hazardous materials; in the category slippery slope technology, local advisory committees, often composed of both medical specialists and lay people, have helped to make difficult decisions about the use of life-support machines in hospitals and in recommending priorities in health care.

The question, "Who should establish and administer controls?" opens up a whole range of important issues which my analysis has so far not touched. This additional set of questions is essentially political and it involves the problem of the motivation of the critics of science. Some people, upon reading the analysis presented in this chapter, would comment that my efforts to analyze concerns about science and to evaluate their legitimacy is misdirected, at least in part, because it fails to take into account that many of the critics of science are not really interested in these analytical categories; what they are really interested in is politics. They would argue that such critics of science believe that the present system under which scientists usually control their own work is fundamentally wrong, and that the critics wish to wrest control of science from the scientists and give it to popular nonscientific groups. Such criticism of my analysis up to this point has some merit, but the political controversy cannot be resolved in a useful way, in my opinion, until the categories of concern about science have been viewed separately and until the varying degrees of legitimacy of these concerns have been discerned. Otherwise, the discussion will degenerate into a shouting match in which each side impugns the motives of the other side.

On the question of motivation, it is clear that the knife cuts both ways. If we ask what are the motives of the critics when they attack science, so also can we ask what are the motives of the defenders of science when they reject the critics. The extreme and entirely useless way of approaching the problem is to fall into opposing camps which might be described as "science for the people" versus "science for the scientists." The first group can then erroneously be written off as political radicals who have no interest in intellectual values; the second group can then inaccurately be dismissed as elitists who are interested only in preserving their privileges while having society pay for their work.

The only way out of this dilemma is to recognize that it is a misleading portrayal of the existing situation. The greatest value of science is not what it does for the scientists, but what it does in both intellectual and material terms for society, of which scientists are one part. It is also clear that society is deeply affected by science and technology in ambiguous and sometimes disturbing ways; some of the public concerns about what might happen if science and technology are left entirely to scientists and engineers are clearly legitimate. In such areas as slippery slope technology and human subjects research it is obvious that the scientists and engineers directly involved in the work have no monopoly of wisdom about the ethical, psychological, and societal impacts of their work. At the same time, we know that the assertion by lay groups of control over the determination of the inherent value of fundamental research could have disastrous results; history provides some rather graphic examples of how destructive that interference can be.

chapter eleven

ATTEMPTS TO PROVIDE A PHILOSOPHICAL OVERVIEW

DURING THE course of the present century there have been three important schools of thought that have recognized the role of science as a formative factor in intellectual history and have taken explicit positions on the question of the relationship of science and values. They have proposed candidates for world views to fill the gaps left by the gradual erosion of the older ethical and natural philosophical conceptions. Since all three go beyond the facts of the individual sciences to make observations about the universe as a whole, they are considered somewhat disreputable by those scientists and philosophers who eschew anything approaching natural philosophy. Nonetheless, individual scientists and philosophers of recognized merit have played important roles in developing each of the three schools.

The labels I use to denote the three schools are 1) the reductionists, 2) the religious natural philosophers, and 3) the holistic materialists. I must conclude that all three have been instructive and interesting failures in the effort to define the relationship between science and values. Yet the failures are only partial, and from the derailments one can extract materials for the construction of competing and supplementing conceptual frameworks that are necessary for the continuing accommodation of science.

The Reductionists

Of all philosophic positions on the question of the relationship between science and values, the ultimate in expansionism is reductionism. By reductionism we mean the belief that all phenomena in the universe can be explained on the basis of physical laws operating at the lowest level of atomic order. A consistent reductionist does not stop at saying that biological processes such as fertilization, cell division, embryological development, birth, and death can be explained in terms of such physical laws. Most biologists today, including many who are not reductionists in the ultimate sense, would agree that physical explanations of these processes are possible. The thoroughgoing reductionist must also maintain that all social, intellectual, artistic, and emotional processes can be, in principle, equally well explained in terms of physical laws governing atomic particles. It should be obvious that a truly successful fulfillment of this program would be the final victory of expansionism, leaving no room at all for continuing debates about the limits of science.

For obvious reasons, few scientists have asserted the maximum claims of reductionism. Probably most of them do not consider the claims valid. And those who, in principle, do, know that they are so far from realizing the reductionist goals that they consider arguments about them pointless. Some of the most successful scientists accept reductionism implicitly as the goal toward which they work, knowing that they will never have complete success in reaching it, but taking satisfaction in any step toward it. Jacques Monod, discussed in Chapter 5, is an example of such a reductionist.

Of all twentieth-century scientists who supported reductionism, perhaps the most enthusiastic and brilliant was the German-American biologist Jacques Loeb (1859–1924). Loeb made explicit what other scientists only hinted at, or, in many cases, shrank from. To him, living organisms were "chemical machines,"[1] and even his devoted admirer and biographer W. J. V. Osterhout remarked that to Loeb mechanism "approached a dogma" for which "his zeal knew no limits."[2] Loeb believed that the "inner life" of man—his wishes, hopes, disappointments, and sufferings—was amenable to physicochemical analysis.[3] He extended the purview of physics and chemistry right up to and including human consciousness and the intellectual creations of man, such as philosophy and ethics.[4] The mechanistic conception of life, he wrote, is "the only conception of life which can lead to an understanding of the source of ethics."[5]

Loeb's expansionism reminds one somewhat of the position of the sociobiologists of later decades. One thinks of E. O. Wilson's call for a biologicization of ethics. But Loeb expressed his view in much sharper reductionist terms than Wilson. Loeb, in contrast to Wilson, was suspicious of attempts to interpret human or animal behavior in terms of Darwinian evolution, for evolution still lacked an exhaustive physical and chemical explanation. Loeb accepted evolution, of course, but he believed that some biologists had exaggerated the importance of natural selection by promoting it into some kind of "law."[6] Loeb thought that the only laws that exist are physical and chemical ones. As he expressed it, "the sum of all life phenomena can be unequivocally explained in physico-chemical laws."[7] He looked forward to the day when evolution would be explained in terms of such laws, and warned his fellow biologists that "we cannot consider any theory of evolution as proved unless it permits us to transform one possible species into another."[8] The proper goals of biologists, he wrote, were the creation of life, the transformation of species, and the molding of life forms.[9]

Loeb grew up in Germany, and as a student his first interest was philosophy. In 1880 he went to the University of Berlin with the intention of pursuing studies in that field, but he was quickly disappointed by the ineffectiveness and pretention of academic philosophy. The issue that consumed him was the philosophical problem of free will, but he soon decided that scientists had a great deal more to say on the subject than philosophers. After turning to science he first thought the study of extirpation of the cerebral cortex was the most promising avenue to an understanding of human will, and he wrote a dissertation on the subject.[10] He then decided that "the method of cerebral operations may give data concerning the paths of nerves in the central nervous system but that it teaches little about the dynamics of brain processes."[11] He then turned to a different topic, the experimental control of behavior of animals; he soon achieved a string of successes that brought him world fame and offers to work in several countries, including the United States, to which he emigrated in 1891.

It was already well known that some caterpillars in the spring climb directly from their winter nests to the tips of the branches of trees where the opening buds provide them with food. People had often wondered how the caterpillars "knew" to climb immediately up the trees and branches to their goal, without wasting time in meanderings. The obvious answer, for many biologists, was that the caterpillars possessed a hereditary instinct which drove them to the correct place for feeding. Loeb, however, did not care for this unrigorous explanation, since it was not based on physicochemical causality, and did not seem to be verifiable. Loeb came up with a beautifully simple explanation of the caterpillars' behavior that met all of his mechanistic criteria. The caterpillars were "slaves of light," and mechanically went toward the brightest portion of their surroundings. This behavior was so automatic and undeviating that if food was placed immediately behind them as they pointed to the light they were incapable of turning around. They starved if going toward food would require going away from light. Loeb announced that he had shown that these animals have no will and no instinct for going toward food. He called them photochemical machines.

In many other experiments, with both animals and plants, Loeb extended the concept of "tropisms," the determination of behavior by physical and chemical influence. Animals or plants, such as the caterpillars, which reacted strongly to light were "heliotropic." Others were "galvanotropic," "geotropic," "chemotropic," and "stereotropic," depending on whether they reacted to electric current, gravity, chemicals, or solid bodies. In all cases, the organisms were described as automatons, reacting to certain forces in accordance with physicochemical laws. And Loeb always saw his approach as a direct attack on the problem of will, that issue before which he thought philosophy had been so impotent; he wrote, "The scientific solution of the problem of will seemed then to consist in finding the forces which determine the movements of animals, and in discovering the laws according to which these forces act."[12]

These experiments were usually carried out on rather primitive organisms, but Loeb never concealed the fact that his ultimate target was man. He predicted, "I believe that the investigation of the conditions which produce tropisms may be of importance for psychiatry." He thought that in the case of man, with his system of communication of ideas, a new type of tropism might be at work:

Let us bear in mind that "ideas" also can act, much as acids do for the heliotropism of certain animals. . . . It might be possible that under the influence of certain ideas chemical changes, for instance, internal secretions within the body, are produced which increase the sensitiveness to certain stimuli to such an unusual degree that such people become slaves to certain stimuli, just as copepods become slaves to the light when carbon dioxide is added to the water.[13]

Loeb was not proposing here a true theory of intellectual causation since he thought that any idea that might have physicochemical effects in the human body was, in turn, the result of earlier physicochemical action. But his view that ideas have both physical effects and sources shows how thoroughly he rejected traditional dualism in his effort to explain all psychic phenomena in reductionist terms.

In his commitment to the concept of man the machine Loeb was certainly not original, but the startling success of his experimental methods on nonhuman material commanded attention. He was an extraordinary scientist, both in his conceptualization of problems and in his skill in carrying out experiments, and some of his work is still regarded as classic. Perhaps the best-known experiments were those that led to artificial parthenogenesis.

Loeb believed that biological development should be looked at from the standpoint of physical chemistry. An obviously crucial moment in the development of an organism is fertilization. What causes an egg to begin to develop? Past biologists often looked upon the uniting of sperm and egg as an inexplicable process involving the interaction of vital cells; Loeb preferred to look at it as a problem in physical chemistry.

Loeb had earlier been interested in disassociation theory, and he was aware that ions can have strong physiological effects. He began experimenting with placing unfertilized eggs in salt solutions of differing concentrations. In 1899 he succeeded in initiating cell division in the unfertilized egg of the sea urchin by giving it a carefully phased treatment in sea water and a solution of magnesium chloride. Subsequent experiments by Loeb and other workers resulted in artificial parthenogenesis by various means in several species. This work resulted in a sensation that went beyond the scientific world. Loeb had shown that the mysterious process of fertilization was an exercise in laboratory chemistry! Newspaper journalists warned virgins not to spend too much time in the ocean surf, and some lay people assumed that what Loeb did for the sea urchin he could do for barren wives.

After Loeb's death, research continued in many of the directions he initiated. Artificial parthenogenesis was achieved in a larger number of species, although there were many failures. Loeb's great achievement here was to establish the principle that laboratory experimentation could extend into an area where earlier it had been considered inappropriate. His theory of tropisms was often criticized, not because his experiments were found faulty, but because he placed so much emphasis on the deterministic nature of animal behavior that he left little room for spontaneous or "trial-and-error" behavior. Here, too, though, Loeb deserves enormous credit for pointing to the effects of physical forces on animal and plant behavior.

In the years between World War I and World War II there was a diminution in the popularity of explicit reductionism and mechanism of the type espoused by Loeb. Biologists continued, of course, to look for the physical and chemical bases of life as industriously as ever. But Loeb's metaphysical commitments to determinism and reductionism were no longer as evident. The causes of this shift were various, and some of them are still unclear. Progress toward an explanation of all of living phenomena in physical terms did not come as rapidly as scientists had hoped; certainly Loeb was wrong in predicting the resolution even of the problems of "human philosophy" in "a few decades" in terms of physics and chemistry.[14] This slowness of development may have led to some disenchantment. The new discoveries in physics—relativity and quantum mechanics—seemed to diminish the persuasiveness of a deterministic and materialistic world view. Politics, also, weighed against the popularity of determinism and materialism as the unattractive features of the Soviet Union, a state in which materialism was an official ideology, became more and more apparent. And, finally, an alternative to materialism and mechanism that still permitted one to use all the advantages of these viewpoints became available to scientists. This viewpoint was the philosophy of "as if," which Donald Fleming has emphasized in his interpretation of the decline of Loeb's form of mechanism.[15] Increasingly, scholars said that all the "isms"—materialism, mechanism, reductionism, as well as idealism and vitalism—were metaphysical positions inappropriate to scientific thought. One should not waste time either supporting or defending them, but instead ignore them. What was necessary in order for a scientist to do productive work was simply "to build a model" or to act "as if" living phenomena performed in accordance with a certain hypothesis. When one compared a complicated molecule to a Tinker-Toy construction, one was merely using a heuristic device that could be readily abandoned whenever its usefulness ended. The model contained no ontological message. Commitments to the nature of ultimate reality were not necessary.

This point of view still has many adherents, but its persuasiveness began to diminish when molecular biology achieved brilliant breakthroughs after World War II. As Fleming observed:

> As the metaphysical brand of mechanism had been undercut by as-if, so the ontological evasion of the latter was undercut by the explosion in the biochemistry of genetics after 1944. In studying the organism in being, it might still be necessary to conceive of mechanical models as no more than crude approximations to biological reality. But not if one reverted to the chromosomes with the men who were cracking the genetic code. There no ontological evasion seemed necessary or even permissible. The genes were simply physicochemical units of which the constitution and properties were increasingly known.[16]

In 1966 a distinguished molecular biologist published a small book that in terms of goals and outlook was remarkably similar to Loeb's *The Mechanistic Conception of Life* of 1912. It was Francis Crick's *Of Molecules and Men.* Flushed with the victory, jointly with James Watson and Maurice Wilkins, of the identification of the structure of DNA—probably the most important event in the whole history of biology—Crick renewed Loeb's message. To Crick, the questions of vitalism and mechanism were not inappropriate for a scientist to address. He chose for the prefatory quotation on the first page of his book the statement, "Exact science is the enemy of vitalism."[17] He said that the ultimate aim of the new biology "is in fact to explain *all* biology in terms of physics and chemistry." He noted that many people were entering biology from physics and chemistry because they felt that this goal was becoming more realistic than it had been. And, like Loeb, he was confident that the victory in explaining the life processes of elementary organisms—such as *E. coli*—was a preliminary step to the attack on human consciousness. He rejected outright the traditional dualism between "mind" and "brain": "What we call our minds is simply a way of talking about the functions of our brains."[18] He predicted that the next conquest of biologists would be the human brain, and agreed with the already current statement that "the neuron bids fair to become the phage of the future."[19] Crick believed that the result of this new expansionism of science would be a transformation in human values, a revolution that nonscientists did not appreciate. He lamented the fact that "people with training in the arts" still feel that modern science "has little to do with what concerns them most deeply." He warned these people, however, that science will soon "knock their culture right out from under them."[20]

Although reductionists believe that all phenomena can in principle be explained in terms of physical and chemical laws operating on the most elementary level, it does not follow that all of them think that only studies on that level are justified. At certain moments in the development of a science an attack on the molecular or atomic level may not promise success because the field is too undeveloped. Many reductionists would, therefore, grant that in fields such as psychology and animal behavior useful work can be done on the level of the study of the whole organism even though they hope eventually to explain such fields in terms of elementary physical and chemical laws. Crick acknowledged this position when he wrote:

> It is an old idea that a biological system can be regarded as a hierarchy of levels of organization, the "wholes" of one level being the parts of the next. . . . In my view a simultaneous attack at more than one level will, in the long run, pay off better than an attack at a single level, even though in the short run one may concentrate on one level at a time.
>
> Thus eventually one may hope to have the whole of biology "explained" in terms of the level below it, and so on right down to the atomic level.[21]

To Crick the decision about which level to address is a question of tactics, not of principle. While he thought that all of biology will eventually be explained in terms of the atomic level, studies at higher levels can be useful in blocking out the main outlines of the phenomena until the time arrives that a reductive explanation can be given.

From this standpoint we can understand more fully why the term "reductionist" can also be applied to behaviorists such as B. F. Skinner, even though they do not concern themselves with phenomena on the atomic level. Skinner's main goal was to explain the behavior of organisms in the causal terms of reinforcement, and he accepted the reductionist principle that such explanations must be identical with those of physics. The strongly reductionist flavor emerged in Skinner's comparison of a human being with a stone; he noted that Aristotle argued that a falling stone "accelerated because it grew more jubilant as it found itself nearer home," and he equated this view with the one that says "that a person carrying good news walks more rapidly because he feels jubilant."[22] Skinner maintained that just as modern science has decided that the stone has no internal mental state, so should one rule out discussions of a person's internal state.

Skinner's analogy between a person and a stone is like all analogies: its strength depends on the number of ways in which a stone and a person are similar. A powerful case can be made that there is more difference between a stone and a person than just about any two entities in nature that might be compared, and that the analogy may therefore not be very helpful. One doubts that even Crick would agree with Skinner here. After all, Crick observed, "Let us concentrate on what I regard as the crucial difference between a virus and a rock. A virus has a very large amount of ordered complexity, all at the atomic level. A rock is much less ordered. . . . The distinction becomes much more striking when we consider higher organisms." He then noted that the DNA molecules in just one human germ cell contain as much information as in "about five hundred large books, all different—a fair-sized library." And he estimated that if we string out all the DNA molecules from every cell in a person's body the distance would be comparable to the diameter of the solar system.[23] Some rock, that jubilant man!

It is clear that one can be totally opposed to talk of vitalism and of human "souls" and still be unimpressed with Skinner's reductionist analogy. As Noam Chomsky commented in a review of Skinner's *Beyond Freedom and Dignity*:

> For Skinner's argument to have any force he must show that people have wills, impulses, feelings, purposes, and the like no more than rocks do. If people do differ from rocks in this respect, then a science of human behavior will have to take account of this fact.[24]

REDUCTIONISM AS A STANCE

Reductionists are those thinkers who believe that the world is a unified physical entity potentially explicable by the laws of physics. The most important part of their intellectual stance is the principle that all complicated entities and dynamic processes can be ultimately reduced to simple interactions. To a reductionist, physics is the preeminent science since the most elementary interactions of nature are described by the physicist. Chemistry and biology describe entities and processes on a more gross level than physics, but the reductionist's goal, stated or unstated, is to provide physical explanations for phenomena on all levels. Ultimately, conscious humans must be accessible to reductionist explanations. A reductionist who is worth his or her salt will not dodge the question of the potential explanation of human values. Whether he or she emphasizes the genetic inheritance of humans, or the environmental influences on them, or some combination of both, the reductionist believes that all human values, all human behavior, all human intellectual and artistic creations can be, in principle, accounted for in terms of elemental interactions.

Conclusions about the value of reductionism as a stance depend heavily on whether one views it primarily as a research program or a philosophy. As a research program it is extremely valuable; its record in the history of science has been brilliant. It is probably the most fruitful of all approaches to nature available to a scientist. Indeed, at the crucial moments in the history of biology it is very nearly the only approach that works. Even when a biologist approaches phenomena at the organismic level, he simplifies his task enormously if he makes no assumptions that contradict that of reductionism.

Almost all of the researchers who openly opposed the reductionists in the past are now justly disregarded as vitalists and romantics. Furthermore, there is every reason to believe that the record of achievements of the reductionists will continue indefinitely. There are no absolute barriers to their research program—not human consciousness, not human ethics. Even more, every success of the reductionist increases man's power and freedom to change the world, for good or ill, and therefore the continuation and even acceleration of the rate of success of the reductionists is the mark of a civilization that has not lost self-confidence.

If, on the other hand, we view reductionism not as a research program but as a philosophy, i.e., as a pursuit of wisdom that helps a person in his or her permanent state of relative ignorance, we must judge it to be a failure. Reductionism makes predictions about how values will some day be explained; it says little about how one should view values before that day arrives. And since we have good reason to believe that the "some day" predicted by reductionists will retreat into the future forever as the reductionists continue to march toward it on a path of continuing successes, we must take for ourselves a world view that is not predicated on our possession of the reductionists' ultimate victory.

Reductionism, then, produces explanations that are, and always will be in differing ways, incomplete, while human beings constantly have to make decisions in precisely the areas where reductionism has not given a persuasive answer. But reductionism as a philosophy is not merely incomplete; it is positively dangerous, for it dismisses the worth of older value systems as clearly unscientific while it provides no adequate substitute for them. Reductionism fails to recognize that value systems that are obviously flawed and gradually must be modified are nonetheless indispensable; unscientific though they are, based on religion or superstition though they may be, they act as temporary restraints that play critical roles while the consequences and options of the next potentialities of science provided by the reductionists are examined. Genetic engineering and behavior modification, to name only two applications of science, are already too potent as possibilities for us to rely on reductionism as a world view, for it provides no safeguards. The older value systems are not adequate for the task either, and carry the danger of becoming dogmatic resistance to all scientific advance, but at least they give a workable set of values that, one hopes, can be modified at appropriate moments—they give a framework, and a target. The inadequacies of older value systems usually become evident over time, and the historical record is one of concession. The inadequacies of the older value systems, then, are relative. The inadequacy of reductionism in handling such challenges is very nearly total.

The disrespect that reductionists have for older value systems is a part of their general attitude toward history. Most reductionists are not only ahistorical, but antihistorical. Jacques Loeb harbored misgivings about Darwinism precisely because it was descriptive and historical, without enough emphasis on mechanistic causality. Loeb liked to tell a joke at the expense of the method of work of palaeontologists, who often date fossils on the basis of the succession of geologic strata. Loeb remarked that of two different species of dinosaurs which ventured forth on a given day, it was possible for one to fall into a deeper hole than the other without being a more primitive organism.[25] Francis Crick shared Loeb's view that evolution was nonscientific, messy, and unordered. Speaking of evolution, he said, "It may be history rather than science."[26] B. F. Skinner noted that "some historians have made a virtue of the unpredictability of history" and complained that "historical evidence is always against the probability of anything new; that is what is meant by history."[27] In his utopian novel *Walden II*, Skinner has his hero Frazier expound, "History is honored in Walden II only as entertainment. It isn't taken seriously as food for thought."[28]

In their devaluation of history these men have, of course, a point: if one's goal is scientific explanation, then describing any phenomenon in a quantitative way that is sufficiently accurate to permit prediction is superior to describing it in a qualitative, nonpredictive way. But their antipathy to history also reveals a flaw in their analysis, for it is likely that historical descriptions are a necessary part even of science. The crucial moments in human evolution, origin of life and origin of consciousness, can probably never be explained in totally rigorous ways, simply because they occurred in open systems with intersecting chains of causality in ways that are not amenable to rigorous description. The development of biological organisms contains such large elements of adventitiousness and contingency that a dogmatic reductionism is simply wrong-headed as an approach. One can easily recognize the importance of nonreductive historical analysis here without surrendering to vitalism.

If historical analysis is necessary for biology, how much more necessary it is for the study of society, where the interplay of science and value occurs! Here contingency and adventitiousness are the normal order. This flat side of reductionism, its lack of appreciation of social history, is one of the reasons that reductionists fail to understand that an obviously unscientific value system can nonetheless play, at certain moments, an indispensable role in helping human beings meet a given challenge.

In the next section, dealing with the religious natural philosophers, we will see examples of writers on science and values who give full recognition to the importance of history and tradition. However they have usually committed an even more serious mistake: in their eagerness to meld science with religion, they have sacrificed scientific rigor to religious aims.

The Religious Natural Philosophers

A number of thinkers in the twentieth century have attempted to combine the evolutionary ideas of natural science with religious values, believing that such a synthesis would place both religion and philosophy on a firmer base. The ideas of Bergson on creative evolution, considered in chapter 5, were important in this effort. Other influential writers in this vein were Samuel Alexander, Conway Lloyd Morgan, and Jan Christiaan Smuts.[29] Central to their views was the concept of "emergence," the rise during the cosmic evolutionary process of qualities and structures of behavior that never before existed. The existence of God and the appropriateness of religious belief were often justified on the basis of this cosmic scenario. Thus Morgan defined "deity" as "an emergent quality of the highest natural systems that we know, i.e., some human persons."[30] And Alexander waxed:

> Within the all-embracing stuff of Space-Time, the universe exhibits an emergence in Time of successive levels of finite existences, each with its characteristic empirical quality. The highest of these empirical qualities known to us is mind or consciousness. Deity is . . . the next higher

empirical quality to mind, which the universe is engaged in bringing to birth. That the universe is pregnant with such a quality we are speculatively assured.[31]

Teilhard de Chardin

The most interesting effort in this century to synthesize science and religion was made by the unorthodox Jesuit priest Teilhard de Chardin (1881–1955), and I wish to look at his views in some detail. We will note similarities between his ideas and several of the earlier emergent evolutionists.

Teilhard's justification of religion did not accord with that of the officials of his order, and they forbade him to publish his most important writings. After his death, however, these works were edited and published by his followers, and by the early sixties Teilhard became one of the best-known names among contemporary French writers; his *The Phenomenon of Man* quickly became a best seller, and Teilhard was celebrated across the spectrum of French political and religious thought. Although slivers of public opinion in France on both the extreme left and the extreme right continued to criticize his views, he gathered support among a very mixed group of individuals in several countries, including Arnold Toynbee, André Malraux, Julian Huxley, Theodosius Dobzhansky, Leopold Senghor, and even the unorthodox Marxist Roger Garaudy. The 1964 edition of the *Grand Larousse Encyclopedia* stated that Teilhard had caused a "revolution in philosophy as important as that of Kant."[32]

Although Teilhard continues even today to attract some attention, it is clear that the original outburst of enthusiasm has not been sustained. Scientists who turned to his writings found them much more religious than scientific, and furthermore, despite Teilhard's achievements as a paleontologist, often considered some of his scientific views to be erroneous. Biologists such as Peter Medawar and George Gaylord Simpson brought Teilhard under their critical view and judged him unscientific.[33] Professional philosophers found his writings maddeningly ambiguous and unrigorous, and, consequently, proceeded to ignore him. More and more, the followers of Teilhard could be described as a cult of uncritical admirers who celebrated rather than analyzed him. By the time the 1976 Larousse encyclopedia appeared, the favorable comparison with Kant as a revolutionary in philosophy had disappeared; Teilhard now represented a "very classical type of philosophy" which had gained renewed strength because of his time.[34]

Teilhard failed to achieve his goal, and his form of biological philosophy contained many intellectual flaws. He was, nonetheless, a remarkable man who carried his analysis of the relationship of science and values far beyond the degree necessary if his only goal had been an apology for his faith. His truly radical views on biological engineering, for example, offended not only the Church but most secular intellectuals who learned of them.

THE EVOLUTIONARY SCHEME OF TEILHARD

The concept of evolution was the key to Teilhard's thought, governing not only the biological world, but all the universe, from the first physical atoms to the ultimate social and normative activities of man. Evolution was a method by which Teilhard studied every phenomenon. In the end, he said, the concept of evolution would lead "to a profound reshaping of planetary values."[35]

Teilhard renounced the body-mind or body-soul dualism in Western thought stretching back to Aristotle and codified for the Church by Saint Thomas Aquinas. The warnings of the Church against Teilhard emphasized the "dangers" of this renunciation.[36] Teilhard thought values emerged from matter, that the way to ultimate political and religious values was *through* and not *outside* matter. He observed that to "Christify matter . . . therein lies the whole adventure of my inner life."[37]

The Catholic Church considered this approach unorthodox, but Teilhard thought he was the Church's modernizer and defender; he once described the Church's oppression of him (by prohibiting him to publish and by sending him in exile to China) as "the Church's shelling of its own front lines."[38] Teilhard pleaded that "what I am doing is . . . simply to substitute for the juridical terms in which the Church has couched her faith terms borrowed from physical reality."[39] But mere camouflage was not Teilhard's goal.

He wanted to change the Church's conception of the place of man in nature. Both Teilhard and Rome were correct in their uneasiness toward each other, for Teilhard's system of thought did, indeed, undermine the classical dualistic teaching of the Church.

Teilhard based his evolutionary scheme on the affirmation that material particles should be considered from two different perspectives, the *without* and the *within*. Scientists contented themselves only with the *without* of things, he said, and the types of energy they studied were entirely external. In order to understand evolution, however, it is necessary, according to Teilhard, to look upon the internal nature of things, and to see another type of energy there. Out of the interaction of these two types of energy emerges one of the "laws" of evolution of which Teilhard was very fond: the Law of Increasing Complexity and Consciousness. As the material particles of the universe evolve in a process of "corpuscularisation" they become increasingly complex on the outside and increasingly "psychic" on the inside.

Teilhard liked to plot the ascending curve of evolution on axes depicting, in rough approximation, the coefficients of size and complexity in order to illustrate the major events of evolution. Figure 1 illustrates such a graph.

On the curve there are certain nodal points, at which new qualities of being emerge, and the two most important are points "a" and "b", the origin of life and the origin of consciousness, respectively (points of "Vitalisation" and of "Hominisation"). As is the case for all Emergent Philosophies (Alexander, Sellars, Smuts, Morgan, the dialectical materialists), these nodal points are crucial for an understanding of the structure of Teilhard's philosophy. Do they mark something totally new which emerges (life, consciousness) or are they merely indicating the enlarged expression of something already there? Teilhard admitted the difficulty he had in describing these moments. At times they seemed to be true qualitative leaps: "Critical points have been reached, rungs on the ladder, involving a change of state—jumps of all sorts in the course of development." Thus, the "degree of interiority" of a cosmic element "rises suddenly on to another level."[40] At other times, he described these points on his graph as the mere becoming visible of what was there all along. One of his analogies stressed

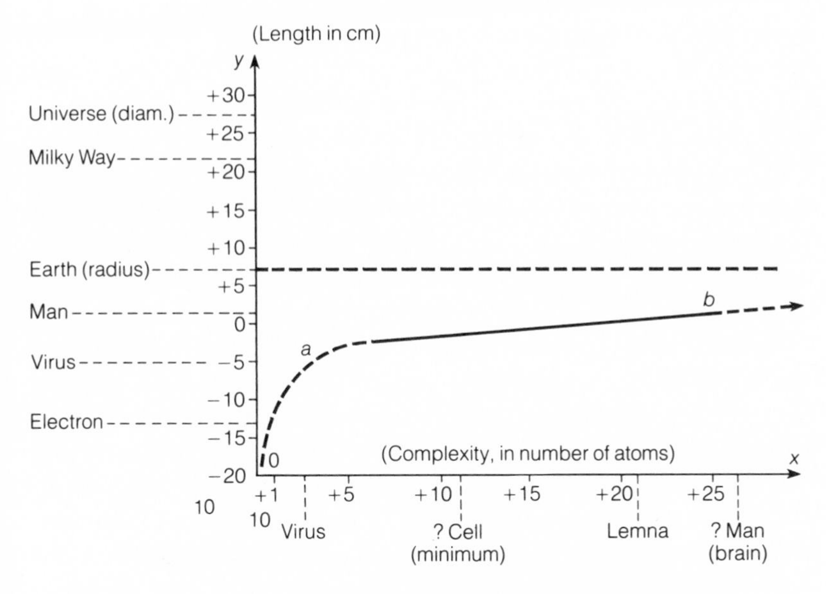

Figure 1. Natural curve of complexities
a, point of Vitalization
b, point of Hominization

From Pierre de Chardin, *Man's Place in Nature*, p. 22.

this latter interpretation; he said that in physics we had learned with the advent of relativity theory that at velocities approaching that of light certain effects (increase in mass, decrease in length) became apparent that were earlier imperceptible, and he likened the emergence of life or of human consciousness to such effects.[41] Relativistic effects exist in physics at all velocities, but only at very high ones are they significant. Similarly, he said, that "psychic energy" exists at all levels of physical reality—even at the molecular level—but only at the level of complexity of living things and especially of man does this energy take on perceptible effects. Graphic as Teilhard's analogy here is, it should be seen that there is a tension between his two descriptions of emergence; the one occurring by "leaps" or "transformations," the other by the gradual enlarging of effects that earlier had been negligible. Teilhard never solved this problem, and the two available solutions lie at the heart of the mind-body problem.

Teilhard's scheme of evolution was based on the concept of orthogenesis, the belief that there is a privileged axis of evolution. The evolutionary and morphological "tree of life" that leads up from one-celled organisms through the chordate line to fishes, amphibia, mammals, primates, and man is the privileged path of evolution, while the branches off to the side (arthropods, molluscs, crustacea, insects) are all of lesser merit, dead-ends of sorts. Evolution is not accidental, Teilhard assured his readers, and man did not develop "by chance." Indeed, life was always "groping forward" toward "irreversible perfection;" the changes that occur during evolution "*add*, one to the other, and their sum increases *in a pre-determined direction*" (emphasis in original).[42] Those two words "pre-determined direction," repeated in many variations, are the parts of Teilhard's biological scheme most difficult for modern biologists to accept, and these scientists have excellent reasons for their objections. As the American George Gaylord Simpson wrote in a review of Teilhard:

> History is inherently unrepeatable, so that any segment of a historical sequence (such as that of organic evolution) begins with one state and ends with another. It therefore necessarily has a direction of change, and if orthogenesis is merely that direction, it explains nothing and only applies a Greek term to what is obvious without the term. However, when Teilhard says that the direction is "pre-determined" . . . it is no longer trivial.
>
> Now it is easy enough to show that, although evolution is directional as a historical process must always be, it is multidirectional; when all directions are taken into account, it is erratic and opportunistic. . . . Teilhard was well aware of the consensus to that effect, but he brushed it aside and refused to grapple with it in terms of the detailed evidence.[43]

Teilhard was, indeed, well aware, as he put it, that "nine of ten biologists" rejected his description of evolution. He disagreed with that consensus not nonchalantly but because he had no other choice if he wished to preserve what to him was the most important element of his whole scheme: a privileged path of evolutionary progress along which he wished to lead his readers, a path which would end in an affirmation of religious faith.

To Teilhard, the indicator of a "precise orientation" in evolution was the "increasing cerebralisation" in the prehominids and the ultimate origin of man:

> The impetus of the world, glimpsed in the great drive of consciousness, can only have its ultimate source in some *inner* principle which alone could explain its irreversible advance toward higher psychisms.[44]

Teilhard granted that animals "think," but said that only "man knows that he thinks." Once this property of reflection emerged, a whole new era had been reached. If the lifeless earth could be called the geosphere and the living layer the biosphere, then the thinking layer was the "noosphere."

But the path of evolution continued on after the arrival of man and continues even today. Just as the "interior psychic energy" always in the past increased, so also does it today, and a new nodal point is approaching. Because of the limited dimensions of the earthly globe, mankind "turns in on itself" and accumulates in ever more complex and dense forms, leading to a "socialisation" around a common set of ideals, a "forced coalescence." Man gains control of his environment, and even of his own evolution. At that point the "pluralisms" of human society disappear, and mankind unites around a new product of emergence, the Point Omega, which is described as the "Prime Mover Ahead," the "Soul of Souls," the principle which explains "the persistent march of things toward greater consciousness." Teilhard made it evident that Omega is Christ, and described the last phase of his cosmogenesis as "Christogenesis." Just as man is the "axis and leading shoot of evolution," so also through Christianity "the principle axis of evolution truly passes."[45]

At this point in the development of his scheme, which has brought us to the Point Omega as a sort of ultimate product of emergence, Teilhard turns around and announces that Omega is *already in existence,* and that it "animates" evolution even in its lower stages, although only "in an impersonal form and under the veil of biology."[46] We then see that even in the earlier chapters, when we had not yet heard of Omega, this "Prime Mover Ahead" was really behind orthogenesis along the privileged axis of evolution all the time; the occasional reader can be excused for feeling a bit misled in not having been informed of this fact much earlier.

THE STRUCTURAL WEAKNESS OF TEILHARD'S SCHEME

A major flaw in Teilhard's biological philosophy from the standpoint of the biologist is his view of a predetermined path of evolution, as already mentioned. That concept is rejected by almost all modern biologists, but it is difficult to see how Teilhard could retain his overall evolutionary model without that particular factor.

Another structural problem in Teilhard's argument concerns the question, "What are his premises and what are his conclusions?" [47] Teilhard presented his most influential writings in a form that made it appear that he started on a strictly scientific basis and that religious conclusions flowed from this analysis. The premise allegedly was science, the conclusion religion. As Teilhard said in the very first sentence of his *The Phenomenon of Man*: "If this book is to be properly understood, it must be read not as a work on metaphysics, still less as a sort of theological essay, but purely and simply as a scientific treatise."[48] A few sentences later he affirmed that although he *could* have engaged in philosophical and theological speculations, "I have at all times carefully avoided venturing into that field." And yet what follows is three hundred pages of biological philosophy filled with speculations, and ending with an essay on Christian theology. The disparity between his promise and his deed is so great that Peter Medawar called *The Phenonemon of Man* a "bag of tricks" and he chastised Julian Huxley for being taken in by it.[49] And George Gaylord Simpson, in a more gentle fashion, observed that despite what Teilhard said, his "major premises are in fact religious and, except for the conclusion that evolution has indeed occurred, most of his conclusions about evolution derive from those and not from scientific premises."[50]

An explanation for the strange, even apparently deceitful, organization of Teilhard's book can probably be found in its history. *The Phenomenon of Man* was published after Teilhard's death from a manuscript copy that Teilhard had unsuccessfully attempted to have published with the permission of the Jesuit order and the Vatican. For over fifteen years Teilhard had modified the manuscript in order to try to mollify his religious critics. In 1947 his Jesuit colleagues and friends Bruno de Solages and Henri de Lubac had suggested almost three hundred "major revisions" and he received other suggestions from less congenial quarters in the Church.[51] He made a serious effort to incorporate many of these suggestions. On several occasions, Teilhard had been told by his Church superiors that it was all right for him to publish on scientific topics, but not on religion or philosophy. It seems likely, then, that Teilhard's affirmation in the beginning of the book that it was "purely and simply a scientific treatise" was an attempt to allay the suspicions of his censors. As it was, Teilhard's book was not placed on the Index (still in existence in 1955 at the time of publication of *The Phenomenon of Man*), but the Vatican instead issued an official warning ("Monitum") against his views in 1962.[52]

The drama of Teilhard's unsuccessful struggle with the Church and the popularity of the issue of conflict between science and religion has deflected the public's attention away from some of the most interesting and controversial aspects of Teilhard's views. Teilhard not only attempted an *apologia* for Christianity; he went far beyond that effort and sketched a scenario of the future which is based on his interpretation of the relationship of science and social values. Oddly enough, the political or religious liberal who has only scorn for the Vatican's treatment of Teilhard may sense a growing discomfort as he delves into Teilhard's views on the future of society.

THE SIGNIFICANCE OF TEILHARD'S SCHEME FOR VALUES

Teilhard described the current phase of human history as "socialization," and he divided this phase into two contrasting periods. He called the first period "the socialization of expansion" and the second "the socialization of compression."

Teilhard believed that once man began to evolve into ever more complex systems of social organization he began to spread out into all available space. In this stage the great voyages of exploration of the early modern period occurred, and a variety of forms of political organization arose. Accompanying the expansion period of socialization was a growing sense of the importance of the individual and the development of human rights. The eighteenth and nineteenth centuries were the "age of individuation," and

Teilhard saw it as being filled with "pluralisms" of all sorts; one sees rather clearly that Teilhard did not think too highly of this period (he was more attracted to the collective and to the aspiring for an absolute):

This was the age of the rights of man (i.e., of the "citizen") against the community: the age of democracy, naively conceived as a system in which everything is for the individual and the individual is everything: the age of the superman, envisaged and awaited as standing out in isolation above the common herd.[53]

This period peaked around World War I, with its rash of petty nationalisms and autonomous movements. In the interwar period there came "the promise of good things to come, the massive and yet undreamt of forces of totalisation."[54] Teilhard certainly had no wish for the victory of Germany in World War II, yet he still saw the advent of that war as a harbinger of a new age; he wrote in Peking in 1939:

At the roots of the great trouble in which the nations find themselves engaged, I think I can see the signs of a new age for Humanity. . . . Whether we wish it or not, the era of tepid pluralisms has definitely come to an end. Either a single people will succeed in destroying all others, or all people will come together in a common soul in order to be more human. This, if I make no mistake, is the dilemma posed by the current crisis.[55]

In his book *Human Energy*, written before World War II, Teilhard wrote that the coming political order "must be international and in the end totalitarian."[56] His followers were quick to point out that Teilhard criticized both National Socialism and Communism, but they were not quite able to erase the impression that Teilhard sympathized with the striving for the absolute inherent in both doctrines. And Teilhard was explicit: "Monstrous as it is, is not modern totalitarianism really the distortion of something magnificent and thus quite near to the truth?"[57]

The reason for the end of the phase of pluralisms and of emphasis on human rights, said Teilhard, was the fact that the terrestrial globe is of limited size, and mankind, after reaching its limits, turns in on itself and reaches ever more complexly organized units with greater and greater "psychic energy," reaching toward a more progressive stage of evolution. Shortage of space and limits on natural resources dictate that man must organize in order to survive. The whole world becomes a sort of superorganism, and the process of continuing cerebralization means that the world must develop an intellectual center. Teilhard saw this process as one that people would join voluntarily.

The key to survival in the new stage of "compressive socialization" is science, for only through science can man learn how to control his own fate. Teilhard produced an apotheosis of science that is surely unmatched in the history of the priesthood. Scientific research had earlier been regarded, he said, as a luxury or an eccentricity; now we must recognize it as "the highest of human functions."[58] Science would even conquer war, because "giant telescopes and atom smashers would absorb more money and excite more spontaneous admiration than all the bombs and cannons put together." Teilhard seriously asked his readers to believe that the day would come when "the man in the street" would center his ideals on "the wresting of another secret or another force from corpuscles, stars, or organized matter."[59]

Teilhard believed that science would show how man could control himself in every way possible: eugenically, psychologically, physiologically. Man would grab "the tiller of his own fate." Teilhard called for the improvement of individual human brains in a process of "auto-cerebralisation" in which the natural evolutionary process of cerebralization would be accelerated. Only in this way could the "ultra-hominisation" of man occur. He lamented the fear that some people have of turning science and technology on the human body:

> For a complex of obscure reasons, our generation still regards with distrust all efforts proposed by science for controlling the machinery of heredity, of sex-determination, and development of the nervous system. It is as if man had the right and power to interfere with all the channels in the world except those which make him himself. And yet it is eminently on this ground that we must try *everything* to its conclusions.[60]

Teilhard called for the creation of a "Center for the Study of Human Problems" in which eugenics would be an important concern, a "nobly human eugenics" which would, nonetheless, include both individual and racial measures.[61] By "racial eugenics" Teilhard said he meant "the grouping or intermixing of different ethnic types being not left to chance but effected as a controlled process in the proportions most beneficial to humanity as a whole."[62] Accused of being a racist, Teilhard flatly denied it.[63] Yet his letters from China revealed that his attitudes on race were, at least in the 1920s and 1930s, of the type that can be described charitably as typical of many white male European anthropologists of his time.[64] It is with a sense of unease, therefore, that we learn that Teilhard asked, "What attitude should the advancing sector of humanity adopt towards static and decidedly unprogressive ethnic groups? . . . Up to what point should the development of the stronger—always supposing that this can be clearly defined—take precedence over the conservation of the weak? . . . In what does the true charity consist?"[65]

Teilhard had assigned himself the lifelong task of reconciling science and religion by creating a synthesis of both. In assuming this task he took a number of steps which even yet command our attention. Renouncing the Thomistic and Cartesian dualism of matter and spirit, he decided that he would seek values through nature rather than outside it. He became an outspoken expansionist. He realized that in the final analysis the grounding of values and ethics entirely outside the material world was unsupportable:

> Many people, I am convinced, still regard the higher morality which they look for and advocate as no more than a sort of compensation or *external* counter-balance, to be adroitly applied to the human machine from outside in order to offset the overflow of Matter within it. But to me the phenomenon seems to go far deeper and to have wider implications. The ethical principles which hitherto we have regarded as an appendage, superimposed more or less by our own free will upon the laws of biology, are now showing themselves—not metaphorically but literally—to be a condition of *survival* for the human race. In other words, Evolution, in rebounding reflectively upon itself, acquires *morality* for the purpose of its further advance. In yet other terms, and whatever anyone may say, above a certain level, technical progress necessarily and functionally adds moral progress to itself. All this is surely proof that the two events are interdependent.[66]

And yet Teilhard was so convinced that morality was becoming an integral part of the upward curve of evolution that he literally lost his critical sense when he looked at science. He had devoted his career to portraying science as harmless to Christian theology, and in the process he overlooked how easily science can be used for inhumane purposes. He believed that as man becomes more conscious and rational, the chances of his making mistakes as he remakes himself and the world are becoming less and less. When critics challenged this belief by pointing to the abuses of an ever more powerful science, Teilhard was undeterred. In 1946 Gabriel Marcel, in a debate with Teilhard on "Science and Rationality," asked the priest if the behavior of the doctors at Dachau did not shake his belief in the inevitability of human progress. Teilhard shocked everyone in the room by replying, "Man, to become fully man, must have tried everything." Man would eventually learn what was good and what was bad, and eliminate the bad. When Marcel cried "Prometheus" to this reply, Teilhard answered, "No, only man as God has made him."[67]

In yet another illustration of his uncritical surrender to science, Teilhard in a 1946 essay entitled "Some Reflections on the Spiritual Repercussions of the Atomic Bomb" expressed only undiluted elation for this victory for science. He had little patience for people who were troubled by the advent of the atomic age. To them Teilhard caustically remarked, "As though it were not every man's duty to pursue the creative forces of knowledge and action to their uttermost end! As though any force on earth is capable of restraining human thought from following any course upon which it has embarked!" Teilhard thought that man's control of atomic energy heralded his future control of his own heredity and psyche. He saw little danger in these new potentialities. The only real danger, he said, was "boredom," which he called "Public Enemy No. 1." So long as man remained interested in science by "pursuing even further, to the very end, the forces of life," this enemy could be conquered.[68]

It is a great irony of our age that one of the most ardent expansionists in the name of science should have been a Jesuit. One wishes Teilhard had had success in his effort to force the Vatican finally to recognize the fact of evolution. One applauds him for seeing that in the future we will not be able to keep our values in a compartment separate from our knowledge of nature. One recognizes that he was correct in seeing the potential good results in man's greater control over his heredity and evolution. But the uncritical scientific optimism of Teilhard, united with an ethnocentric religious commitment to Christianity, was a dangerous path out of the dilemma he saw so clearly. Such an approach to the immensely difficult problems of biomedical ethics is an invitation to disaster.

The Holistic Materialists

In the early part of the twentieth century a school of natural philosophy developed which has often been described as "organicism" or "holism." Early exponents of this point of view included J. H. Woodger, E. S. Russell, and Ludwig von Bertalanffy.[69] Although there were important differences among these writers, they shared the belief that the old positions of vitalism and mechanism were both false, and should be replaced by a third. They agreed that there was no special "life force" in biological and social systems, and that, in principle, such systems were explicable by science. They were, then, expansionists. But they rejected the mechanistic point of view, so well represented by Loeb, that all biological and social systems can be explained in physicochemical terms. The organicists advanced the idea that the whole organism is greater than the sum of its parts; as Russell said, "No part of any living unity and no single process of any complex organic activity can be fully understood in isolation from the structure and activities of the organism as a whole."[70] In order to achieve a scientific understanding of living organisms, it is necessary, thought the organicists, to include purely biological concepts (i.e., not ones taken from physics or chemistry).

To this idea of irreducibility the organicists often united the further concept of a hierarchy of levels of being. In multicellular organisms different types of analysis were deemed appropriate at the levels of the cell, the tissue, the organ, the organ system, and the organism as a whole.

Most of the debates about the validity of organicism concerned the relationship between physics and chemistry, on the one hand, and biology on the other. The equally appropriate issue about the relationship between biology and social behavior was much more rarely raised. The debate about the reduction of biology to physics and chemistry became particularly intense after the identification of the structure of DNA in 1953, an event which was a startlingly new advance in the ability of scientists to explain biological processes in physicochemical terms.

Some writers extended the holistic approach above the biological level; they spoke of a hierarchy of levels that included the physical, the biological, and the social. Such an approach was directly relevant to the question of the relationship between science and human values, since it brought all of the normative concerns of human society within the purview of a purportedly scientific approach. One such author was Alex B. Novikoff, who wrote in 1945 in *Science* a frequently cited article on the subject:

> The concept of integrative levels describes the progress of evolution of the inanimate, animal, and social worlds. It maintains that such progress is the result of forces which differ in each level and which can properly be described only by laws which are unique for each level. Since higher level phenomena always include phenomena at lower levels, one cannot fully understand the higher levels without an understanding of the lower level phenomena as well. But a knowledge of the lower levels does not enable us to predict *a priori* what will occur at a higher level. Although some may have validity for the higher level, laws of a lower level are inadequate to describe the higher level. The laws unique to the higher level can be discovered by approaches appropriate to the particular level; to do otherwise is invalid scientifically and, in some instances, dangerous socially.[71]

For all its somewhat abstract and even metaphysical overtones, this approach has considerable heuristic value. It provides a justification of the autonomy of each discipline (e.g., physics, chemistry, biology, sociology) by granting each a separate realm of jurisdiction. It is expansionist for science in general, but restrictionist for particular scientific disciplines. The integrative or holistic approach protects its adherents from gross extrapolations from one level to another, such as attempts to explain human behavior in terms of animal behavior. This warning has ethical and political impact, especially in view of the fact that several of the most ambitious efforts in history to explain society in biological terms have been politically conservative or even fascist. Novikoff was referring to such episodes when he said that to ignore the differences between levels can be "dangerous socially."

But the holistic approach also has some obvious drawbacks.[72] If the concept of autonomous realms of jurisdiction for the various sciences is taken as literally and absolutely true, it can lead back to a rigid separation of humans and animals, almost in a pre-Darwinian or even religious sense. Furthermore, we know that biological and physiological studies *do* often inform us in important ways about human beings. How does one decide, within the holistic framework, which of these extrapolations are legitimate and which are not? And, last, if mechanism and reductionism can be "dangerous socially," so, too, can a dogmatic holism, as I will try to illustrate below.

Among holistic materialists, the best known are the dialectical materialists, who have developed a Marxist philosophy of nature. The dimensions of this effort in the Soviet Union have been astounding; tens of thousands of books, articles, pamphlets, and brochures on Marxism and science have appeared there in the decades since the 1920s. Every graduate of a Soviet university in the last fifty years has been required to take at least one course in dialectical materialism. Although the majority of Soviet students and scientists have apparently not paid much attention to this educational experience, some—particularly in the earlier, more ideological years of the Soviet period—have been intellectually affected by it. As I have illustrated in an earlier book, several outstanding Soviet scientists (e.g., V. A. Fock, V. A. Ambartsumian, A. I. Oparin, L. S. Vygotsky, S. L. Rubinshtein, P. K. Anokhin) have so closely tied their scientific ideas to Marxism that a thorough discussion of their work must include the influence of Marxist philosophy, just as an analysis of the views of scientists in other cultures must often include references to their religious or philosophical viewpoints.[73] Furthermore, the writings of these Soviet scientists are philosophically and scientifically significant and interesting.

In my opinion, the attempt of the holistic materialists to find a solution to the question of the relationship of science to sociopolitical values has been a failure, just as the efforts of the reductionists and the religious natural philosophers have also failed. In this section, I will discuss several of the particular reasons for the lack of success of the Soviet Marxists. Before doing so, however, I would like to summarize dialectical materialism as an intellectual scheme, at the same time illustrating some of its genuine strengths.

Dialectical materialism is an effort to find an interpretation of the natural world (including man) that steers between reductionism and vitalism. Like the reductionists, the dialectical materialists believe that only matter (or, since relativity, matter-energy) exists in the universe. But unlike the reductionists they maintain that on different levels of this material reality, different principles or regularities obtain. With reference to the biological and social sciences in particular, they are quite skeptical of the possibility of explaining life, thought, and values in elementary physicochemical terms. The long-term evolutionary development of matter during the history of the earth from the simplest nonliving forms up through life and eventually to man and his social organizations is regarded by them as a series of quantitative transitions involving correlative qualitative changes. Thus, there are "dialectical levels" of natural laws, or what J. D. Bernal once called "the truth of different laws for different levels, an essentially Marxist idea."[74]

According to this view, social laws cannot be reduced to biological laws, and biological laws cannot be reduced to physicochemical laws. The person who speaks of a possible future identity, or even a close parallel, between the behavior of a computer and the human mind is ignoring those very significant differences, however valuable the comparison may be in certain limited ways; the popular ethologists who speak of humans as "naked apes" are committing the same error. As the American Marxist and Nobel Prize laureate H. J. Muller once observed, such scientists are "forgetting the special laws which apply to social processes and structures that put the latter on a different level, as it were, from the simple biological non-social, non-intelligent organisms."[75]

This emphasis upon the uniqueness of biological and social phenomena is a valuable safeguard in dialectical materialism against the vulgar simplicity of sociobiology, behaviorism, and other reductionist explanations of humans and their values. In recent years dialectical materialists in the Soviet Union have often played the role that humanists assume in Western countries, opposing scientistic efforts to reduce human behavior and values to explanations based on physical and chemical research, whether it be behavioral genetics or psychopharmacology.[76] And while dialectical materialism insists on the irreducible uniqueness of human beings, it also emphasizes that no transcendent principles are at work in nature and, as an expansionist doctrine, it avoids the antiscientific restrictionism of religious viewpoints. As a nonreductionist materialism, dialectical materialism manages to preserve a place for uniquely human values while at the same time basing itself on the principle that nothing exists in nature but matter-energy. It is therefore compatible with science without surrendering to the reductionist sciences of physics and chemistry. These characteristics of dialectical materialism qualify it for much more serious attention from intellectuals in the West than it has so far received.

Dialectical materialism has been discussed in countless books, and I do not need to examine it fully here. The form of dialectical materialism developed in the Soviet Union can be summarized in the following tenets:

1. The world is material, and is made up of what current science would describe as matter-energy.
2. The material world forms an interconnected whole.
3. Man's knowledge is derived from objectively existing matter.
4. The world is constantly changing, and, indeed, there are no truly static entities in the world.
5. The motion in matter and the changes which occur in it are self-generated, and no transcendent principles are needed to explain them.
6. An understanding of the world must include a knowledge of its historical development, and not just its functional behavior.
7. The changes in matter occur in accordance with overall regularities or laws.
8. The laws of the development of matter exist on different levels corresponding to the different subject matters of the sciences; social phenomena cannot be reduced to biological ones, and biological ones cannot be reduced to physical ones.

9. Matter is infinite in its principles and therefore man's knowledge will never be complete.
10. Man's knowledge grows with time, as is illustrated by his increasing success in applying it in practice, but this growth occurs through the accumulation of relative—not absolute—truths.

Presented in this intellectual form, dialectical materialism is superior both to reductionism and religious natural philosophy as a heuristic approach to the world.

But dialectical materialism as it exists today suffers from debilitating flaws, both political and intellectual. On a political level, dialectical materialism has been damaged—probably beyond the point of salvage—by the fact that it is the political doctrine of an oppressive, nondemocratic state. It has become, in effect, a state religion. There is little chance that intellectuals in the politically free parts of the world will give dialectical materialism the serious attention it deserves when its strongest exponent is a state which denies its intellectuals elementary civil rights such as the freedoms of expression, political choice, and travel. Sciences such as psychiatry continue to be perverted in the Soviet Union by being made the instruments of political authorities. Dissidents have been declared insane on the grounds that any healthy person would recognize the virtues of the Soviet system. Scientists who defend democracy and human rights, of whom Andrei Sakharov is the best known, have been persecuted.

The authoritarian policies of the USSR have caused the great majority of intellectuals there to lose interest in dialectical materialism and in Marxism in general. What was once a liberating and innovative doctrine has become a scholastic and orthodox one. Rather than opening up vistas, it closes them down.

Ironically, the principle of nonreductionism in dialectical materialism, which provides valuable safeguards against scientistic exaggerations based on biological or behavioral explanations of social values (sociobiology, behaviorism), also strengthens the hand of the authoritarian rulers of Soviet intellectual and political life. According to dialectical materialism, each level (physical, biological, social) of nature has its own laws; it therefore seems logical that each level of nature should be authoritatively interpreted by an individual class of specialists. Physicists should stick to physics, while particularly qualified experts evaluate and direct the social organism. Thus, the Marxist philosophy of science has become useful as a state doctrine for rationalizing an authoritarian government in which a political party representing a minority of the population bears special responsibility for discerning and applying socioeconomic regularities. Nonreductionism, in the Soviet context at least, has strengthened elitism.

Reductionism, for all its vulgarities, has the healthy feature of being vulnerable to scientific criticism. Its commitment to the explanation of all of reality by one set of laws automatically opens up the discussion of that reality to all the sciences. Since reductionism attempts to reduce complex phenomena to simple ones, the question can always be raised, "Are the postulated simple mechanisms adequate for the task?" Examples of reductionist systems which are now discredited because of such criticism are Social Darwinism in the late nineteenth century and eugenics in the early twentieth. Critics of these movements were able to convince increasing numbers of people that Darwinian biological evolutionary theory is not adequate for an explanation of social and economic life, nor are Mendelian genetics adequate for explanations of such phenomena as crime, prostitution, or unemployment. Sociobiology and behaviorism are now being subjected to similar questions.

The critic of a social or political theory based on the nonreductionist principle that "each level of reality has its own laws" has a more difficult task. By definition, these laws cannot be reduced logically to sets of simpler principles, so how does one decide if they are coherent or persuasive? In the Soviet Union, a physicist or biologist is granted no authority in criticizing political or social theories, because he is "out of his area." Nonreductionism can act as a buttress for dogmatism, and has done so in the Soviet Union. Nonreductionism reduces the vulnerability of social theorists to criticism from other disciplines. In the end, of course, the reality principle still holds: if the postulated social theory does not seem to fit well with actual events as they unfold, its persuasiveness will decline. But the questioning of nonreductionist theories is less direct than that of reductionist ones.

A further weakness in dialectical materialism is that its opposition to reductionism brings it (especially in its Stalinist form) dangerously close to vitalism or animism. Such critics of dialectical materialism as Jacques Monod have, for this reason, termed it a form of animism, as we have seen. Monod's criticism is exaggerated. Dialectical materialists distinguish themselves from vitalists and animists by insisting that biological and social phenomena are just as amenable to scientific investigation and explanations as are physical phenomena, even though the principles which reign on these upper levels are different from those on the lower levels. In fact, Soviet Marxists maintain that they have found these upper-level principles in historical materialism, which purportedly describes the principles of Soviet socialism (thus reinforcing the arguments of their critics who correctly condemn the violations of human rights that abound in Soviet socialism and therefore conclude that any philosophical system that supports such a political system must be corrupt itself).

Dialectical materialism is a universalistic philosophy that attempts to bring all of nature—including living phenomena, human beings, society, and even ethical and value systems—under the sway of scientific study. It is, therefore, clearly an expansionist doctrine. And herein lies one of its differences from animism and vitalism. Animists and vitalists are usually restrictionists, people who exclude, in principle, such social phenomena from the possibility of scientific examination. In the Soviet Union the scientific study of society (e.g., sociology, anthropology) has been heavily restricted out of fear that the research might reveal weaknesses in Soviet society and lead to embarrassing discussions, but in principle the scientific study of society is just as legitimate there as is the scientific study of inanimate nature. And here is a great paradox: the Soviet Union, which is officially based on a materialistic and scientific expansionism, actually restricts research on society and its values more than countries in the West that—to the degree that they still have reigning ideologies—are based on philosophical and religious restrictionism.

We see, then, that all three of the recent major attempts to describe the relationship of science to values—those of the reductionists, the religious natural philosophers, and the holistic materialists—have been failures. Yet we still need a means of coping with this relationship—if not a true world view, then at least a set of suggestions and warnings. In the next chapter I attempt to give a few of these guides. The framework within which I pose my suggestions is closer to that of the holistic materialists than to that of the reductionists or religious natural philosophers; nonetheless, I grant reductionism great strength as a scientific research program and I grant the religious viewpoint a worth-while role when one approaches areas where one's ignorance is still great and when older value systems may be our best temporary resort.

PART FOUR

WHAT KIND OF EXPANSIONISM DO WE WANT?

chapter twelve

SCIENCE AND VALUES: ATTEMPTS AT HISTORICAL UNDERSTANDING

THE MAIN subject of this book has been the relationship of twentieth-century developments in the various fields of science to different sorts of values. I recognized from the beginning the impossibility of covering this subject in a comprehensive way, and I therefore decided to concentrate my attention on what I saw as the most significant recent interrelationships between science and values. I introduced priorities into my inquiry by asking the question, "What has been the fundamental impact of science upon values during the twentieth century?" My answer was that of the myriad of influences of science upon values, the most important could be grouped around two transformations in thought and attitudes. The first was the "epistemological transformation" which grew out of the new physics—relativity theory and quantum mechanics—born in the first decades of the century, and the second was the "ethical transformation" that resulted from the development of new biological theories and practices in the last half of the century. I examined the works of authoritative writers on those subjects, as well as several specific episodes of controversies over science and values connected with these developments in physics and biology, and concluded that most of the authors could be classified as either restrictionists or expansionists of various sorts. Expansionists were people who believed that science could be related, either directly or indirectly, to value questions and who thought, therefore, that our values should be informed and perhaps even changed by developments in science. Restrictionists were people who contended that science and values belong to separate realms, and who resisted any indication that our search for, or defense of, values should turn to science for help. We also saw that some restrictionists found it impossible to remain true to the principle of restrictionism; they sometimes even used the principle of restrictionism to defend value preferences, in an approach that may have been illogical, but that was comprehensible in social terms.

It is obvious that some fields of science approach social values much more closely than others. Cultural anthropology or psychology excite emotions among more people than do theoretical physics or organic chemistry. The word "science" is clearly too broad a term to serve as a reliable reference point in discussions about whether science is "value-free" or "value-laden." Instead of a bipolar relationship between science and values we need to think in terms of a spectrum, or a series of spectra, between concepts and goals taken from different scientific fields, on the one hand, and various types of values (political, moral, religious, etc.), on the other. Stephen Toulmin described such an alternative to a value-free picture of science in the following way:

> The basic concepts of the sciences range along a spectrum from the effectively "value-free" to the irretrievably "value-laden"; the goals of the scientific enterprise range along a spectrum from a purely abstract interest in theoretical speculations to a direct concern with human good and ill; the professional responsibilities of the scientific community range along a spectrum from the strictly internal and intellectual to the most public and practical.[1]

The relationship between values and a certain point on the spectrum of science does not remain constant over time; atomic physics is rather far from human concerns, but when the advent of quantum theory in the early decades of this century seemed to hold implications for such human values as free will, atomic physics became a subject of heated debate among philosophers and lay people. That controversy has now become quiescent, and physics currently excites people much more through its technological applications (e.g., atomic weapons, nuclear power) than as a result of its philosophical implications, but it is entirely possible that new developments in physics will revive the older type of debate. Fields at the far "value-free" end of the spectrum of science, such as physics and astronomy, can affect our world views, including their normative components.

Linkages Between Science and Values

Let us examine more closely the types of linkages between various fields of science and various types of values that can be identified. Wherever possible, I will give examples taken from discussion in previous chapters of this book.

LINKS BASED ON VALUE TERMS WITHIN SCIENCE

In the body of core concepts of some fields of science—the social and biological sciences—there are terms that are probably inextricably linked with values. Thus, in physiology, psychology, and psychiatry terms such as "normal," "abnormal," and "deviant" carry value implications. In biology, the term "adaptation" may also be value-laden, as it is often based on the assumption that survival and propagation are the good results which distinguish adequate from inadequate adaptation.

Many researchers in these fields of the social and biological sciences have done their best to eliminate value connotations. Some psychologists try to avoid ethical or moral considerations by saying, for example, that "normal" is a statistical term; "normal behavior" is behavior displayed or practiced frequently by a majority of the members of a society. Thus, what is normal for one society (homosexuality, polygamy) may not be normal for another.

Yet this approach, leading in its extreme form to pure anthropological relativism, is hardly satisfying when examined carefully. If an anthropologist of this statistical bent should encounter a society where a group decision had been made that it was all right to eat young children in order to increase the food supply, then that anthropologist would be forced to conclude that the society was acting normally as it ate itself to extinction.[2]

Increasingly, social scientists have recognized the impossibility and immorality of a complete divorce of value judgments from their science, and have termed those kinds of behavior normal or even superior which help ensure the survival and prosperity of the human species and the fulfillment of the potentials of its individual members. (Although this is clearly an ethical decision, it is also informed by a biological notion of survival, and shows how closely science and values are becoming intertwined in some fields). In approaching their studies from this standpoint, the social scientists are unapologetically placing a human value at the base of their science. The Society for Applied Anthropology in 1951 adopted a code of ethics which called for its members to have "respect for the individual and for human rights and the promotion of human and social well-being."[3] Many other professional societies have adopted similar resolutions, usually quite general in wording, but clear enough in intent. In taking such positions these professional societies are not only recommending a code of ethics for the guidance of their members in conducting research projects, but also postulating an implicit value scale in the societies and individuals whom they study.

In the social sciences the inclusion of carefully considered value commitments in the theoretical conceptions of the field is not only unavoidable but also commendable. In the physical and mathematical sciences, however, the case for keeping such considerations separate from science to the maximum degree possible is much more persuasive. Here, if one centers one's eyes firmly on the theoretical terms of the science itself (and not on the applications of science or the behavior of its practitioners, where ethical considerations obviously come into play) value links are difficult to discern. Attempts to insert such value considerations into the terms of physical theory may even lead to damage to science and, eventually, to society.

After modern physical science emerged as an experimental and rigorous science, shedding its earlier teleological and anthropomorphic elements, an effort was made to rid it of all terms that contained human references. Physicists of the nineteenth century sometimes lamented the fact that the same word "force" had currency in both mechanics and human behavior, fearing that students might confuse the meanings.

In recent years an interesting reversal of this tendency has occurred. In high energy physics scientists now speak, with obvious relish, of "quarks," "charm," "the eightfold way," and even of properties of quarks called "truth" and "beauty." It would be interesting to explore the reasons this change has occurred, but a detailed examination would take us too far afield. An explanation of this development would undoubtedly refer both to sociological and intellectual factors. Physicists today are often youthful, puckish, and decidedly unsolemn. The nature of the profession has changed dramatically in this century. The science of physics has changed even more markedly. Newtonian physicists of the nineteenth century talked about events close enough to the everyday experience of ordinary citizens that the possibility of confusion of similar terms was realistic; high-energy physicists of today deal with phenomena so remote from ordinary events that "contamination" is excluded. And by adopting terms taken from human characteristics or the Buddhist religion, physicists can have the personal satisfaction of appearing more congenial and universalistic, and can have the pleasure of confusing those critics who would describe them as unknowledgeable about the humanities and uninterested in humanistic questions.

At any rate, the roles of such terms as "normal" in psychology and "charm" in physics are totally different, despite the superficial appearance that both make reference to human characteristics usually considered desirable. The first term is unavoidable in science, and covers what is probably an inherent linkage between science and values; the second is a convention that, while picturesque, has no direct relationship to the science-value question.

LINKS BASED ON SCIENTIFIC THEORIES OR HYPOTHESES ALLEGED TO IMPINGE ON VALUES

In the previous section we considered links between science and values based on terms used in various fields of science as a part of the common vocabulary, and gave as examples words like "normality" and "adaptation." Our conclusion was that examples of such value-laden terms can be found in the social and perhaps biological sciences, but that the significance of such terms in the physical sciences in modern times has been slight indeed.

There exists, however, another category of links between science and values that is much more common and significant than the one just considered. This category consists of value attributions given by individuals to theories or hypotheses found in the sciences. Of course, the source of the value here may not be so much in the theory itself as in the mind of the person who is trying to find significance for human values in science. An example would be the Social Darwinist of the nineteenth century (an "indirect expansionist") who tried to justify a certain political and economic order by drawing an analogy between Darwin's biological survival of the fittest and the economic struggle of unregulated capitalism. Another, from astronomy, would be the attempt to justify divine creation by referring to expanding theories of the universe. Yet we should not relegate all efforts to relate scientific theories to social values to this level of simplicity. In other instances, the relationship is more complex and interesting.

As we have seen in an earlier chapter, Niels Bohr liked to refer to nuclear physics in explaining human attributes. Sometimes he used analogies (as when he described "love" and "justice" as complementary abstractions analogous to "corpuscle" and "wave") and sometimes he spoke in much more direct and concrete terms, as when he indicated that freedom of will may be related to quantum phenomena in the human nervous system.

In biology, the frequency of alleged links between scientific hypotheses or theories and values is much greater than in physics. The Lamarckian hypothesis, now judged invalid, was connected to social and political questions by many writers in Germany and Russia in the 1920s. And even today we must admit that *if* humans could genetically inherit acquired physical and behavioral characteristics in a controllable and significant way we would have to look at our educational and social institutions in a completely new light.

The fractious controversies over ethology and sociobiology discussed in chapters 6 and 7 are examples of how debates about hypotheses and theories in the study of animal behavior almost directly lead to value conflicts. Some sociobiologists have maintained that all cultural behavior in the human animal, including the development of ethics, can be explained in terms of "inclusive fitness maximizing;" this hypothesis is probably the most ambitious direct expansionist argument to be advanced in this century. It goes, of course, much beyond the more restricted claim that some behavior believed to be ethical, such as "altruism," can be explained in accordance with the sharing of genes through kinship.

In some instances the participants in the ethology and sociobiology discussions brought a priori values with them into the debates and merely sought evidence in biological theory for those values, ignoring counterevidence; in other instances, the participants were not seeking value confirmation by science, but were willing to follow science wherever it seemed to take them. In almost all cases in which researchers had broad interests in ethology and sociobiology, however, value disputes arose. Thus, the relationship between scientific theories and human values is too intimate in these areas for us to dismiss its importance by observing that "it is the people who bring the values into the controversy; the science is neutral." The statement, supported by some sociobiologists, that "ethics derives from inclusive fitness maximization among humans" is a scientific hypothesis that brings values squarely to the middle of scientific discussion. The testing of the claims of the sociobiologists is a socially and politically important task that still has not been accomplished.

In a historical and social sense, areas of scientific investigation such as sociobiology are inextricably linked to values. A researcher who stays in the field over the long term, who becomes a full participant in its controversies, will eventually have to commit himself or herself on value issues, either implicitly or explicitly.

LINKS TO VALUES BASED ON THE EMPIRICAL FINDINGS OF SCIENCE

We have so far discussed links between science and values that have been based on the terms contained in the scientific vocabulary of different disciplines, and links that are based upon the content of scientific theories. We now turn to the question of the value significance of empirical findings or data produced by science.

The empirical data of science can have significant influence on values because they can support or contradict scientific hypotheses that impinge upon existing social values. The Copernican hypothesis of a heliostatic universe, for example, caused an uproar in seventeenth-century Europe because it ran counter to an alternative description of the universe, the Ptolemaic system, which was deeply imbedded in European culture and religious teachings. In 1543, however, when Copernicus presented the hypothesis, there was little empirical evidence that permitted one to choose between his hypothesis and the Ptolemaic variant.

Against this background, the data later presented by Galileo had great significance for the discussions about man's dignity, and his place in the eyes of God, for Galileo introduced evidence which made the Copernican description, in the minds of many astronomers at least, no longer an "alternative" or a "hypothesis," but a physical fact. Before Galileo a rational astronomer could still defend Ptolemaic astronomy if he chose; after Galileo produced a mountain of evidence in favor of Copernicus, our rational astronomer would probably end up on Copernicus' side. Thus, the empirical findings of science in this instance had significance for value beliefs, although it would, of course, be a mistake to say that "values were contained in the data." The value transformation occurred because Galileo's data supported a view of the universe that contradicted an existing view interwoven with societal values.

The possibility of new scientific data having an impact on social values continues today. At the moment, the viewpoint that there are many civilizations in the universe is only a hypothesis. If in the next few years, scientists would produce recordings of radio transmissions from an extraterrestrial civilization that had a level of culture and technology far superior to ours, the value impact would be significant, although it is difficult for us to predict how great. Some scientists say we might well be as demoralized as were primitive tribes overwhelmed by European culture in recent centuries. Certainly there is a limit to the rapidity with which humans can absorb change; if it turned out that the extraterrestrial civilization not only possessed a vastly superior technology, but a radically different religious and moral ideology which seemed to provide an underpinning for its technology, then the disruptive effects upon us might be large indeed. In this instance we could justly say that "values *are* contained in the data," the values of another civilization. If we wished to catch up, we might not be able to do so merely by learning new technical tricks, but might be forced to compare our deepest values with theirs.

Other writers doubt that the discovery of extraterrestrial civilizations would have such an impact; they cite the enormous distances involved, the impossibility or unlikelihood of true exchanges of information in a short enough time to be meaningful, and the unlikelihood of actual physical contact. Yet "question and answer exchanges" or physical contact are probably not necessary for disruption; we could broadcast into space at random the whole *Encyclopedia Britannica* in a week or two, repeating it in cycles, and another civilization might already have chosen to do something similar. If we had such a huge amount of information about a superior civilization, we would be sure to find some of it disturbing to our ordinary way of looking at things, even if we could not get answers to specific questions or could not make physical contact with the other civilization.

These speculations serve to illustrate the point that the empirical data supplied by science has the potential to cause value conflicts. As discussed more fully below, the origin of these conflicts is not necessarily in the data, but usually in the relationship between the data and values already possessed by society.

LINKS BASED ON THE METHODS AND SOURCES OF SCIENCE

Another type of link sometimes discerned between science and values is based on the methods and sources of science. People who emphasize this type of link do not assert that value statements reside in the core of scientific theory, nor are they talking about the attribution of value implications to the facts or findings of science. Rather, they believe that the source of scientific creativity and the methods by which science proceeds have value meanings.

Einstein was a person who, as we have seen, frequently emphasized this type of relevance of science to ethical domains. In all other regards, Einstein was a restrictionist, and he usually prefaced his comments on this topic by disavowing any direct connection between science and values, agreeing with Hume that facts and values are incommensurable; as he liked to point out, "Science can only ascertain what *is*, but not what should be." Yet Einstein went on to maintain that science as a model of rigor and deductive clarity could be helpful to ethics once the goals and value content of the ethical system had already been acquired on a nonscientific basis.

Beyond the usefulness of science as a model for nonscientific reasoning, Einstein stressed the similarity of goals of scientists and those of persons striving for justice and harmony in the world. In his linking of science, ethics, and religion by their mutual striving for perfection, Einstein was voicing a view that can be traced back to the Pythagoreans of classical Greece. One of them, Philolaus, wrote:

> The nature of number is a standard capable of leading and teaching the one who wonders at everything or the one who is ignorant of everything. Were it not for number and its essence, nothing, either of itself or in relation to other things, would be clear to anyone. . . .
>
> The nature and harmony of number do not admit any falsehood because that is foreign to them. Falsehood and envy are innate only in the unlimited, the unintelligible, and the irrational.[4]

Plutarch echoed this view in his discussion of Plato's opinions on geometry; Plutarch noted that Plato "sang the praise of geometry for drawing us away from the world of sense to which we cling and turning us toward the intelligible and eternal level of existence."[5]

Science as a value choice symbolizing rationality and harmony is probably particularly attractive to a person who believes that he or she is living in a society in which disorder, or falseness and hypocrisy, affect almost all other areas of activity. We have seen that Werner Heisenberg was very much drawn to Platonism and that he found in the world of geometric forms and number a kind of harmony that reassured him during a life which contained an unusually large measure of political turmoil and the necessity for political compromise. We remember Heisenberg's description of the "confusion, pillage, and robbery" in Munich in 1919, and then his observation that reading Plato's *Dialogues* brought him a sense of peace. In the last decades of Heisenberg's life, whether he was making adjustments to Hitler's government, or obeying the camp regime brought by his imprisonment by the Allies after World War II, or discussing the philosophical principles of physics with East German Marxists in the 1960s, the permanence and harmony contained in Platonic idealism continued to have great appeal.

Some scientists have even gone so far as to maintain that the rationality of science contains a moral lesson. One of the most explicit statements of this view was given in 1961 by the Soviet mathematician A. Ia. Khinchin, who contrasted the objective reality of scientific arguments with the pettiness of social and political quarrels:

> In common lawsuits each of the disputing sides usually aims toward the solution of the issue that would be desirable or profitable from its standpoint; with the greatest inventiveness possible, each side devises arguments for the resolution of the case in its favor. Depending on the epoch, the social environment and the content of the dispute, the sides appeal to various higher authorities—common morals, "natural" law, holy writ, and the juridical code, the current rules of internal order, and often even the opinions of authoritative scholars or recognized political leaders. All of us have many times observed the passion with which such disputes are waged. . . .
>
> Only mathematics is fully spared from this. . . . Every mathematician learns early that in his science any attempt, whatever the reason, to act in a tendentious way, inclining toward a certain solution of the problem, favoring only those arguments which speak in favor of that solution—any such attempt is doomed to fail and to bring nothing but disappointment. Therefore, the mathematician quickly learns that in his science only correct, objective arguments free from all tendentiousness have benefit for his cause. . . .
>
> I have always been interested in these features of mathematics and many times have observed how they affect people seriously working in science, and especially those who study mathematics. It is joyous and morally elevating to see a person in this way gradually overcome his repulsive and narrow-minded habit of subordinating the laws of thinking to his own personal, petty, or mercenary interests.[6]

It would be easy to show that Khinchin has vastly simplified mathematics and its history in this passage, ignoring its grubbier moments. The birth of radical innovations in mathematics, such as non-Euclidean geometry, has been surrounded by controversy and even personal attacks. Khinchin's countryman, Nikolai Lobachevskii, never received recognition during his liftime for his version of non-Euclidean geometry, even though we now recognize it as "objective and correct." But my goal here is not to assess the validity of Khinchin's obviously exaggerated opinion of the moral virtues of mathematics, but merely to emphasize that to Khinchin and others these virtues seemed real and important. And Khinchin's viewpoint is one of the reasons that in countries where hypocrisy in politics is particularly rampant people with good minds and moral sensitivity often go into science.

The links between science and values which we have been discussing in this section are derived from the methods and sources of science: the link between the permanence and harmony in geometric forms and the hungering for such qualities in hectic social and political life; the striving for completion or perfection that can be seen in both scientific and religious systems; and the link between truth-seeking in science and similar pursuits in other fields. One other value of this type frequently mentioned in discussions of the relationship between science and esthetics is the term "elegance." In mathematics, a certain proof or procedure may be termed "elegant," while another variation, equally accurate, may be considered much less elegant. The term "elegance" has thus become one of the criteria by which a mathematical achievement may be legitimately judged.

This type of link between some fields of science and certain types of values—the commitments to harmony, perfection, truth-seeking, and elegance—has been very important in explaining the motivation of certain scientists and the criteria by which scientific achievements may be judged; this link does not, however, say very much about the content of science. The link concerns the sources and methods of science, not information contained in scientific theories or statements. Furthermore, the actual effects of such motivational and methodological factors are very uncertain; if commitment to harmony and perfection has driven some scientists to embrace rationality as the most important goal, it has on occasion led others, including the Pythagoreans, into a form of number mysticism that eventually became an obstacle to further scientific work. Some of the critics of Eddington believed the last phase of his work was marred by this exaggerated approach, and similar things have been said about the last decades of Einstein's career.

Platonic or Pythagorean idealism has also on occasion created an intellectual refuge leading to political conservatism in times of social stress, or reassurance when political concessions were necessary, as it did for Werner Heisenberg.

LINKS BASED ON TECHNOLOGICAL CAPABILITIES

Scientific theory in the physical sciences is distant from considerations of value. In the biological sciences, the gap is narrower, and, as we have seen, there are even a few direct links, but the significance of the gap should still not be underestimated. Technology, however, affects human values enormously and in a great variety of ways. Most people who speak of the impact of science upon ethics mean technology, not science. We have seen in chapter 9 that the issues being discussed in recent years under the rubric of "biomedical ethics" are actually issues of medical technology and ethics. Of course, biological and medical science stand behind medical technology, providing the necessary fundamental knowledge which eventually is translated into techniques and apparatus; however, at the stage of technology such a large number of socioeconomic factors come into play (Will the potential technology actually be built and distributed? What will it cost and who will pay for it? Who will be treated and under what rules? Where will it be placed? Who will control it?) that any attribution of primary causation for developments in biomedical ethics to science rather than to the society-technology interaction is likely either to go entirely awry or to be highly simplistic.

Nonetheless, science definitely plays a role in the changes in ethical views that often follow the introduction of new technologies—whether in medicine or in other areas. In earlier centuries, particularly before the eighteenth, it was possible to speak of technology that was quite separate from science. At that time the traditions of the engineers and the craftsmen were to a remarkable degree independent from those of the natural philosophers (scientists). In the nineteenth century these two traditions began to come much closer together, intellectually and institutionally, and today it is sometimes impossible in advanced areas of research and development to unravel them. The period of time between a basic discovery and its widespread application is still usually longer than many lay people believe, but the transition does occur on a regular basis. No matter how complicated that transition may be, with all of the intersections of options in which social and economic considerations often play determining roles, the fact still remains that science was often a necessary early step in the chain of links that eventually led to actions changing the ways in which humans regard themselves and in what they consider to be acceptable behavior. We saw in chapter 9, for example, how the biology standing behind the technology of amniocentesis has played a role in modifying some social values. Therefore, the common observation that "science is changing our values"—a statement that would be more accurate if it read "technology is changing our values"—contains a kernel of truth, even though the relationship is usually far more indirect than is often thought to be the case.

Technology changes our values by making possible actions which were earlier either impossible or extremely difficult, and by radically altering the speed and scale of these actions. We are then sometimes forced to see the results of our actions in a new light, and we may decide we would like to stop doing some things we earlier have often done, or we may decide that we wish to do something which we earlier considered improper. The case of the couple who changed their minds on abortion after amniocentesis gave them an option earlier unavailable is an example of people deciding to do something they had earlier considered impermissible. Examples of people who decided to try to stop doing something they had earlier done can be found in areas such as military technology and environmental pollution. Although wars continue all over our planet, the public notion of the acceptability of major war has unquestionably been altered by the advent of modern military technology. A similar pressure to change past behavior is evident in the current movement to protect the environment.

The ethical dimensions of a problem involving technology are so highly influenced by the specific characteristics of the given technology that it is impossible to decide in advance what would be proper to do in all future contingencies involving as yet undeveloped technologies. Let us look briefly at the history of trying to use knowledge of genetics to improve the genetic characteristics of human beings. When the eugenics movement of the early twentieth century flourished, there was no way of influencing the genetic characteristics of the human species except by selection among phenotypes, i.e., living human beings. The eugenic movements soon became expressions of racial and political prejudices. However, even one of the leaders of a conservative eugenics movement, Alfred Ploetz, managed to see that he was caught in a historical moment when the only available option was not nearly as desirable as later ones might be; he observed that if the time ever came when medical technology would make selection possible not among living human beings but among the sperm and ova of the legal parents, the whole eugenics movement might be transformed into a more humane and acceptable enterprise. We have not even yet reached that point, but we now make selection, in some instances, among fetuses, and the possibility of selection among sex cells is close enough to reality to interest medical researchers. While it would clearly be mistaken for us to embrace whatever medical technology makes possible—each potentiality needs criticism and caution—it would be equally incorrect to take a dogmatic position refusing to consider the new options presented by medical technology.

REVERSE LINKS OF VALUES UPON SCIENCE AND TECHNOLOGY

Most of the discussion to this point has been devoted to the influence of science and technology upon values, but the existence of reverse influences is quite clear; our values have influence on science and technology. This influence is of different types; I will briefly discuss here two: 1) the influence of values upon the cognitive structure of science; 2) the influence of values upon the procedures, limits, and applicability of science and technology.

The first category is one which is of primary concern to external historians of science, as I mentioned in the introduction to this book. Many historians of science believe that among the factors that influence the growth and character of scientific knowledge are social and economic ones; values of various sorts are usually included among "social influences." Philosophical, religious, and ethical influences upon scientific growth have frequently been cited.

Among the examples often given are the influence of Puritan religion on seventeenth-century English science; of German *Naturphilosophie* on field theory; of number mysticism on Greek geometry and the work of Johannes Kepler; of Weimar cultural antideterminism on quantum mechanics; of Malthusian political and economic ideas on Darwin; of philosophical monism on Ernst Haeckel; of racism on late nineteenth- and early twentieth-century anthropology and human genetics; of Marxism on the Soviet psychologist Vygotsky and the Soviet physicist Fock; of Mach's philosophy of science on the early Einstein; and of Kierkegaard's philosophy on Niels Bohr.

Although the external historians of science have made genuine progress in establishing their case that such influences can affect scientific development, the individual instances cited above are still, in almost every case, controversial. Many scientists (and quite a few historians and philosophers) are reluctant to admit that values can play important roles in the internal development of science. Evidently they think that attributing influence to extrascientific factors diminishes in some way the prestige of science (although just why this should be so is unclear).

Equally important in explaining the controversy over external history of science is the fact that it is never possible to prove that a certain religious, ethical, or philosophical influence was a necessary causal factor for a given scientific development. Indeed, it is not even clear what such a "proof" would look like. One can always say—even if it is obvious that a certain value commitment was enormously important to an individual scientist—that the scientist could have developed the same scientific theory, or made the same scientific discovery without the value commitment.

Careful external historians of science admit that the above point is correct, and they advance their cases not as logical proofs, but as parts of the overall background of a scientist's work without which the scientist simply cannot be understood as an intellectual and as a scientist. They are content to make persuasive arguments in favor of the existence of genuine links between philosophical and value systems and the internal character of a scientist's work.

In addition to influencing the internal development of science, social factors obviously impinge on the procedures, limits, and applicability of science and technology. This influence was particularly clear in the discussions of biomedical issues in chapter 9.

It is mistaken to believe that the recent debates over biomedical ethics are solely the result of the advance of science, or, as some people put it, "the fact that our science and technology advance more rapidly than our ethical systems." Of course the development of science has been an important factor in these disputes. If we look closely, however, we will see that the discussion was sometimes initiated by rising ethical awareness about medical practices that had earlier gone on for long periods of time without questions; at other moments the discussion was initiated as a result of changes in the way certain groups of people are viewed by the majority of society. The new element that caused the debate to break out was sometimes a heightened sense of human rights rather than an advance in medical technology.

Quite a few examples of such discussions could be taken from the area of human experimentation. Research on certain categories of subjects, such as children, pregnant women, prisoners, the mentally ill, the poor, and racial minorities was conducted in earlier decades without the same degree of sensitivity that is now usually expected. The now notorous Tuskegee Project, in which two groups of black men in Macon County, Alabama, were for several decades subjects of investigation of the results of syphilis serves as an example. One group was composed of several hundred syphilitic men; another somewhat smaller group of uninfected men served as controls. The goal of the project was to observe the "natural course" of syphilis in untreated black males, and the directors of the project therefore prevented the men from obtaining treatment, even after penicillin became available in the 1950s. The subjects of the experiment were never told of its purposes. Untreated patients suffering the ravages of syphilis were observed from 1932 to 1972, when the Department of Health, Education, and Welfare halted the project. At least twenty-eight subjects died from advanced syphilitic lesions.

The public outcry against such experiments was primarily a result of heightened awareness of human rights, not of an advance in science; this shift in attitudes toward rights was very broad indeed, involving the whole social history of a nation. The same kind of observation can be made about many other biomedical issues, such as abortion.

In other instances, however, changes in medical science and technology have played important roles in intiating discussions about values. Some of the recent debates about the prolongation of life and definitions of death were initiated by developments in medical technology—the advent of life-support machines in hospitals. Once the debates began, however, they soon absorbed older problems which had nothing, or very little, to do with new technology. The question of whether a horribly deformed child should be saved at the moment of birth is as old as civilization, as the ancient practice of exposure indicates. The questions of the criteria and attributes of "personhood" and of "the life worth living" are immensely deeper than the specific issues raised by the individual new medical technologies. We see once again that we should be careful about interpreting all these controversies as the results of advances in science and technology; in many instances the reverse phenomenon of changing human values having impact upon medical practice is also at work.

LINKS BETWEEN FACTUAL STATEMENTS AND VALUE STATEMENTS

As I indicated at the beginning of chapter 1, many scholars who have explored the relationship of science and values have concentrated their attention on the question of whether or not science is "value-free." In its most common academic form, this discussion is a careful philosophical analysis of the relationship between factual and normative statements, between "is" and "ought." Historical and social factors are not important to such analyses; logic is what counts.

It should be rather clear by now to the reader of this book that I am not primarily concerned with such analyses, since I think they miss the main point about the historical relationship of science, technology, and values. I am willing to grant that without the introduction of a value commitment somewhere along the line, a factual statement cannot logically entail a normative one. And yet history shows us innumerable examples of science and technology—supposedly based on facts—interacting with social values. How do we explain this? It is not hard to explain if we see that a normative statement was introduced somewhere in the chain of reasoning. Sometimes the original scientific statement already contained normative elements. Sometimes the scientific statement interacted with previously existing societal value positions, and caused a shift to new value positions. And sometimes the value interaction was simply based on bad logic, but, if so, we still need to know why the bad logic seemed persuasive.

From the standpoint of the philosopher who is studying the question of whether science is really value-free or not, the most interesting instances of science-value interactions I have cited are probably the ones based on "value terms" in science itself—those words like "normal," "deviant," and perhaps "adaptation." If these terms are really value-laden, and if some sciences find these terms necessary, then we can say that these sciences are in turn value-laden.

Although I find this analysis interesting, it is, in my opinion, still not directed at the most important cluster of issues; it is still based on an effort to "get at the essence of science" without consideration of its relationship to the society in which science is situated. And it is in this *relationship* between science and society where the historically interesting value conflicts arise. Society at any given moment always already possesses a large number of value commitments; new developments in science and technology may buttress or contradict some of these existing commitments. Then a dispute arises. The question, "At what point did extra-scientific normative considerations get introduced?" is, seen from this standpoint, almost elementary. Society was never without such commitments and cannot exist without them; when new science or new technology seems to affect those commitments, controversies will arise and value transformations may occur. Seen in its social context, science is far from value-free. The value transformations in which it plays a crucial role are of immense intellectual and social importance. The two major transformations considered in this book are examples: when quantum mechanics undermined the nineteenth-century scientific goal of a reductive, rigidly causal explanation of all the universe, including, eventually, man, the effects were widespread. When advancing biological science and technology today present new options in such a way that people revise their opinions about what kinds of actions or behavior are desirable, the results are, again, of great significance.

The Importance of Situational Analysis

The cardinal feature of an analysis of the relationship of science to social values is the *interface* of science and society.[7] This relationship might be helpfully represented in a schematic way (see figure 2).

If we briefly consider the controversy which has so often served

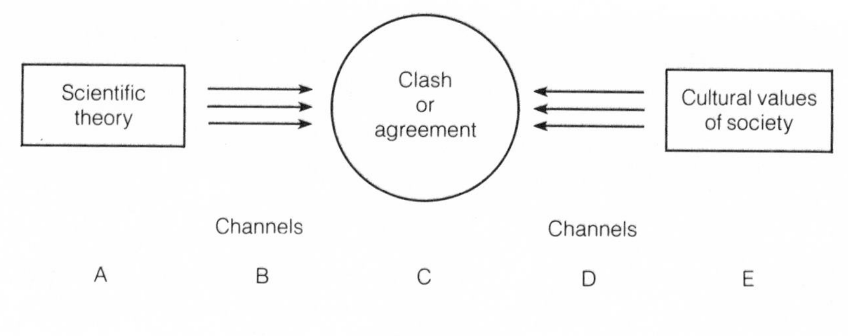

Figure 2.

as a model for our discussions—Galileo's espousal of the Copernican theory—we can view the development of the dispute on the basis of this scheme. We will see that the social context in which this discussion occurred, and the mechanism by which the issues were joined, were as important as anything in the theory itself in influencing the dimensions and outcome of the controversy.

The kinematic scheme of the solar system known as the Copernican theory (A in the diagram) did not contain inherent value statements. It was a highly technical exposition which at first caused no outcry. In order for a dispute to occur, the descriptive theory had to be transmitted to the area of public discussion through some channel (B), receiving a certain interpretation at the same time. The first channel by which the theory came to the public—the book *De Revolutionibus*—already had an interpretation appended to it, and one with which Copernicus probably would not have agreed. (We cannot be sure, for Copernicus died before he could comment on the Lutheran theologian Andreas Osiander's introduction to his book). Osiander maintained that the theory was merely a method of convenient calculation.

It was not until Galileo, much more of a showman than Copernicus, presented the Copernican theory in a dramatic way under a different interpretation and through a distinctly different channel that serious disagreements began. Galileo's *Dialogue Concerning the Two Chief Systems of the World* was directed to literate people outside the miniscule community of astronomers, and although it purportedly presented both sides of the argument, it clearly favored the Copernican view. Furthermore, Galileo's bias was toward Copernicanism as a true description of the world, the way it "really is." Galileo spoke the language of the educated nonscientific community, literally and metaphorically, since his work was in Italian (Copernicus' had been in Latin) and he kept the text simple (Copernicus' had been quite technical). All these features of Galileo's approach (the characteristics of the channel B) heightened the chance of conflict. Galileo might even be called an early "authoritative popularizer of science," similar to scientists such as Heisenberg and Eddington, described in this book, since he both did creative work himself and interpreted it for the lay public. These authors are of signal importance in shaping clashes or resonances between science and cultural values because they not only make scientific theory accessible, but they also mold its interpretation according to their own preferences.

Just as the scientific theory must reach the arena of public discussion through some channel, and acquire additional characteristics in the process, so also must social values be brought into contact with the scientific theory through one or more channels (D in the diagram). This process is far from automatic, and it can also have great influence on the type of clash or resonance which occurs. If a great authoritative institution like the Catholic Church responds, that is one thing; if an individual poet and divine like John Donne in England responds, that is another. Both objected to Copernican theory, but we now remember only the action of the Church.

It would be interesting to view the reception of Darwinian theory in detail on the basis of the same scheme, but I will take space here only for a few comments. The channels by which Darwinism came to public debate helped shape the discussions (e.g., the significance of Huxley in England and of Haeckel in Germany); the channels by which cultural values were brought into contact with Darwinism were also significant (the objections of Bishop Wilberforce in England; the outcries of fundamentalists in the United States, as well as the endorsements by businessmen such as Andrew Carnegie). Once again, we see that although we are entitled to speak of the impact of science on social values, an adequate description of that impact must give as much attention to the channels by which the scientific theory and the social values were brought together in agreement or disagreement as is given to the theories and the values themselves.

In this book I have noted that animal behaviorism and sociobiology are contemporary fields of science which have provoked considerable controversy over values. The channels by which these topics became public issues would need to be studied in order to understand the discussions more fully. Writers like Robert Ardrey and Desmond Morris presented the findings of animal behaviorism in a highly popularized and even vulgarized form. Konrad Lorenz, in his popular writings, interpreted events in terms of a clash of world views, and saw civilization as being in danger for its failure to adopt his particular conservative ideology. Nikolaas Tinbergen usually stayed clear of such rank popularization. E. O. Wilson at first concentrated primarily on scholarly analysis, but later made forays into popular literature. He was criticized not only in academic journals but also in political publications.

I have been emphasizing that the interrelationship between a scientific theory, on the one hand, and the society in which it appears, on the other, can have significant importance in determining what the value impact of the theory will be. The significance of such relationships is also important for technology.

A rather simple example can be seen in the varying response in the United States and the Soviet Union to certain technical innovations. The same technology is, in several instances, viewed by the respective governments in radically different ways. In the United States, with its tradition of the importance of individual privacy, the introduction of computers for the collection and centralization of information on the lives of private citizens has been highly controversial. The creation of central governmental computer centers having access to police, military, business, and tax records has been stoutly opposed. In the Soviet Union, no such controversy has occurred. On the other hand, the introduction of widespread photoduplication facilities (Xerox, etc.), has been resisted by authorities in the Soviet Union, and such facilities are still under tight controls. In the United States, such photoduplication equipment has spread rapidly. We see here an example of how the same technology can be treated in strikingly different ways in two different societies: computer-based personnel files are recognized by the United States government as a political issue, but not by the Soviet government; photoduplication facilities are recognized by the Soviet government as a political issue, but not by the American government. Such differences may complicate the lives of the salespeople who shuttle across the oceans trying to sell equipment!

What Kind of Expansionism Do We Want?

The natural reaction of a Western intellectual hearing discussions about expansionism and restrictionism is likely to be a sympathy for restrictionism. Restrictionism is, after all, based on the philosophical dualism that has been such an essential ingredient of Western civilization for several centuries. Furthermore, the abuses that have resulted from occasional expressions of expansionism are quite sobering: the condemnation of Galileo, the censures of Darwin, the Nazi support of Aryan science, and the Soviet suppression of genetics can all be described as episodes in the history of expansionism, i.e., the belief that the realms of science and values overlap. Not pleased by the apparent implications of individual cases of overlapping, political authorities have supported some scientific theories or interpretations to the detriment of others. Science has suffered as a result, and, ultimately, so has society.

And yet what does restrictionism offer, and how persuasive is its case? In its pure form restrictionism states that science and values simply have nothing to do with each other, that they talk about entirely different realms. Seductive as that prospect may be in terms of the safety it offers, it is no longer tenable. If one considers the massive influence that science and technology are currently having on our values—and only a few of those influences have been discussed in this book—and then adds the knowledge that these influences will be even greater in the future, it becomes clear that some new approach apart from classical restrictionism is needed if we are to have a means of coping with the changes.

We are now in a new era in our understanding of science-value relationships, and this new period brings with it both novel opportunities and novel dangers. The task of finding a method of living with the growing expansionist tendencies without again scuttling to the illusory safety of restrictionism—or, even worse, yielding to a violent backlash against all of science—is one of the most significant assignments of the next generation. At moments, one doubts the assignment can be fulfilled. But we must try. We now must live in the middle range of the science-value spectrum, recognizing the erroneousness of the value-free conception of science so prevalent in the previous generation, and the equal erroneousness of the countering view that "all of science is value-laden."

We acknowledge today more openly than before that some of the concepts of science, especially those of the social and behavioral sciences, contain value elements. We also know that the findings of science in areas such as psychology, genetics, neurophysiology, and animal behavior can have important value effects. It seems, furthermore, increasingly likely that some of the aspects of human behavior usually assigned to the ethical realm are influenced by genetic and physiological bases. As we learn more about what sociobiologists and others have called the "emotive centers of the hypothalamic-limbic system" we will probably see more clearly that genetics and physiology are relevant to discussions of ethics. The claims of science to provide increasingly successful explanations of human behavior are not limited, however, to genetic interpretations. It should be noticed that both sides of the famous nature-nurture debate are operating on expansionist assumptions. The nature side would expand the explanation of human beings on the basis of physiology and genetics; the nurture side would expand the explanation of humans on the basis of social conditioning and the environment. Both approaches are in principle amenable to scientific explanation and conceptualization. Both types of analysis are expansionist. And as our knowledge of human genetics, neurophysiology, and psychology increases, our power of intervention also grows.

While there are many reasons to recognize the untenability of restrictionism, the worst course we could follow as we turn away from it would be to embrace an uncritical or unexamined expansionism. That is the path toward a repetition of past expansionist abuses. One hopes that a recognition of the validity of expansionist claims *in principle* will give us the intellectual room we need in order to be critical of expansionist claims *in practice*. Expansionists should be questioned closely about the scientific validity of their intellectual claims and the ethical implications of their suggested practices. If they are concealing a personal ideology in their views, that ideology should be revealed and discussed. Such an approach might protect us from regarding the popular writings of a Robert Ardrey, a Desmond Morris, or a Konrad Lorenz solely as "scientific entertainment." In the past, we almost never recognized the intellectual legitimacy of expansionism, and, therefore, when a wave of expansionism swept over us it was often either trivialized, or, in the worst cases, accompanied by irrational abuses and the abandonment of critical debate. Critical debate is now the order of the day.

In chronological terms the most dangerous period of the development of a science is when enough is known to advance the first fruitful speculations and to try a few interventions, but not enough is known to bring discipline to these speculations or to predict the possible side effects or after effects of intervention. When the science of human genetics first began to develop at the end of the nineteenth century and the first decades of the twentieth it contained so many flawed conceptions (e.g., belief in single-gene determination of behavioral and psychological characteristics) that it allowed room for a rash of pseudoscientific eugenic theories and practices in which social and political prejudices played large roles. The most recent areas of science-value interactions such as the still fairly new field of sociobiology may also contain flawed conceptions; conservatism about accepting all the claims advanced by advocates of the field is entirely warranted.

Maintaining a moderate and skeptical position toward the value claims of advancing science, while admitting that science has an increasing justification for making such claims, will be difficult unless we simultaneously recognize the continuing validity of our traditional value systems. The apparent contradiction here is not genuine. Our traditional systems of ethics and values doubtlessly perform valuable functions of which we are even yet not fully aware. As a result of the testing of these systems over a very long history, human beings are probably wiser than they know.

Throughout the development of civilization the nongenetic cultural and ethical frameworks of human behavior have been subjected to selection as well as the genetic ones. The results which we possess today as cultural heritages contain reserves and functions that no anthropologist or ethicist fully understands. We should not give away or alter some aspect of this heritage without good reasons for doing so. The choices that have to be made cannot always be made explicit in discussions, because we do not always know what it is that needs to be preserved. Therefore, the tension between traditional ethics and advancing science is not best described as a tension between "irrational tradition" and "rational conceptual schemes." We should keep in mind that cultural traditions conceal unknown useful functions similarly to the way in which we assume our genes do.

In recent decades we have come to appreciate the concepts of flexibility and reserve in our genetic endowments. We know that sickle-cell trait, at first glance only a biological "defect," is based on genetic and physiological characteristics that protect people from malaria. In some environments it is an adaptive genetic characteristic; in others, maladaptive. This example, taken from genetic inheritance, can serve as an analogy when we examine cultural inheritance. We should be willing to consider the possibility that "units" in our cultural and religious heritages, at first glance only superstitious and irrational, may similarly serve hidden functions, perhaps still unknown. Until we learn what those hidden functions are, it would be premature to jettison those heritages.

Anthropologists tell us that the beliefs and superstitions of primitive peoples, at first glance irrational, often serve very practical goals in preserving the security of the particular primitive society, although the society itself may not be aware of the value of their customs. Our contemporary traditional value systems undoubtedly still work in some similar ways. Despite the injustices of contemporary civilization, it works fairly well; or, at least, we can say that the scientist who proposes changing a fundamental value in the light of recent science runs the risk of causing his or her civilization to work less well. One does not have to sanctify nature or fall into rank conservatism, then, to see that we should not alter the underlying assumptions of human society in a wholesale way just because a piece of scientific evidence indicates that some changes in those assumptions are in order. It makes little difference whether the piece of knowledge comes from a B. F. Skinner who decries the nonscientific foundation of our belief in human dignity or an E. O. Wilson who asks for our ethics to be informed and perhaps modified by biology. We have time to ask many questions and to proceed slowly, preserving our social equilibrium.

Perhaps a simple analogy on the theme of equilibrium will help a bit here. A grade-school child is often superb at riding a bicycle even though he or she knows nothing of the principles of physics and human physiology that permit one to ride a bicycle and govern what can be done with it. Later he or she may learn some of the necessary science at school. If the youth would on one fine day decide to relearn how to ride a bicycle on a scientific basis, applying the principles learned at school, the riding would at first gain nothing, and he or she might even have a wreck. Accumulated experience is much more important here than science. In a similar way, it is highly probable that some of the values necessary for the maintenance of the equilibrium of human society were learned on a nonscientific basis and are now encased in nonscientific or even irrational beliefs; as we discover what the scientific explanations of some of these values and ethical systems are, we should be intelligently cautious about attempting a sudden new way of keeping our equilibrium.

What, then, has the recent history of science-value interactions taught us? It has taught us, first of all, that restrictionism is dead as an intellectual option. Indeed, there always has been something quite strange about the assumption that our values, ethics, and other norms exist outside of space, time, and matter, and that they cannot be explained in terms of our material culture. We also have left behind the view that science is value-free, not because all of science is value-laden (it is not), but because the value-free approach to science is simply inadequate to describe what is currently happening in the relationship between science and values. Some aspects of science and some scientific disciplines are much more intimately involved with social values than others, and we therefore need a differentiated approach that will reveal these complex, reciprocal relationships. We now recognize the links that exist between many areas of science and our values and ethics. We are ready to benefit from the insights that science can bring to our understanding of these values. We know that we must live in the middle range of science-value interactions, seeing that the pure poles of "value-free science" and "science-free values" are both, in similar measure, diminishing in strength. But living on this particular slippery slope will require extreme caution. The major flaw in the view of the past generation was the refusal to see where science was affected by values; we should guard against a possible future period in which we might fall into one of two possible different errors: the premature attempt to explain cultural values exhaustively in terms of science, or the attempt to define all science as intrinsically value-laden.

Avoiding these extremes, there is a lot of important work to be done. We need to examine the internal concepts of science to find out how we might analyze the connections of science with values. We need to study more thoroughly the importance of genetic evolution, physiological structure, and psychological conditioning for understanding our social behavior and values. We also need to reexamine the history of science to see where science-value interactions have occurred with important social effects in the past, including those instances in which some of the reasoning was, from our present point of view, faulty. And, of course, we need to explore the ethical dimensions of present technologies and scientific research procedures. By pursuing these different approaches we will learn much more about the great variety of ways in which science and values can interact.

NOTES

Introduction

1. Charles Coulston Gillispie, *Genesis and Geology*.
2. B. F. Skinner, *Beyond Freedom and Dignity*, p. 104.
3. Stephen Toulmin, "Can Science and Ethics Be Reconnected?" See also my reply, "The Multiple Connections Between Science and Ethics."
4. Toulmin, "Can Science and Ethics Be Reconnected?" p. 28.
5. A. S. Eddington, *The Nature of the Physical World*, p. 333.
6. *Ibid.*, p. 353.
7. A. S. Eddington, *Science and Religion*, pp. 7–8.
8. Everett W. Hall, *Modern Science and Human Values*, p. 4.
9. Marshall Clagett, *Nicole Oresme and the Medieval Geometry of Qualities and Motions*, p. 44.
10. *Ibid.*, p. 45.
11. A. C. Crombie, *Medieval and Early Modern Science*, 1:16.
12. Augustine, *De Genesi ad Litteram*, 1:18.
13. Quoted in Crombie, *Medieval and Early Modern Science*, 1:26.
14. Copernicus, *On the Revolution of the Heavenly Spheres*, p. 50.
15. Quoted in Gillispie, *Genesis and Geology*, p. 3, from Robert Boyle, *Seraphick Love; Some Motives and Incentives to the Love of God, Pathetically Discours'd of in a Letter to a Friend* (4th ed.; London, 1665), pp. 53–55.
16. M. C. Jacob, "The Church and the Formulation of the Newtonian World-View," p. 148.
17. *Ibid.*, pp. 130–33; 135.
18. Gillispie, *Genesis and Geology*, pp. 3–4.
19. Keith Michael Baker, *Condorcet*, p. 183. I am grateful to Peter Buck, historian of science at MIT, for suggestions on the topic of eighteenth- and early nineteenth-century applications of mathematics to social and moral issues. See his essay, "From Celestial Mechanics to Social Physics: Discontinuity in the Development of the Sciences in the Early Nineteenth Century," in H. N. Jahnke and Michael Otte, eds., *Epistemological and*

Social Problems of the Sciences in the Early Nineteenth Century (Boston: Reidel, 1981), pp. 19–33.

20. Baker, *Condorcet*, p. 228.

21. *Ibid.*, pp. 223–24.

22. "Human Decency is Animal," *New York Times Magazine*, October 12, 1975, p. 39.

23. Charles Darwin, *The Descent of Man*, p. 126.

24. J. D. Y. Peel, *Herbert Spencer*, p. 132.

25. I do not consider the growth of science to be linear or cumulative in any simple sense; the account of paradigm shifts given by Thomas Kuhn in his well-known *Structure of Scientific Revolutions* effectively counters such a view. But I do think that in the long run succeeding paradigms are more comprehensive (i.e., they explain everything the older ones did, and more) than earlier ones, and I think that the gains in application derived from newer science provide evidence of its progressive quality. That progress, however, must be seen as a very indirect and halting one, and one subject even to reverses. Science is not made up of cumulative absolute truths, but rather of a system of relative truths, any part of which may be fundamentally revised by later science.

26. Quoted in Gillispie, *Genesis and Geology*, p. 3.

27. T. H. Huxley, *Evolution and Ethics*, p. 51.

28. Quoted in Ronald W. Clark, *The Huxleys*, p. 118. Spencer, as a Lamarckian, saw an escape from the social implications of natural selection that was not persuasive to Huxley. See Peel, *Herbert Spencer*, pp. 151–53.

29. T. H. Huxley, p. 80.

30. Samuel Eliot Morison, *Three Centuries of Harvard, 1636–1936*, pp. 24–25.

31. *Johns Hopkins University*, pp. 15–16.

32. Morison, *Three Centuries of Harvard*, p. 279.

33. Douglas Sloan, "The Teaching of Ethics in the American Undergraduate Curriculum, 1876–1976," p. 205.

34. Peel, *Herbert Spencer*, p. 87.

35. Sloan, "The Teaching of Ethics," p. 228.

1. The New Physics

1. For an interesting discussion of ideological currents surrounding the early days of quantum mechanics, see Paul Forman, "Weimar Culture, Causality, and Quantum Theory, 1918–1927."

2. Nell Boni, Monique Russ, and Dan H. Laurence, eds., *A Bibliographical Checklist and Index to the Published Writings of Albert Einstein.* Einstein's collected works are not yet conveniently available, but there is a microprint edition: Albert Einstein, *Collected Works, 1901–1956.* The first edition of Einstein's published works was in Russian: Albert Einstein, *Sobranie*

nauchnykh trudov. An earlier incomplete collection of his works up to 1922 was published in Japan, 1922–24. See Herbert S. Klickstein, "A Cumulative Review of Bibliographies of the Published Writings of Albert Einstein," pp. 141–49. All these will be superseded eventually by the publication of a new multivolume edition of both published and unpublished materials. A biography which puts perhaps too much emphasis on the shifts in Einstein's thought, particularly on topics such as pacifism, is Ronald W. Clark, *Einstein: The Life and Times*. A more intellectually persuasive account, particularly of the thought of the young Einstein, is Philipp Frank, *Einstein: His Life and Times*. By far the best analysis of the shifts in Einstein's epistemological views is Gerald Holton, "Mach, Einstein, and the Search for Reality." See also the other essays on Einstein in the latter source.

3. The Einstein Archives are at the Institute for Advanced Study, Princeton, New Jersey, under the care of Helen Dukas, Einstein's secretary. Einstein's works will be published in a multivolume edition under the editorship of Professor John Stachel.

4. Quoted in Holton, "Mach, Einstein, and the Search for Reality," p. 226.

5. The possibility of a link between these two forces was not, of course, original to Einstein. It had been a recurrent theme in the mechanistic world picture dating from Newton's time and had received new impetus with the strengthening of the atomic hypothesis in the early nineteenth century.

6. Einstein concluded his article with a recognition that he had not solved the problem he set out: "Die Frage, ob und wie unsere Kräfte mit den Gravitations-kräften verwandt sind, muss also nach vollkommen offen gelassen werden." A. Einstein, "Folgerungen aus den Capillaritätserscheinungen," p. 523.

7. Holton, "Mach, Einstein, and the Search for Reality," p. 219.

8. Einstein, "Remarks on Bertrand Russell's Theory of Knowledge," p. 289.

9. The division made by the bibliographers was quite arbitrary, and gives only the most general impression. As an example of the inadequacy of the classification, notice that the author and lecturer James Murphy held two interviews with Einstein on approximately the same subjects, yet one interview is included under "Scientific Writings," the other under "General Writings;" see entries nos. 212 and 309 in Boni, Russ, and Laurence, eds., *Bibliographical Checklist*.

10. Einstein, "Science and Religion," p. 210.

11. Einstein, "Science and God: A Dialog," p. 375.

12. Quoted in R. S. Shankland, "Conversations with Albert Einstein," p. 50.

13. A particularly sensitive discussion of Einstein's attitude toward social and political issues, including the Zionist movement, is in Frank, *Einstein: His Life and Times*, especially pp. 147–53.

14. Einstein, "Dialog über Einwände gegen die Relativitätstheorie," p. 697.

15. Einstein, "Epilogue: A Socratic Dialogue."

16. Einstein, "The Laws of Science and the Laws of Ethics," p. vi. It is likely that Einstein had Spinoza's ethics in mind when he discussed the role of logical thinking in ethics.

17. *Ibid.*

18. *Gelegentliches . . . zum fünfzigsten Geburtstag . . . dargebracht von der Soncino-Gesellschaft der Freunde des jüdischen Buches zu Berlin*, p. 9.

19. Einstein, "Science and God," p. 377.

20. For a discussion of these passions, see Joseph Haberer, *Politics and the Community of Science*, pp. 103 ff.

21. *Gelegentliches . . . zum fünfzigsten Geburtstag*, p. 13.

22. Einstein, *The World as I See It*, pp. 20–21. This essay first appeared as "Motiv des Forschung," in *Zu Max Plancks 60 Geburtstag: Ansprachen in der Deutschen physikalischen Gesellschaft* (Karlsruhe: Muller, 1918).

23. B. Kuznetsov, *Einstein*, p. 15.

24. Einstein, "Science and God," p. 373.

25. Einstein, "Religion and Science: Irreconcilable?" p. 19.

26. In his article "Why Socialism?" Einstein observed that both astronomers and economists "attempt to discover laws of general acceptability for a circumscribed group of phenomena in order to make the interconnection of these phenomena as clearly understandable as possible." But he went on to warn that there are essential methodological differences between astronomy and economics and he noted that "socialism is directed toward a social-ethical end. Science, however, cannot create ends." Einstein, "Why Socialism?" p. 9.

27. One of the best short biographical articles on Bohr, complete with bibliographical materials, is Leon Rosenfeld "Bohr, Niels Henrik David." Also see Rosenfeld's *Niels Bohr: An Essay Dedicated to Him on the Occasion of His Sixtieth Birthday*; S. Rozental, ed., *Niels Bohr: His Life and Work as Seen by His Friends and Colleagues*; and the Niels Bohr Memorial Session recorded in *Physics Today* (October 1963), 16:21–62. A bibliography of Bohr's works is in *Nuclear Physics* (1963), 41:7–12.

28. Quoted in Klaus Michael Meyer-Abich, *Korrespondenz, Individualität und Komplementarität*, p. 133, from C. F. von Weizsäcker, *Zum Weltbild der Physik*, p. 378, and A. Petersen, *The Bulletin of Atomic Scientists*, p. 9.

29. Niels Bohr, *Atomic Physics and Human Knowledge*, p. 96.

30. Quoted by Niels Bohr in *Atomic Physics and Human Knowledge*, p. 96.

31. Meyer-Abich, *Korrespondenz*, especially pp. 104, 133–40.

32. For an example of references to Indian and Chinese thought, see N. Bohr's *Essays 1958–1962 on Atomic Physics and Human Knowledge*, pp. 14–15.

33. Leon Rosenfeld, "Niels Bohr: Biographical Sketch," in his edited *Niels Bohr: Collected Works* (Amsterdam: North-Holland, 1972), 1:xix.

34. The original article was N. Bohr, H. A. Kramers, and J. C. Slater, "The Quantum Theory of Radiation." The retreat was signaled in N. Bohr, "Über der Wirkung von Atomen bei Stössen," especially pp. 156–57.

35. N. Bohr, *Atomic Physics and Human Knowledge*, p. 5. The essay from which the quotation is taken, "Light and Life," was written in 1932.

36. N. Bohr, "Über die Anwendung der Quantentheorie auf den Atombau," p. 141.

37. N. Bohr, "Atomic Theory and Mechanics," p. 848.

38. N. Bohr, "The Quantum Postulate and the Recent Development of Atomic Theory," pp. 580, 584.

39. *Ibid.*, pp. 580, 590.

40. N. Bohr, *Atomic Physics and Human Knowledge*, p. 41.

41. Bohr later described these conversations in "Discussion with Einstein on Epistemological Problems in Atomic Physics." The article was reprinted in N. Bohr, *Atomic Physics and Human Knowledge*, pp. 32–66.

42. N. Bohr, *Atomic Physics and Human Knowledge*, p. 47.

43. N. Bohr, *Atomic Theory and the Description of Nature*, p. 92.

44. *Ibid.*, pp. 1, 4.

45. *Ibid.*, p. 96.

46. *Ibid.*, p. 100.

47. *Ibid.*, p. 101.

48. N. Bohr, "On the Notions of Causality and Complementarity," *Dialectica*, p. 54. See also his "On the Notions of Causality and Complementarity, *Science*."

49. N. Bohr, "Science and the Unity of Knowledge," p. 58.

50. N. Bohr, "Medical Research and Natural Philosophy," p. 971.

51. N. Bohr, "Natural Philosophy and Human Cultures," p. 271.

52. N. Bohr, "Science and the Unity of Knowledge," pp. 58–59, 53.

53. *Ibid.*, pp. 60–61.

54. N. Bohr, "Physical Science and the Study of Religions," p. 389.

55. N. Bohr, "Light and Life," p. 457.

56. *Ibid.*

57. *Ibid.*, p. 458.

58. *Ibid.*

59. N. Bohr, "Analysis and Synthesis in Science," p. 28.

60. N. Bohr, "Medical Research and Natural Philosophy," p. 970.

61. Francis Crick, *Of Molecules and Men.*

62. N. Bohr, *Essays 1958–1962 on Atomic Physics and Human Knowledge*, pp. 26, 28.

63. N. Bohr, "Natural Philosophy and Human Cultures," p. 271.

2. Eddington and the English-Speaking World

1. The best biography of Eddington is A. Vibert Douglas, *The Life of Arthur Stanley Eddington*; the Douglas book includes a bibliography of

Eddington's works on pp. 193–98. There is no central collection of Eddington's papers; items are scattered in several places. I know of no central listing. The Trinity College Library has a collection of several of his schoolboy papers, his *Journal* (1905–1914, with later occasional additions), a few letters, the manuscript of *The Nature of the Physical World*, and some miscellaneous items; the Cambridge University Library has several papers; the Institute of Astronomy of the University of Cambridge has some papers; the Einstein Collection in Princeton has about a dozen letters; the Fitzwilliam Museum, Cambridge, has four letters from Einstein to Eddington; the Eidgenössische Technische Hochschule has about a dozen Eddington letters to H. Weyl, 1918–1944; the Royal Society in London has letters to Sir J. Larmor and A. Schuster; the Library of the Université Libre of Brussels has a few letters to T. de Donden, 1921–1936; the Royal Astronomical Society in London has a few letters to Eddington; the H. N. Russell correspondence in Princeton has about ten letters from Eddington; other items exist in the Bohr and Shapley correspondences, and doubtlessly elsewhere as well. According to the Douglas biography (p. 184), Eddington willed "many of his manuscripts and letters to the Royal Astronomical Society." When I visited there in 1978, I was told that a collection of Eddington's papers that were "merely personal" was burned at an unspecified date, apparently in accordance with the wishes of surviving relatives and friends.

Among the secondary works, a valuable study is J. W. Yolton, *The Philosophy of Science of A. S. Eddington.* A number of studies of Eddington's life and work have been published in the "Eddington Memorial Lectures" of the Cambridge University Press, including: A. D. Ritchie, *Reflections on the Philosophy of Sir Arthur Eddington*, 1948; L. P. Jacks, *Sir Arthur Eddington: Man of Science and Mystic*, 1949; E. T. Whittaker, *Eddington's Principle in the Philosophy of Science*, 1951; Martin Johnson, *Time and Universe for Scientific Conscience*, 1952; and H. Dingle, *The Sources of Eddington's Philosophy*, 1954. Other sources are C. W. Kilmister, *Men of Physics: Sir Arthur Eddington*; C. W. Kilmister and B. O. J. Topper, *Eddington's Statistical Theory*; Noel B. Slater, *The Development and Meaning of Eddington's Fundamental Theory*; and Edmund Whittaker, *From Euclid to Eddington.*

The *Reader's Guide to Periodical Literature* shows an almost continuous stream of articles by and about Eddington between 1907 and 1947, with the heaviest flow in the years 1929 to 1937.

2. Arthur Stanley Eddington, "A Total Eclipse of the Sun," manuscript dated June 30, 1898, and evidently based on a talk given by Eddington at the Brymelyn Literary Society Meeting on March 7, 1898; manuscript now on deposit in Eddington Papers, Trinity College Library, Cambridge.

3. Douglas, *Life*, pp. 40–41.

4. Quoted in Douglas, *Life*, p. 185.

5. Quoted in Douglas, *Life*, p. 104.

6. Eddington, *The Nature of the Physical World*, p. 270.

7. Eddington, *Science and Religion*, p. 8.

8. Eddington, *The Nature of the Physical World*, p. 275.
9. Eddington, *New Pathways in Science*, p. 12.
10. *Ibid.*
11. *Ibid.*, p. 22.
12. Eddington, *The Nature of the Physical World*, pp. 230–31.
13. *Ibid.*, p. 244.
14. Eddington, *The Philosophy of Physical Science*, pp. 56–57.
15. Eddington, *The Nature of the Physical World*, p. 335.
16. *Ibid.*, p. 322
17. *Ibid.*, pp. 288–89.
18. Eddington, *Science and the Unseen World*, p. 49.
19. Eddington, *The Nature of the Physical World*, pp. 324, 327.
20. Eddington, *New Pathways in Science*, p. 322.
21. See the discussion of Fox in D. Elton Trueblood, *The People Called Quakers*.
22. Eddington, *The Nature of the Physical World*, p. 321.
23. Trueblood, *People Called Quakers*, p. 244.
24. "Eddington Journal," Trinity College Library, Cambridge, p. 56.
25. Eddington, *Science and Religion*, pp. 15–16.
26. Eddington, *Science and Religion*, pp. 9–10.
27. Eddington, *The Nature of the Physical World*, p. 254.
28. Eddington affirmed, "I repudiate the idea of proving the distinctive beliefs of religion either from the data of physical science or by the methods of physical science." Eddington, *The Nature of the Physical World*, p. 333.
29. Eddington, *New Pathways in Science*, p. 305.
30. Eddington, *The Nature of the Physical World*, p. vi.
31. Manuscript copy, "The Nature of the Physical World," Trinity College Library, Cambridge, pp. 309–311.
32. *Ibid.*, p. 317.
33. *Ibid.*, p. 339.
34. Herbert Dingle, *The Sources of Eddington's Philosophy*, p. 19.
35. Jeremy Bernstein, *Einstein*, p. ix.
36. J. G. Crowther, *British Scientists of the Twentieth Century*, p. 165.
37. Trueblood, *People Called Quakers*, pp. 243–44. Swarthmore, incidentally, had a long tradition of close relations with the Quakers.
38. Crowther, *British Scientists*, p. 142.
39. *Time* (December 4, 1944), 44:50.
40. *Current Biography* (1941), pp. 254–56.
41. Eddington, "The Decline of Determinism."
42. Dingle, *Sources of Eddington's Philosophy*, p. 64.

3. Fock and the Soviet Union

1. Detailed discussions of Shmidt and Fock can be found in my *Science and Philosophy in the Soviet Union*.

2. A. A. Friedmann, "Über die Krümmung des Raumes," A. Einstein, "Bemerkung zu der Arbeit von A. Friedmann, 'Über die Krümmung des Raumes,'" p. 326. A. Einstein, "Notiz zu der Arbeit von A. Friedmann 'Über die Krümmung des Raumes,'" p. 228. Friedmann also wrote a popularization of relativity, *Mir kak prostranstvo i vremiia.*

3. O. Iu. Shmidt, *Zhizn' i deiatel'nost'*; O. Iu. Shmidt, *Izbrannye trudy.*

4. S. I. Vavilov, *Sobranie sochineniia.*

5. M. G. Veselov, P. L. Kapitsa, and M. A. Leontovich, "Pamiati Vladimira Aleksandrovicha Foka," p. 375.

6. A bibliography of Fock's works from 1923 to 1956 is in *Vladimir Aleksandrovich Fok.*

7. V. A. Fock, "Velikii fizik sovremennosti"; V. A. Fock, "Protiv nevezhestvennoi kritiki sovremennykh fizicheskikh teorii."

8. Fock, "Protiv nevezhestvennoi kritiki sovremennykh fizicheskikh teorii," p. 172.

9. Fock, "K diskussii po voprosam fiziki," p. 151.

10. Fock, "Sovremennaia teoriia prostranstva i vremeni," p. 13.

11. V. A. Fock, "K diskussii po voprosam fiziki," pp. 153, 155.

12. See Jehoshua Yakhot, "The Theory of Relativity and Soviet Philosophy."

13. Fock, "K diskussii po voprosam fiziki," p. 15.

14. "Es ist nicht überflüssig, zu unterstreichen, dass das Verhältnis von Körpern oder Prozessen zum Bezugssystem ebenso objektiv ist (d.h. unabhängig von unserem Bewusstsein) wie überhaupt alle physikalischen und anderen Eigenschaften der Körpern." Fock, "Über philosophische Fragen der modernen Physik," p. 742.

15. Planck commented, "The concept of relativity is based on a more fundamental absolute than the erroneously assumed absolute which it has supplanted." Max Planck, *The New Science*, p. 146. Planck expressed the same idea in earlier publications, e.g., *Das Weltbild der neuen Physik*, p. 18.

16. Fock, "Protiv nevezhestvennoi kritiki sovremennykh fizicheskikh teorii," p. 172.

17. Fock liked to put his viewpoint in French: "(1) La relativité physique n'est pas général; (2) la relativité générale n'est pas physique." V. A. Fock, "Les principes mécaniques de Galilée et la théorie d' Einstein," p. 12.

18. See V. A. Fock, "Sistema Kopernika i sistema Ptolemeia v svete obshchei teorii otnositel'nosti," pp. 180–86; also his "Nekotorye primeneniia idei Lobachevskogo v mekhanike i fizike," pp. 48–86.

19. Fock, "Osnovnye zakony fiziki v svete dialekticheskogo materializma," especially 44 ff., and Fock, "Sovremennaia teoriia prostranstva i vremeni," p. 18.

20. V. I. Lenin, *Materialism and Empirio-Criticism: Critical Comments on a Reactionary Philosophy*, p. 271.

21. Fock, "Protiv nevezhestvennoi kritiki sovremennykh fizicheskikh teorii," p. 171.

22. Fock, "Osnovnye zakony fiziki v svete dialekticheskogo materializma," p. 36.
23. V. A. Fock, "Protiv nevezhestvennoi kritiki sovremennykh fizicheskikh teorii," p. 171.
24. For a detailed discussion of this point, see Loren R. Graham, *Science and Philosophy in the Soviet Union*, pp. 128–35.
25. See V. A. Fock, *The Theory of Space, Time and Gravitation*, esp. pp. xv–xvi.
26. *Ibid.*, p. 351; see also Fock, "Poniatiia odnorodnosti, kovariantnosti i otnositel'nosti," p. 133.
27. See my brief discussion in *Science and Philosophy in the Soviet Union*, p. 135.
28. Leopold Infeld, *Why I Left Canada*, p. 88.
29. Fock, "Otnositel'nosti teoriia"; and I. Iu. Kobzarev, "Otnositel'nosti teoriia."
30. Infeld, *Why I Left Canada*, pp. 75–77.
31. Fock, "Ob interpretatsii kvantovoi mekhaniki," p. 235.
32. Jehoshua Yakhot, "Theory of Relativity and Soviet Philosophy."
33. See the discussion in my *Science and Philosophy in the Soviet Union*, *passim*, and especially pp. 116–221.
34. M. G. Veselov, P. L. Kapitsa, M. A. Leontovich, "Pamiati Vladimira Aleksandrovicha Foka," pp. 375–76.
35. For a strong defense by Fock of dialectical materialism as a way of viewing physics, see his "Comments," *Slavic Review*; see also Loren R. Graham, "Quantum Mechanics and Dialectical Materialism" and "Reply" and Paul K. Feyerabend, "Dialectical Materialism and Quantum Theory," pp. 381–417.
36. In the preface to the 1955 edition of his *The Theory of Space, Time, and Gravitation*, Fock commented: "The philosophical side of our views on the theory of space, time and gravitation was formed under the influence of the philosophy of dialectical materialism, in particular, under the influence of Lenin's 'Materialism and Empirio-criticism.' The teachings of dialectical materialism helped us to approach critically Einstein's point of view concerning the theory created by him and to think it out anew. It helped us to understand correctly, and to interpret, the new results we ourselves obtained."
37. I am grateful to Professor David Noble of the Massachusetts Institute of Technology for comments on this section.

4. Heisenberg and Germany

1. The most complete bibliography of Heisenberg's works I have seen is that compiled after his death by Annemarie Giese of the Max-Planck-Institut für Physik und Astrophysik, München, and kindly mailed to me

by W. Huber of that institute. Including journal articles, books, and recordings, it includes 220 items. Newspaper articles, however, are not included, an omission of some significance; an example of an important newspaper article, from a political standpoint, is his defense of physics against National Socialist attacks, "Zum Artikel 'Deutsche und jüdische Physik,'" p. 6.

2. The dates of the various editions are as follows: first, 1935; second, 1936; third, 1941; fourth, 1943; fifth, 1944; sixth, 1945; seventh, 1947; eighth, 1949; ninth, 1959; tenth, 1973.

3. Between the time of his first publication in 1922 and the appearance of the first version of his popular essays going into the 1935 book, Heisenberg published fifty articles and one technical book. Almost all of them were for specialists in quantum mechanics, and most of them (29) were in one journal (*Zeitschrift für Physik*). All in all, ten journals were involved. In later years he published in a much greater variety of places.

4. A year later Heisenberg told Einstein that "the idea of observable quantities was actually taken from his relativity." Quoted in Max Jammer, *The Conceptual Development of Quantum Mechanics*, p. 198.

5. Heisenberg, "Über quantentheoretische Umdeutung kinematischer und mechanischer Beziehungen."

6. Jammer, *Conceptual Development of Quantum Mechanics*, p. 198.

7. *Ibid.*, p. 180. Wittgenstein's statement, "Wovon man nicht sprechen kann, därüber muss man schweigen," was in his *Tractatus Logico-Philosophicus*, p. 284.

8. Heisenberg, "Quantenmechanik."

9. In a particularly eloquent passage, he proclaimed, "Das Programm der Qu. M. musste daher sein sich zunächst von diesen anschaulichen Bildern freizumachen und an Stelle der bisher benutzten Gesetze der klassischen Kinematik und Mechanik einfache Beziehungen zu setzen zwischen experimentell gegebenen Grössen. Während also die frühere Theorie den Vorteil der unmittelbaren Anschaulichkeit und des Gebrauchs bewährter physikalischer Prinzipien mit dem Nachteil verband, im allgemeinen mit Beziehungen zu rechnen, die principiell nicht prüfbar waren und daher zu inneren Widersprüchen führen konnten, sollte die neue Theorie auf die Anschaulichkeit zunächst ganz verzichten, dafür aber nur ganz konkrete Beziehungen enthalten, die einer unmittelbaren experimentellen Prüfung zugänglich sind und deswegen kaum in die Gefahr inneren Widersprüch kömmen." *Ibid.*, p. 990.

10. In another place, in 1930, Heisenberg observed: ". . . one should particularly remember that the human language permits the construction of sentences which do not involve any consequences and which therefore have no content at all—in spite of the fact that these sentences produce some kind of picture in our imagination; e.g., the statement that besides our world there exists another world, with which any connection is impossible in principle, does not lead to any experimental consequence,

but does produce a kind of picture in the mind. Obviously such a statement can neither be proved nor disproved. One should be especially careful in using the words 'reality,' 'actually,' etc., since these words very often lead to statements of the type just mentioned." W. Heisenberg, *The Physical Principles of the Quantum Theory*, p. 15.

11. Heisenberg, "Kausalgesetz und Quantenmechanik." He had briefly raised the same question in an offhand manner in his earlier "Über den anschaulichen Inhalt der quantentheoretischen Kinematik und Mechanik."

12. Heisenberg, "Kausalgesetz und Quantenmechanik," pp. 175, 181, and *passim*.

13. Heisenberg, *Wandlungen* (1935). For the original journal articles, see Heisenberg, "Zur Geschichte der physikalischen Naturerklärung"; Heisenberg, "Wandlungen der Grundlagen der exakten Naturwissenschaft in jüngster Zeit."

14. The essay was presented at a public meeting of the *Sächsische Akademie* on November 19, 1932.

15. Heisenberg, *Wandlungen* (1935), pp. 42–43.

16. *Ibid.*, p. 28.

17. *Ibid.*, p. 5.

18. *Ibid.*, p. 16.

19. The best discussion of Stark is in Alan D. Beyerchen, *Scientists Under Hitler: Politics and the Physics Community in the Third Reich*; also see Armin Hermann's biographical article in *The Dictionary of Scientific Biography*.

20. Beyerchen, *Scientists Under Hitler*.

21. *Ibid.*, p. 116.

22. *Ibid.*, p. 144.

23. Heisenberg, "Zum Artikel 'Deutsche und jüdische Physik.'"

24. Although Stark did not sign this article, Beyerchen attributes it to him. Beyerchen, *Scientists under Hitler*, p. 158.

25. Heisenberg gave the date of publication of Stark's *Nationalsozialismus und Wissenschaft* as 1932, while the edition I was able to locate was published in 1934. Perhaps he saw it as an earlier brochure. In any case, the publication was prior to Heisenberg's *Wandlungen*, and known to Heisenberg when he published *Wandlungen* in 1935.

26. Stark, *Nationalsozialismus und Wissenschaft*, p. 11.

27. *Ibid.*, p. 14.

28. Heisenberg, *Wandlungen* (1935), p. 15.

29. *Ibid.*, p. 7.

30. Heisenberg, "Die Entwicklung der Quantentheorie, 1918–1928." The care which Heisenberg gave his treatment of Lenard is shown by the fact that the sentence referring to him is one of the few spots where the text was changed between the 1934 "Wandlungen" article and its 1935 book form. In the 1934 article, Heisenberg spoke of the " . . . durch die Untersuchungen von Lenard und Einstein gewonnene Einsicht. . . ;" in the 1935 book, he spoke of " . . . durch die Untersuchungen von Lenard

und ihre Deutung durch Einstein gewonnene Einsicht." Heisenberg was making sure that he gave independent credits accurately to Lenard and Einstein in this controversial area. Compare Heisenberg, "Wandlungen der Grundlagen der exakten Naturwissenschaft in der jüngsten Zeit," p. 670, with his *Wandlungen* (1935), p. 9.

31. Heisenberg, *Wandlungen* (1935), pp. 18–19.

32. Heisenberg, *Physics and Philosophy.*

33. Beyerchen, *Scientists under Hitler, passim.*

34. *Ibid.*, p. 163.

35. "An die Leser der 'Antike,'" *Die Antike* (1937), 13(2):77–78. The three new editors were Wolfgang Schadewaldt, Bernhard Schweitzer, and Johannes Stroux.

36. Werner Heisenberg, "Ideas of the Natural Philosophy of Ancient Times in Modern Physics," pp. 54–55, 56.

37. *Ibid.*, p. 59.

38. Heisenberg in later editions of the *Wandlungen* gave the date of the lecture as May 5, 1941, but this is surely an error, since newspapers of the time tell us the lecture was held at 6:00 p.m., Monday, April 28, 1941. Compare Heisenberg, *Wandlungen* . . . (9th ed.), (Stuttgart: S. Hirzel, 1959), p. 85, with *Pester Lloyd*, Budapest, April 27, 1941, p. 9 and April 29, 1941, p. 6. The lecture was first printed in *Geist der Zeit* (May 1941), 19(5):261–75. *Geist der Zeit* was an organ of the German academic exchange office, and by this time was a voice of Nazi propaganda. I am grateful to Professor Istvan Deak of Columbia University for helping me with Hungarian sources.

39. For conflicting views, see Wolfgang Leppmann, *The German Image of Goethe*, especially pp. 196–99; Heinz Kindermann, *Das Goethebild des 20. Jahrhunderts*, especially pp. 514–20, 544–46; August Raabe, *Goethes Sendung im Dritten Reich*; Hans Brandenburg, *Goethe und Wir*; Hans Carossa, *Wirkungen Goethes in der Gegenwart*; Franz Koch, *Goethe und die Juden*; Walter Rauschenberger, *Goethes Abstammung und Rassenmerkmale*; Mark Waldman, *Goethe and the Jews: A Challenge to Hitlerism.*

40. Baldur von Schirach, ed., *Goethe an Uns: Ewige Gedanken des grossen. Deutschen.*

41. *Pester Lloyd*, Budapest, April 27, 1941, p. 9; April 29, 1941, p. 6.

42. Heisenberg quoted Helmholtz to this effect. Heisenberg, "The Teachings of Goethe and Newton on Colour in the Light of Modern Physics," p. 70.

43. *Ibid.*, pp. 70–71.

44. *Ibid.*, p. 71.

45. *Ibid.*, p. 75.

46. Heisenberg wrote: "Der Vortrag kommt von einem Vergleich der Grundlagen der Goethe'schen and der Newton'schen Farbenlehr zu einem Vergleich der Wirklichkeitsbereiche, denen die beiden Farbenlehren

gelten. Es wird darauf hingewiesen, dass schon die moderne Atomphysik die Wirklichkeitsbereich der Newton'schen Physik überschreitet, dass also auch die konsequente Verfolgung des von Galilei und Newton eingeschlagenen Weges schliesslich in Gebiete führt, die sich von denen der klassischen Physik grundsätzlich unterscheiden. Aus diesen Umstand kann die Hoffnung geschöpft werden, dass sich in nicht allzufern Zeit auch ein Verständnis für die Verbindung zwischen den naturwissenschaftlichen und den geisteswissenschaftlichen Bereichen anbahnen wird." Heisenberg, "Goethe és Newton szinelmélete a modern fizika megvilágitásában."

47. Heisenberg, "The Teachings of Goethe and Newton," p. 60.

48. The English translation of Heisenberg's interpretation here is incorrect, and, in fact, gives the opposite meaning to what Heisenberg intended. The translation reads: "The scientist must reconcile himself to the idea of directly linking the fundamental concepts on which his science rests with the world of the senses." The German original reads: "Der Naturforscher muss also hier darauf verzichten, die Grundbegriffe, auf die er seine Wissenschaft baut, mit der Sinnenwelt unmittelbar zu verknüpfen." An accurate translation would be: "The scientist must renounce the idea of directly linking the fundamental concepts on which his science rests with the world of the senses." Compare *Philosophic Problems of Nuclear Science* (1952), p. 70, with *Wandlungen* (1973), p. 98.

49. Heisenberg, *Philosophic Problems*, pp. 73, 75.

50. In a 1967 paper, Heisenberg returned to the subject of Goethe's views of science, and although his respect for Goethe's genius was still very evident, he was less laudatory of Goethe's scientific method. Instead, he saw Goethe as prophetic of the evil that science could do. See "Das Naturbild Goethes und die technisch-naturwissenschaftliche Welt," in Heisenberg, *Zwei Vorträge*, pp. 5–17.

51. Heisenberg, "On the Unity of the Scientific Outlook on Nature," *Philosophic Problems of Nuclear Science*, p. 79.

52. *Ibid.*, p. 82.

53. *Ibid.*

54. *Ibid.*, pp. 77, 93, 94.

55. Heisenberg, "Science as a Means of International Understanding," *Philosophic Problems of Nuclear Science*, p. 118.

56. *Ibid.*, p. 109.

57. *Ibid.*, p. 113.

58. Heisenberg, "Die gegenwärtigen Grundprobleme der Atomphysik," *Wandlungen* (1973), p. 146.

59. Heisenberg, "Fundamental Problems of Present-day Atomic Physics," *Philosophic Problems of Nuclear Science*, p. 107.

60. He wrote, "Damit tritt also von neuem der Gedanke Platons in die Naturwissenschaft ein, dass der atomaren Struktur der Materie letzten Endes ein mathematisches Gesetz, eine mathematische Symmetrie zu-

grunde liege. Die Existenz der Atome oder der Elementarteilchen als Ausdruck einer mathematischen Struktur das war die neue Möglichkeit, die Planck mit seiner Entdeckung aufgezeigt hatte, und hier berührt er Grundfragen der Philosophie." Heisenberg, "Die Plancksche Entdeckung und die philosophischen Grundfragen der Atomlehre," *Wandlungen* (1973), p. 164.

61. *Ibid.*, p. 183.

62. Heisenberg, *Physics and Beyond: Encounters and Conversations*, pp. 7–9.

63. See p. 113 above and Heisenberg, *Wandlungen* (1935), p. 15.

64. For example, he wrote, "I think . . . modern physics has definitely decided for Plato. For the smallest units of matter are in fact not physical objects in the ordinary sense of the word; they are forms, structures, or—in Plato's sense—Ideas, which can be unambiguously spoken of only in the language of mathematics." *Across the Frontiers*, p. 116.

65. *Ibid.*, p. 165.

66. See, for example, his chapter "Goethe's View of Nature and the World of Science and Technology," *ibid.*, pp. 122–41.

67. *Ibid.*, p. 229 and *passim.*

68. Heisenberg, *Physics and Beyond*, p. 213.

69. Heisenberg, *Physics and Philosophy*, pp. 140–141.

70. Heisenberg, *Philosophic Problems of Nuclear Science* (New York: Pantheon, 1952), p. 75 (from the 1941 essay).

71. Heisenberg, *Physics and Beyond*, 1971, p. 82.

72. Heisenberg, *Zwei Vorträge*, pp. 11, 17.

73. Heisenberg, *Physics and Philosophy*, p. 200.

5. *Bergson, Monod, and France*

1. Quoted in Irwin Edman, "Foreword," to Bergson, *Creative Evolution*, pp. ix–x.

2. Edouard Le Roy, *Une Philosophie Nouvelle*, p. 3.

3. Quoted in J. Maritain, *Bergsonian Philosophy and Thomism*, p. 43, from Albert Thibaudet, *Le Bergsonisme* 2:101.

4. Thomas Hanna ed., *The Bergsonian Heritage*, p. 16.

5. Teilhard read Bergson's earliest works before meeting Le Roy, so it would be incorrect to think that Le Roy was the channel through which he learned about Bergson. Nonetheless, Teilhard and Le Roy were close friends who maintained a correspondence between 1925 and 1934 and who influenced each other's views. See Madeleine Barthélemy-Madaule, *Bergson et Teilhard de Chardin*, especially appendix 3, "Pierre Teilhard de Chardin et Édouard Le Roy," pp. 655–59.

6. Bergson, *Creative Evolution*, p. 14.

7. *Ibid.*, p. 335.

8. *Ibid.*, pp. 12, 12–13.
9. *Ibid.*, pp. 52–53.
10. *Ibid.*, pp. 274, 277.
11. *Ibid.*, pp. 18, 288.
12. See the analysis on this point in A. N. Chanyshev, *Filosofiia Anri Bergsona*, pp. 34–35.
13. Bergson, *Creative Evolution*, pp. 59–60.
14. *Ibid.*, pp. 14–15.
15. *Ibid.*, p. 182.
16. See, in particular, Milič Čapek, *Bergson and Modern Physics.* Also, P. A. Y. Gunter, ed., *Bergson and the Evolution of Physics.*
17. Bergson, *Creative Evolution*, pp. xxiii–xxiv.
18. *Ibid.*, p. xix.
19. Čapek, *Bergson and Modern Physics*, p. 11.
20. Bergson, *Creative Evolution*, p. 55.
21. De Broglie wrote, "Personally, from our early youth, we have been struck by Bergson's very original ideas concerning time, duration, and movement. More recently, turning again these celebrated pages and reflecting on the progress achieved by science since the already distant time when we first read them, we have been struck by the analogy between new concepts of contemporary physics and certain brilliant intuitions of duration." Louis de Broglie, "The Concepts of Contemporary Physics and Bergson's Ideas on Time and Motion," in Gunter, ed., *Bergson and the Evolution of Physics*, p. 46.
22. Bertrand Russell, *The Philosophy of Bergson*, p. 3.
23. Bergson, *Creative Evolution*, pp. 164, 194, 182.
24. *Ibid.*, pp. 277, 99, 58.
25. *Ibid.*, p. 23.
26. See A. Metz, "Le Temps d'Einstein et la Philosophie"; H. Bergson, "Les Temps fictifs et le Temps réel"; and "Controverse au sujet des temps fictifs et des temps réels dans la théorie d'Einstein: Réplique de M. André Metz; Deuxième réponse de M. Henri Bergson; Un dernier mot de M. A. Metz."
27. André Metz, "Einstein's Time and Philosophy," in Gunter, ed., *Bergson and the Evolution of Physics*, pp. 159–60.
28. Bergson, *Creative Evolution*, p. 95.
29. *Ibid.*, pp. 61–62.
30. *Ibid.*, p. 86.
31. Idella J. Gallagher, "Bergson on Closed and Open Morality," p. 56.
32. Bergson, *The Two Sources of Morality and Religion*, p. 68.
33. *Ibid.*, p. 306. One sympathetic reader concluded that this statement "seems to do no more than expose to full view at last the religious root of Bergson's whole career as a metaphysician." James Street Fulton, "Bergson's Religious Interpretation of Evolution," p. 27.
34. See Antonin Gilbert Sertillanges, *Henri Bergson et le Catholicisme.*

Sertillanges called Bergson an "apologist" for Catholicism, but he recognized that Bergson himself was never a Catholic in any literal sense. Jacques Maritain, however, criticized Sertillanges for trying to "baptize" Bergson; Maritain said that there were a number of Catholic disciples of Bergson who were so eager to reconcile his thought with Catholicism that they had become guilty of "spineless eclecticism." Jacques Maritain, *Bergsonian Philosophy and Thomism*, pp. 19, 22. Also, Madeleine Barthelemy-Madaule, *Bergson et Teilhard de Chardin.*

35. Probably the best known Marxist critique of Bergson is Georges Politzer, *La fin d'une parade philosophique: Le bergsonisme.* Another Marxist analysis is A. N. Chanyshev, *Filosofiia Anri Bergsona.* A social democratic criticism is Paul Szende, "Bergson, der Metaphysiker der Gegenrevolution." Also see Felicien Challaye, "Bergson vu par les Soviets." All the above citations give examples of socialists who considered Bergson to be representing bourgeois interests. For examples of socialists who, on the contrary, saw Bergson as being opposed to bourgeois thought and institutions, see the following syndicalist articles: Georges Sorel, "L'Évolution creatrice"; M. Maurice Legendre, "L'influence de la philosophie de M. Bergson"; M. D. Draghiscesco, "L'influence de la philosophie de M. Bergson"; M. Rene Gillouin, "L'influence de la philosophie de M. Bergson"; J. Vietroff, "L'influence de la philosophie de M. Bergson"—all in various issues of *La Mouvement Socialiste.*

36. For some examples of early concerned Catholic reactions, see Joseph de Tonquédec, "Comment interpréter l'ordre du monde?" Albert Farges, "L'erreur fondamentale de la philosophie nouvelle"; Joseph de Tonquédec, "M. Bergson est-il moniste?" Examples of Catholics praising Bergson are Sertillanges, *Henri Bergson et le Catholicisme*, and Joseph Lotte, who wrote in 1912 in the *Bulletin des professeurs catholiques de l'Université*: "I shall never forget my rapturous emotion when, in the spring of 1907, I read the *Évolution créatrice.* I felt the presence of God in every page. One needs to have lived for years without God to know the joy with which one finds Him once more. Bergson's books have led me to find Him again, and for that I shall be everlastingly grateful to them." Quoted in Jacques Chevalier, *Henri Bergson*, p. 65.

37. Georges Sorel defended Bergson against the Action française in "Sur L'évolution créatrice," *L'independant* (May 1, 1911). Walter Lippmann's comment is in his "Most Dangerous Man in the World."

38. G. V. Plekhanov, *Izbrannye filosofskie proizvedeniia*, 3:316.

39. Etienne Gilson, *The Philosopher and Theology*, p. 107.

40. Plekhanov, *Izbrannye filosofskie proizvedeniia*, p. 316.

41. Gilson, *The Philosopher and Theology*, pp. 119, 123.

42. André Joussain, "Bergsonisme et marxisme."

43. Arthur Oncken Lovejoy, "The Practical Tendencies of Bergsonism."

44. Many of Bergson's disciples unconvincingly defended him against the charge that he was antiscientific. M. Desaymard, for example, asserted,

"This philosophy is not 'anti-scientific;' it is 'ultra-scientific.'" Quoted in Lovejoy, "Practical Tendencies," p. 262.

45. Bergson, *Creative Evolution*, p. 165.
46. Jacques Monod, *Chance and Necessity*, p. 103.
47. Bergson, *Creative Evolution* (1944), p. 63.
48. Monod, *Chance and Necessity*, p. 26.
49. *Ibid.*, pp. 26, 159.
50. *Ibid.*, pp. 116, 112, 169.
51. *Ibid.*, pp. 45, 148.
52. *Ibid.*, p. 171.
53. *Ibid.*, pp. 171, 174–75.
54. *Ibid.*, pp. 177, 176, 180.
55. *Ibid.*, pp. 171, 33ff.

6. *Studying Animals to Learn about Man*

1. Julian Jaynes says in his study of the origin of the term, "While it is possible that the term may have been used earlier in a trivial way, this is the first attempt to found comparative psychology as a new science." Julian Jaynes, "The Historical Origins of 'Ethology' and 'Comparative Psychology,'" p. 602.
2. Robert B. Lockard, "Reflections on the Fall of Comparative Psychology."
3. Frank A. Beach, "The Snark Was a Boojum."
4. Lockard, "Reflections," p. 172.
5. *Ibid.*, pp. 173, 176.
6. In *Walden Two* the affairs of the community were governed by the elites—the "Planners" and "Managers." When Burris, a visitor to Walden Two, asked Frazier, the founder of the community, whether constitutional questions were decided by vote of all the members of the community, Frazier replied negatively, and continued, "You're still thinking about government by the people. Get that out of your head."

The most ambitious effort to duplicate Walden Two in real life was probably the community Twin Oaks in Virginia. After a few years the members of the community decided to place constitutional questions before the whole membership, not just an elite. Thus, Skinner's antidemocratic principles were abandoned. Furthermore, one of the founding members of Twin Oaks observed that if the "Planners" there tried to decide important issues unilaterally, "resentment over their blatant presumption and authority would topple the government." B. F. Skinner, *Walden Two*, p. 254; Kathleen Kinkade, "A Walden Two Experiment."

7. Quoted in B. F. Skinner, *Beyond Freedom and Dignity*, pp. 37–38.
8. See Harvey Wheeler, ed., *Beyond the Punitive Society*—especially the essays by Wheeler and by John R. Platt.
9. Skinner, *Beyond Freedom and Dignity*, p. 211.

10. *Ibid.*, p. 213.
11. *Ibid.*, p. 104.
12. *Ibid.*, pp. 102–3.
13. *Ibid.*, pp. 113–14.
14. Max Black, "Some Aversive Responses to a Would-be Reinforcer," in Wheeler, ed., *Beyond the Punitive Society*, p. 133.
15. Skinner, *Beyond Freedom and Dignity*, p. 175.
16. Arnold Toynbee, "Great Expectations," in Wheeler, ed., *Beyond the Punitive Society*, p. 117.
17. Black, "Some Aversive Responses," p. 126.
18. Skinner, *Walden Two*, p. 281.
19. Carl R. Rogers and B. F. Skinner, "Some Issues Concerning the Control of Human Behavior," p. 1062.
20. Skinner, *Beyond Freedom and Dignity*, p. 215.
21. *Ibid.*, p. 214.
22. *Ibid.*, pp. 182, 150.
23. *Ibid.*, pp. 164–65.
24. See N. Tinbergen, "Ethology," in his *The Animal in Its World*, 2:130.
25. In conversation on the origins of his views, Konrad Lorenz commented in 1973, "My father wanted me to study medicine, and, being a nice, obedient boy, I did study medicine. That turned out to have been my good luck because the professor of anatomy was a really great comparative anatomist and an even greater comparative embryologist. . . . Even then I knew a lot about animal behavior, particularly about ducks—I was crazy about ducks from early childhood. . . . I would have been very stupid indeed not to have seen that this comparative method could be applied to behavior as well. . . . I became fascinated with this kind of study of behavior, and thought I had invented it, discovered it. Much later, I found that old Oskar Heinroth knew all about it, but neither of us at the time knew that Charles Otis Whitman had applied exactly the same methods ten or fifteen years before Heinroth. So the first real comparative ethological study of the phylogeny of behavior by comparative methods was a paper published by Charles Otis Whitman in Woods Hole, Massachusetts, in the year 1898." Quoted in Richard I. Evans, *Konrad Lorenz*, p. 6.
26. Lorenz often counterposed ethology to behaviorism in his writings, emphasizing the hereditarian aspects of the first approach, the environmentalism of the second. He wrote, "Liberal and intellectual Americans, attracted to a sound, simple, easily intelligible, and, above all, mechanistic teaching, accepted it [behaviorism] almost without exception, particularly because here was a doctrine proclaiming itself as a liberating and democratic principle." Lorenz, *Civilized Man's Eight Deadly Sins*, p. 86. In an interview with Paul Ferris in 1974, Tinbergen observed that Americans and communists were hostile to ethology because they were inclined to see man as infinitely moldable. Ferris described Tinbergen's views in the

following way: "Ethology, he argues, with its softer, Old World origins, leans naturally toward the view that human behavior is a compromise—that key word in Tinbergen's vocabulary—between what we are born with and what we acquire afterward." Paul Ferris, "Nikolaas Tinbergen Seagull," p. 102.

27. Mary Alice Evans and Howard Ensign Evans, *William Morton Wheeler: Biologist*, pp. 213–15.

28. Lorenz, "Companions as Factors in the Bird's Environment," 1:127.

29. Lorenz, "Instinctive Behavior Patterns in Birds," 1:72.

30. Lorenz, *On Aggression*, p. 241.

31. *Ibid.*, pp. 255–56, 277.

32. *Ibid.*, pp. 253–54.

33. Noam Chomsky, *Reflections on Language*, p. 132.

34. See Theo J. Kalikow, "Konrad Lorenz's 'Brown Past': A Reply to Alec Nisbett," pp. 173–79 and Alec Nisbett's "Reply," pp. 179–80.

35. Lorenz, "Systematik und Entwicklungsgedanke im Unterricht," *Der Biologe* (January–February 1940), 9:24.

36. Lorenz, "Durch Domestikation verursachte Störungen arteigenen Verhaltens," p. 71. Alec Nisbett's discussion of Lorenz' ideas and behavior in this period can only be called a whitewash, although he does correctly point out a translation error in Leon Eisenberg's earlier discussion of the same topic. But Eisenberg's point remains the valid one, despite his error. See Alec Nisbett, *Konrad Lorenz*, pp. 81–88, and Leon Eisenberg, "The 'Human' Nature of Human Nature."

37. Lorenz, *Civilized Man's Eight Deadly Sins*, pp. 58, 54–58.

38. *Ibid.*, pp. 51, 59.

39. Leon Eisenberg, "The 'Human' Nature of Human Nature," p. 169.

40. Stephen Jay Gould, "Biological Potential vs. Biological Determinism."

7. *Primatology and Sociobiology*

1. Stuart R. Altman, ed., *Social Communication among Primates*, p. xi.

2. See, for example, George B. Schaller, *The Mountain Gorilla*; Sherwood L. Washburn and David Hamburg, "The Study of Primate Behavior," Irven De Vore and K. R. L. Hall, "Baboon Ecology," K. R. L. Hall and Irven De Vore, "Baboon Social Behavior," Phyllis Jay, "The Common Langur of North India," George B. Schaller, "The Behavior of the Mountain Gorilla," Jane Goodall, "Chimpanzees of the Gombe Stream Reserve," George B. Schaller, "Behavioral Comparisons of the Apes," and Sherwood L. Washburn and David A. Hamburg, "The Implications of Primate Research," all in Irven De Vore, ed., *Primate Behavior: Field Studies of Monkeys and Apes*; Jane van Lawick-Goodall, "A Preliminary Report on Expressive Movements and Communication in the Gombe Stream Chimpanzees," S. L. Washburn and D. A. Hamburg, "Aggressive Behavior in

Old World Monkeys and Apes," Phyllis Jay, "Primate Field Studies and Human Evolution," all in Phyllis C. Jay, ed., *Primates: Studies in Adaptation and Variability*; Irven De Vore, "Mother-Infant Relations in Free-Ranging Baboons," in Thomas E. McGill, ed., *Readings in Animal Behavior*; and S. L. Washburn and Phyllis C. Jay, eds., *Perspectives on Human Evolution.*

3. Harry F. Harlow, "The Heterosexual Affectional System in Monkeys"; G. L. Arling and H. F. Harlow, "Effects of Social Deprivation on Maternal Behavior of Rhesus Monkeys"; H. F. Harlow et al., "Maternal Behavior of Rhesus Monkeys Deprived of Mothering and Peer Associations in Infancy."

4. An example of a work critical of sociobiology is Marshall D. Sahlins, *The Use and Abuse of Biology.*

5. The best-known work of Desmond Morris (mentioned in the previous chapter as a student of Tinbergen), is *The Naked Ape.*

6. William Etkin, *Social Behavior from Fish to Man.*

7. S. A. Barnett, *Instinct and Intelligence*, pp. 191–92.

8. S. L. Washburn and C. S. Lancaster, "The Evolution of Hunting," in S. L. Washburn and Phyllis C. Jay, eds., *Perspectives on Human Evolution*, p. 217.

9. The best collection of articles is Arthur L. Caplan, ed., *The Sociobiology Debate.*

10. Thus, Wilson says, "Scientists and humanists should consider the possibility that the time has come for ethics to be removed temporarily from the hands of the philosophers and biologicized" (p. 562). And, at the very beginning of the last chapter he wrote: "Let us now consider man in the free spirit of natural history, as though we were zoologists from another planet completing a catalog of social species on Earth. In this macroscopic view the humanities and social sciences shrink to specialized branches of biology; history, biography, and fiction are the research protocols of human ethology; and anthropology and sociology together constitute the sociobiology of a single primate species" (p. 547). E. O. Wilson, *Sociobiology: The New Synthesis.*

11. Edward O. Wilson, *On Human Nature*, p. 175.

12. Edward O. Wilson, *The Insect Societies*, p. 1.

13. Quoted in Wilson, *The Insect Societies*, p. 317.

14. *Ibid.*, pp. 318, 319.

15. Quoted in Mary Alice Evans and Howard Ensign Evans, *William Morton Wheeler, Biologist*, p. 265 from E. O. Wilson, "The Superorganism Concept and Beyond," in *L'effet de Groupe chez les Animaux*, Colloques internationaux du Centre National de la Recherche Scientifique, pp. 1–11.

16. Quoted in Wilson, *The Insect Societies*, p. 319.

17. Wilson, *Sociobiology*, p. 22.

18. Wilson, *The Insect Societies*, p. 224.

19. *Ibid.*, p. 229.

20. *Ibid.*, pp. 318–19.
21. *Ibid.*, p. 319.
22. Arthur L. Caplan, ed., *The Sociobiology Debate*, p. 7.
23. Wilson, *Sociobiology*, p. 118.
24. *Ibid.*, p. 562.
25. *Ibid.*, p. 550.
26. *Ibid.*, p. 575.
27. Wilson, *On Human Nature*, p. 200.
28. *Ibid.*, pp. ix–x, 195, 5–6.
29. *Ibid.*, pp. 7, 19.
30. *Ibid.*, p. 13.
31. *Ibid.*, pp. 13, 11.
32. *Ibid.*, p. 196.
33. Max Black, "Some Aversive Responses to a Would-be Reinforcer," p. 126.
34. Wilson, *On Human Nature*, p. 196.
35. *Ibid.*, pp. 196 ff.
36. *Ibid.*, p. 199.
37. *Ibid.*, pp. 96–97.
38. *Ibid.*, pp. 135, 138.
39. See *Le Figaro* (June 30, 1979); *New York Times* (July 8, 1979); *Time* (August 13, 1979). An example of a right-wing British publication on the sociobiology theme is: Richard Verrall, "Sociobiology: The Instincts in Our Genes," pp. 10–11. A radical critique of some of the right-wing abuses of biology can be found in Martin Barker, "Racism: The New Inheritors." Another critique from the left, one which is much more attentive to the history and structure of science, is Donna Haraway, "The Biological Enterprise: Sex, Mind, and Profit from Human Engineering to Sociobiology."
40. Wilson, *On Human Nature*, p. 6.

8. Eugenics: Weimar Germany and Soviet Russia

1. There is a very large literature on Darwinism and Social Darwinism in Germany. See, in particular, Hans-Günter Zmarzlik, "Der Sozialdarwinismus in Deutschland als geschichtliches Problem," *Vierteljahrshefte für Zeitgeschichte* (1963), 11:246–273, and "Der Sozialdarwinismus in Deutschland—Ein Beispiel für den gesellschaftspolitischen Missbrauch naturwissenschaftlicher Erkenntnisse," in Günter Altner, ed., *Kreatur Mensch: Moderne Wissenschaft auf der Suche nach dem Humanum* (Munich: Moos, 1973), pp. 289–311. Also see William Montgomery, "Germany," pp. 81–116; Gerhard Heberer and Franz Schwanitz, eds., *Hundert Jahre Evolutionsforschung*; and Hedwig Conrad-Martius, *Utopien der Menschenzüchtung*. For interesting but somewhat simplified accounts, see Daniel

Gasman, *The Scientific Origins of National Socialism*; and Niles R. Holt, "Monists and Nazis, pp. 37–43.

2. Otto Ammon, *Die Gesellschaftsordnung und ihre natürlichen Grundlagen*, pp. 25, 26.

3. See the description in Conrad-Martius, *Utopien der Menschenzüchtung*, pp. 74–75.

4. Schallmayer, *Vererbung und Auslese im Lebenslauf der Völker*, p. 100.

5. *Ibid.*, esp. ch. 7, pp. 174–211.

6. For the United States, for example, see Rudolph Vecoli, "Sterilization: A Progressive Measure?" and Donald K. Pickens, *Eugenics and the Progressives*.

7. Alfred Ploetz, *Die Tüchtigkeit unserer Rasse und der Schutz der Schwachen*.

8. *Ibid.*, p. 212.

9. *Ibid.*, pp. 224–225.

10. See Fritz Lenz' review of Grotjähn's *Leitsätze sur sozialen und generativen Hygiene*. Also see Karl Kautsky's review of Grotjähn's *Die Hygiene der menschlichen Fortpflanzung*.

11. Alfred Ploetz proposed the term *Rassenhygiene* in 1895. He was not certain whether or not it was a new term. As he observed, "Ich weiss nicht, ob das Wort Rassenhygiene schon ausgesprochen wurde oder nicht; sicher ist, dass der darin enthaltene Begriff längst in vielen Köpfen lebte, und dass er in den Geisteskämpfen unserer Tage eine grosse Rolle spielt." *Die Tüchtigkeit unserer Rasse*, p. 5.

12. Alfons Fischer also insisted on spelling the word "Rassehygiene." See Fritz Lenz' review of Fischer, *Grundriss der sozialen Hygiene*. Also see Lenz' review of W. Schallmayer, "Einführung in die Rassehygiene" (sic).

13. The emerging demagogic element can be seen in Lenz's statement: "Wenn man schon Rücksicht auf politische Richtungen nimmt, so muss man allerdings auch bedenken, dass für die volkische Bewegung, die in Deutschland auch eine Macht darstellt, und zwar eine wachsende Macht, das Wort Rasse eine angenehmen Klang hat. Gerade mit Rücksicht auf die Aussichten praktischer Durchführung rassenhygienischer Reformen tut man daher meines Erachtens nicht gut, das Wort Rassenhygiene ganz zu vertilgen." As quoted in Günter Altner, *Weltanschauliche Hintergründe der Rassenlehre des Dritten Reiches*," p. 38.

14. K. Saller, *Die Rassenlehre des Nationalsozialismus in Wissenschaft und Propaganda*, pp. 16–17, 72. In his 1930 encyclical "On Christian Marriage," Pope Pius XI went beyond condemnation of contraception and abortion to a specific castigation of all those who "put eugenics before aims of a higher order, and by public authority wish to prevent from marrying all those whom, even though naturally fit for marriage, they consider, according to the norms and conjectures of their investigations, would, through hereditary transmission, bring forth defective offspring." "Encyclical Letter of His Holiness Pius XI by Divine Providence Pope," pp. 23–24. Eugen Fischer became the rector of the University of Berlin after the National Socialist takeover.

15. Karl Kautsky, Jr., "Unterbrechung der Schwangerschaft."

16. George Chaym, *Sozialistische Monatshefte*, p. 638.

17. In addition to the articles and reviews specifically discussed in the text, see the following reviews in *Die Gesellschaft*: R. F. Fuchs of A. Basler, *Einführung in die Rassen- und Gesellschafts-Physiologie*; Karl Kautsky of Alfred Grotjähn, *Die Hygiene der menschlichen Fortpflanzung: Versuch einer praktischen Eugenik*; M. Kantorowicz-Kroll of Robert Sommer, *Familienforschung, Vererbungs- und Rassenlehre*; and Miron Kantorowicz of Ernst Neumann, *Individual-Rassen-und Volkshygiene*.

18. Oda Olberg, review of Stavros Zurukzoglu, *Biologische Probleme der Rassenhygiene und die Kulturvölker*, and review of von Bohr-Pinnow, *Die Zukunft der menschlichen Rasse*.

19. Karl Kautsky, review of Oda Olberg, *Die Entartung in ihrer Kulturbedingtheit*.

20. Hugo Iltis, "Rassenwissenschaft und Rassenwahr."

21. *Ibid.*, p. 113.

22. Iltis continued his strong opposition to National Socialist views on race. See, for example, his "Der Schädelindex in Wissenschaft und Politik."

23. For examples of Lenz' attacks on Lamarckism, see his reviews of Hermann Paull, *Wir und das kommende Geschlecht*; of Wilhelm Schmidt, *Rasse und Volk*; and of Friedrich Hertz, *Rasse und Kultur*. An example of Lenz' praise of Hitler before the National Socialists came to power is the following 1931 statement: "Hitler ist der erste Politiker von wirklich grossen Einfluss, der die Rassenhygiene als eine zentrale Aufgabe aller Politik erkannt hat und der sich tatkräftig dafür einsetzen will;" see Lenz, "Die Stellung des Nationalsozialismus zur Rassenhygiene."

24. Fritz Lenz, "Der Fall Kammerer und seine Umfilmung durch Lunatscharsky." Lenz repeatedly made the argument that science was value-free. In 1921, for example, he said, "Die Natur verlangt überhaupt nichts; die Naturwissenschaft kann nur zeigen, was geschieht, nicht was geschehen soll." *Archiv für Rassen-und Gesellschaftsbiologie* (1921), 13:112.

25. Levien, "Stimmen aus dem deutschen Urwalde (Zwei neue Apostel des Rassenhasses)." The article is a criticism of the views of H. F. K. Günther and A. Basler.

26. James Allen Rogers, "Charles Darwin and Russian Scientists," p. 382. There was even an early advocate in Russia of improvement of humans through selection: V. M. Florinskii, *Usovershenstvovanie i vyrozhdenie chelovescheskogo roda*. For a discussion of this book by an early Soviet eugenist, see M. V. Volotskoi, "K istorii evgenicheskogo dvizheniia." Also see George Kline, "Darwinism and the Russian Orthodox Church." N. Danilevskii, *Darvinizm, kriticheskoe issledovanie*; K. A. Timiriazev, *Charlz Darvin i ego uchenie*; Alexander Vucinich, "Russia: Biological Sciences"; and James Allen Rogers, "Russia: Social Sciences."

27. See Mark B. Adams, "The Founding of Population Genetics" and "Towards a Synthesis."

28. N. K. Kol'tsov was president of the Russian Eugenics Society; Iu.

A. Filipchenko was the director of the Bureau of Eugenics of the Academy of Sciences; and A. S. Serebrovskii was a member of the permanent bureau of the Russian Eugenics Society and a contributor to its journal. For his great hopes for the eugenics movement, see A. S. Serebrovskii, "O zadachakh i putiakh antropogenetiki," esp. p. 112. Ironically, N. I. Vavilov, who ultimately suffered the most at the hands of Soviet authorities, apparently steered clear of eugenics; he became a foe of Lysenko in the late 1930s and died in 1940 in Siberian exile.

29. For a good bibliography of Russian eugenics literature, see K. Gurvich, "Ukazatel' literatury po voprosam evgeniki." For a discussion of the eugenics movement in Soviet Russia, see David Joravsky, *The Lysenko Affair*, pp. 256–66.

30. For accounts of the founding and early activities of the Russian Eugenics Society, see "O deiatel'nosti Russkogo Evgenicheskogo Obshchestva za 1921 god" and similar descriptions in succeeding volumes of the same publication. For an interesting but somewhat one-sided Soviet interpretation of these early eugenic interests, see N. P. Dubinin, *Vechnoe dvizhenie*.

31. "O deiatel'nosti Russkogo Evgenicheskogo Obshchestva za 1921 god," p. 101.

32. For examples of these types of literature, see A. S. Serebrovskii, "Genealogiia roda Aksakovykh"; N. Chulkov, "Rod grafov Tolstykh"; Iu. A. Nelidov, "O potomstve barona Petra Pavlovicha Shafirova"; V. Zolotarev, "Rodoslovnye A. Pushkina, gr. L. N. Tolstogo, P. Ia. Chaadaeva, Iu. F. Samarina, A. I. Gertsena, kn. P. A. Kropotkina, kn. S. N. Trubetskogo"; S. V. Liubimov, "Predki grafa S. Iu. Vitte"; Iu. A. Filipchenko, "Nashi vydaiushchiesia uchenye"; T. K. Lepin, Ia. Ia. Luss, and Iu. A. Filipchenko, "Deistvitel'nye chleny byvsh. imperat., nyne Rossiiskoi, Akademii za poslednie 80 l. (1864–1924 gg.)"; L. S. Berg, "Brachnost' rozhdaemost' i smertnost' v Leningrade za poslednie gody"; A. V. Gorbunov, "Vliianie mirovoi voiny na dvizhenie naseleniia Evropy"; and V. V. Bunak, "Novye dannye k voprosu o voine, kak biologicheskom faktore."

33. Bunak, "Novye dannye," p. 231.

34. The chairman of the "Commission for the Study of the Jewish People" was V. V. Bunak. "Evgenicheskie zametki," p. 58. For examples of the commission's work, see S. S. Vermel', "Prestupnost' evreev" and V. V. Bunak, "Materialy dlia sravnitel'noi kharakteristiki sanitarnoi konstitutsii evreev."

35. See, for example, Kol'tsov's generally positive review of E. Baur, E. Fischer, and F. Lenz, *Grundriss der menschlichen Erblichkeitslehre und Rassenhygiene*. Kol'tsov commented that Lenz' section of this textbook had been an important source of his own article, "Vliianie kul'tury na otbor v chelovechestve." Filipchenko chose as a part of the title of one of his articles a quotation from the same book, Lenz's observation that "der Schutz der geistigen Arbeiter, und speziell der hochbegabten, ist eine

Hauptaufgabe der Rassenhygiene." Filipchenko, "Nashi vydaiushchiesia uchenye," p. 22.

36. See, for example, Kol'tsov's "Die rassenhygienische Bewegung in Russland" and Filipchenko's "Die russische rassenhygienische Literatur, 1921–1925." The German race hygienists paid particular attention to research done by Soviet scholars in the Ukraine that purportedly indicated that the cranial measurements of intellectuals are greater than those of workers and peasants—this in a "workers' and peasants'" state! See "Materialen zur Anthropologie der Ukraine."

37. For his irritation, see N. K. Kol'tsov, "Kritika bibliografiia: Trudy 2-go mezhdunarodnogo evgenicheskogo s"ezda," p. 69.

38. Kol'tsov, "Evgenicheskie s"ezdy v Milane v Sentiabre 1924 goda."

39. For a contemporary survey of the international eugenics movement, describing the various legislative programs, see P. I. Liublinskii, "Sovremennoe sostoianie evgenicheskogo dvizheniia."

40. The article was titled "Rukovodiashchie polozheniia nemetskogo obshchestva rasovoi gigieny."

41. "Sovremennoe sostoianie voprosa o sterilizatsii v Shvetsii" and "Programma prakticheskoi evgenicheskoi politiki."

42. "Evgenicheskaia sterilizatsiia v Germanii," p. 82.

43. "Obsuzhdenie Norvezhskoi evgenicheskoi programmy na zasedaniiakh Leningradskogo Otdeleniia R. E. O." The particular eugenic program at the basis of this discussion was that of the Norwegian J. A. Mjöen. This society's discussion is very interesting, for it encompassed consideration of both "positive and negative race hygiene proposals" by speakers with wide disagreements on issues such as racially mixed marriages and mandatory sterilization.

44. For approval by the Commissariats of Health and Education, as well as for announcement of the subsidy, see "Iz otcheta o deiatel'nosti Russkogo Evgenicheskogo Obshchestva za 1923 g.," p. 4. For approval of the society's charter by the Commissariat of Internal Affairs, see "Evgenicheskie zametki," p. 58.

45. Vasilii Slepkov, "Nasledstvennost' i otbor u cheloveka (Po povodu teoreticheskikh predposylok evgeniki)."

46. B. M. Zavadovskii, "Darvinizm i lamarkizm i problema nasledovaniia priobretennykh priznakov."

47. Kol'tsov, "Noveishie popytki dokazat' nasledstvennost' blagopriobretennykh priznakov," esp. p. 167.

48. For a discussion of the impact of Filipchenko's argument, see N. M. Volotskoi, "Spornye voprosy evgeniki."

49. *Ibid.*, p. 225.

50. *Ibid.*, pp. 212–54.

51. For an example of the type of extrapolation from the animal and plant worlds to that of human society that increasingly attracted criticism, note Filipchenko's observation, "Every plant and animal breeder who

graduates from an agronomy institute is familiar in detail with the genetics of domestic animals and plants; can we really say that the latter are more valuable in these regards than man himself?" Filipchenko then went on to call for the inclusion of eugenics instruction in the standard education curriculum for Soviet youth. See his "Evgenika v shkole," p. 32.

52. Kol'tsov observed in a 1924 article, "If we calculate the average number of children in the family of each member of the Russian Communist party, that number will no doubt be far from what Gruber cites as necessary for a population group to preserve itself in the overall population. What would we say about a stock-breeder who every year castrated his most valuable producers, not permitting them to multiple? But in cultured society approximately the same thing is occurring before our eyes!" Kol'tsov, "Vliianie kul'tury na otbor v chelovechestve," p. 15. The reference is to Max Gruber, *Ursache und Bekampfung des Geburtenrückgangs im deutschen Reiche*.

53. Kol'tsov, "Rodoslovnye nashikh vydvizhentsev." For other examples of this type of article, see N. P. Chulkov, "Genealogiia dekabristov Murav'evykh" and V. Zolotarev, "Dekabristy."

54. The difficulty of the position of the academic eugenicists was illustrated by A. S. Serebrovskii when he called for a Socialist eugenics and then observed, "Every class must create its own eugenics. This slogan, however . . . must in no way be understood in the manner of several of our comrades, especially in Moscow, who maintain that the whole base of Morganist-Mendelian theory is an invention of the Western bourgeoisie and that the proletariat, creating its own eugenics, must base itself on Lamarckism." Serebrovskii, "Teoriia nasledstvennosti Morgana i Mendelia i marksisty," p. 113.

55. "Evgenika," vol. 23, cols. 812–19.

56. For examples of Lenz' criticisms, see his reviews of Kurt Gerlach, *Fernstenliebe*, p. 330 and of L. F. Clauss, *Die nordische Seele*, pp. 445–47. Also see his more reserved criticism of the notorious journalistic race anthropologist H. F. K. Günther in "Notizen: Günthers Berufung nach Jena." Lenz criticized in particular the effort of National Socialist race propagandists to make a mythic, irrational ideal of the Nordic spirit. Lenz believed that racial ideas should be based on science.

57. Noam Chomsky, *Reflections on Language*, p. 132.

9. *Biomedical Ethics*

1. Donald Fleming, "On Living in a Biological Revolution," p. 65.

2. Peter Saunders and Philip Rhodes, "The Origin and Circulation of the Amniotic Fluid," p. 1, citing the works of Albano (1934) and Vosburg et al. (1948).

3. D. V. I. Fairweather, "Technique and Safety of Amniocentesis," p. 19.

4. *Ibid.*, p. 26.
5. Tabitha M. Powledge and John Fletcher, "Guidelines for the Ethical, Social and Legal Issues in Prenatal Diagnosis," 300:171–72.
6. *Ibid.*, pp. 168–72.
7. Fritz Fuchs and Lars L. Cederqvist, "Prenatal Diagnosis Based upon Amniotic Fluid Cells," p. 266.
8. Tabitha M. Powledge, "From Experimental Procedure to Accepted Practice," p. 7.
9. J. W. Ballantyne, *Manual of Antenatal Pathology and Hygiene.*
10. Jerry N. Thompson, "Prenatal Detection of Heritable Metabolic Disorders," p. 25.
11. *Ibid.*, p. 30.
12. Carl E. Wasmuth, "Euthanasia," 10:711.
13. Donald Longmore, *Spare-Part Surgery*, p. 59.
14. Longmore, *Spare-Part Surgery*; Tetsuzo Akutsu, *Artificial Heart*; W. J. Kolff, *Artificial Organs*; Robert L. Dedrick, Kenneth B. Bischoff, and Edward F. Leonard, eds., *The Artificial Kidney*; Clark K. Colton, *A Review of the Development and Performance of Hemodialyzers*; May Sherman, *Artificial Organs*; Fred Warshofsky, *The Rebuilt Man*; J. S. Redding, *Life Support*; Harold M. Schmeck, Jr., *The Semi-Artificial Man*; David Bregman, *Mechanical Support of the Failing Heart*; G. H. Myers and V. Parsonnet, *Engineering in the Heart and Blood Vessels.*
15. Tetsuzo Akutsu, *Artificial Heart*, p. 17.
16. Stanley F. Cohen, "The Manipulation of Genes," p. 25.
17. Joshua Lederberg, "DNA Splicing: Will Fear Rob Us of Its Benefits?" p. 33.
18. The best source on the recombinant DNA controversy is the archive at the Massachusetts Institute of Technology, collected by Dr. Charles Weiner and associates. For published accounts, see Richard Hutton, *Bio-Revolution* and June Goodfield, *Playing God.*
19. William Bennett and Joel Gurin, "Home Rule and the Gene," p. 16.
20. *Ibid.*, p. 59.
21. Loren R. Graham, "Why Study Soviet Science?"
22. Paul Ramsey, "Genetic Engineering," p. 14.
23. Theodore Friedmann and Richard Roblin, "Gene Therapy for Human Genetic Disease?" p. 952.
24. *Ibid.*, p. 953.
25. Carl Djerassi, "Probabilities and Practicalities," p. 27.
26. "Human Genetic Engineering," pp. 45–57.
27. *Ibid.*
28. "Rerun of German Genetic Therapy," pp. 30–31.
29. Friedmann and Roblin, "Gene Therapy?" p. 954.
30. Harold M. Schmeck, "Human Gene Splice Tried by American," *New York Times* (October 9, 1980), p. A31; Philip J. Hilts, "Use of Genes is Condemned," *The Boston Globe* (October 17, 1980), p. 3; Anne C. Roark,

"Use of Genetic Engineering on Humans Investigated by U.S., University," *The Chronicle of Higher Education* (October 14, 1980), 21:2; Gina Bari Kolata and Nicholas Wade, "Human Gene Treatment Stirs New Debate," *Science* (October 24, 1980), 210:407; Nicholas Wade, "UCLA Gene Therapy Racked by Friendly Fire," *Science* (October 31, 1980), 210:509–11.

31. Wade, "UCLA Gene Therapy," p. 509.

32. *Ibid.*

33. *Ibid.*, p. 511.

34. Ramsey, "Genetic Engineering," p. 17.

10. Public Concerns About Science and Technology

1. Letter of January 18, 1977, from Professor Robert S. Morison and Gerald Holton, quoting Walter Rosenblith, inviting participants to a planning conference for the MIT faculty seminar, "Limits to Inquiry."

2. Hans Jonas, "Freedom of Scientific Inquiry and the Public Interest." p. 27.

3. See "An Act for the Development and Control of Atomic Energy," *United States Statutes at Large* (79th Cong., 2d sess.), (Washington, D.C.: U.S. Government Printing Office, 1946), 60:755–75. A discussion of the resistance of scientists to the May-Johnson Bill is in Alice Kimball Smith, *A Peril and a Hope*.

4. Carl Djerassi, "Probabilities and Practicalities," p. 27.

5. *Congressional Record* (94th Cong., 1st sess.), April 9, 1975, 6:H2601–H2602; also, "Bauman Amendment's Chances Down," *Science* (July 4, 1975), 189:27.

6. Although Melanchthon clearly disagreed that the earth moved, he used some other aspects of the Copernican system (data for planets, etc.). See Robert S. Westman, "The Melanchthon Circle, Rheticus, and the Wittenberg Interpretation of the Copernican Theory," *Isis* (June, 1975), pp. 165–93.

7. Robert Sinsheimer, "Comments," p. 18.

8. Letter to the author from Harvey Brooks (April 12, 1977).

9. All quotations in this paragraph are from Cheryl M. Fields, "Who Should Control Recombinant DNA?"

10. Theodore Roszak, *Where the Wasteland Ends*, pp. xxiv, 168.

11. A. S. Eddington, *Science and the Unseen World*, p. 49.

12. Dorothy Nelkin, "The Science-Textbook Controversies." See also her *Science Textbook Controversies*.

13. "NSF: Congress Takes Hard Look at Behavioral Science Course."

14. Let me remind people who discount the possibility of such a situation of the following hypothetical scenario mentioned at the end of chapter 8. Imagine that you are a scientist in Nazi Germany and that you have just discovered Tay-Sachs disease, an abnormality based on a genetic defect which is more common among Jews than other population groups.

Is it not possible that you would try to keep this bit of knowledge away from the eyes of Hitler and National Socialist bureaucrats, and that you might suggest to your trusted colleagues that they do the same?

11. Attempts to Provide a Philosophical Overview

1. Loeb wrote, "Living organisms are in the first place chemical machines. . . ." J. Loeb, Assimilation and Heredity," p. 548. In another place, he defined living organisms as "mécanismes chimiques consistant principalement en matière colloidale, at capables de persister, de se développer et de se reproduire automatiquement." Loeb, "La nature chimique de la vie," p. 305. He gave similar definitions in his *The Dynamics of Living Matter.* p. 1, and in many other places.

2. W. J. V. Osterhout, "Biographical Memoir of Jacques Loeb 1859–1924," p. 365.

3. "The contents of life from the cradle to the bier are wishes and hopes, efforts and struggles, and unfortunately also disappointments and sufferings. And this inner life should be amenable to a physico-chemical analysis? In spite of the gulf which separates us today such an aim I believe that it is attainable." J. Loeb, *The Mechanistic Conception of Life*, p. 26.

4. "Et la on peut se demander par suite si, dans quelques décades, la physique et la chimie n'auront pas résolu même les problèmes de la philosophie humaine." Loeb, "La nature chimique de la vie," p. 315.

5. "If our existence is based on the play of blind forces and only a matter of chance; if we ourselves are only chemical mechanisms—how can there be an ethics for us? The answer is, that our instincts are the root of our ethics and that the instincts are just as hereditary as is the form of our body. . . . Not only is the mechanistic conception of life compatible with ethics; it seems the only conception of life which can lead to an understanding of the source of ethics." Loeb, *The Mechanistic Conception of Life*, p. 31. And a few pages later he continued, "A mechanistic conception of life is not complete unless it includes a physico-chemical explanation of psychic phenomena" (p. 35).

6. Loeb maintained that animals display tropisms that cannot be explained by natural selection, and concluded, "We must, therefore, free ourselves at once from the overvaluation of natural selection and accept the concept of Mendel's theory of heredity, according to which the animal is to be looked upon as an aggregation of independent hereditary qualities." Loeb, *The Mechanistic Conception of Life*, pp. 52–53.

7. Loeb, *The Mechanistic Conception of Life* (1912), p. 3.

8. Loeb, *The Organism as a Whole, From a Physicochemical Viewpoint*, p. 6.

9. "I believe it can only help science if the younger investigators realize that experimental abiogenesis is the goal of biology." Loeb, *The Dynamics of Living Matter*, p. 223. And, elsewhere, "The aim of Physiological

Morphology is not wholly analytical. It has another and higher aim, which is synthetical or constructive, that is, to form new combinations from the elements of living matter, just as the physicist and the chemist form new combinations from the elements of non-living matter." Loeb, *The Mechanistic Conception of Life* (1964), p. 103.

10. Loeb, "Die Sehstörungen nach Verletzung der Grosshirnrinde."
11. Loeb, *The Mechanistic Conception of Life*, pp. 35–36.
12. *Ibid.*
13. *Ibid.*, pp. 62–63.
14. Loeb, "La nature chimique de la vie," p. 315.
15. Donald Fleming, "Introduction," in Jacques Loeb, *The Mechanistic Conception of Life*, pp. xxxix–xli. Also see Hans Vaihinger, *The Philosophy of "As-If."*
16. Fleming, "Introduction," pp. xl–xli.
17. Francis Crick, *Of Molecules and Men*, p. ix.
18. *Ibid.*, pp. 10, 87.
19. Quoted by Crick from Gunther Stent, *The Molecular Biology of Bacterial Viruses.*
20. Crick, *Of Molecules and Men*, p. 95.
21. *Ibid.*, pp. 13–14.
22. B. F. Skinner, *Beyond Freedom and Dignity*, p. 8.
23. Crick, *Of Molecules and Men*, pp. 5–6, 59, 58.
24. Noam Chomsky, "The Case Against B. F. Skinner," p. 19.
25. Donald Fleming in Jacques Loeb, *The Mechanistic Conception of Life*, p. xvii.
26. Crick, *Of Molecules and Men*, p. 15.
27. Skinner, *Beyond Freedom and Dignity*, p. 20.
28. Skinner, *Walden II*, p. 2.
29. Samuel Alexander, *Space, Time, and Deity*; C. Lloyd Morgan, *Emergent Evolution*; Jan Christiaan Smuts, *Holism and Evolution.*
30. Morgan, *Emergent Evolution*, p. 25.
31. Alexander, *Space, Time, and Deity*, p. 345–47.
32. *Grand Larousse Encyclopédique*, 10:210–11. Examples of praise of Teilhard are: Sir Julian Huxley, "Introduction"; Theodosius Dobzhansky, "The Teilhardian Synthesis"; Léopold Senghor, *Pierre Teilhard de Chardin und die afrikanische Politik*; Roger Garaudy, *From Anathema to Dialogue: The Challenge of Marxist-Christian Cooperation.*
33. P. B. Medawar, "Critical Notice"; George Gaylord Simpson, "On the Remarkable Testament of the Jesuit Paleontologist Pierre Teilhard de Chardin."
34. *La Grande Encyclopédie*, 55:11725.
35. Pierre Teilhard de Chardin, *The Future of Man*, p. 298.
36. For a discussion of the Church's reaction to Teilhard's criticism of dualism, see Mary Lukas and Ellen Lukas, *Teilhard*, pp. 63, 160.

37. Quoted in N. M. Wildiers, *An Introduction to Teilhard de Chardin*, p. 27.
38. Quoted in Lukas and Lukas, *Teilhard*, p. 290.
39. Quoted in Wildiers, *Introduction to Teilhard*, p. 152.
40. Teilhard, *The Phenomenon of Man*, pp. 78, 89.
41. Teilhard, *Man's Place in Nature*, p. 24.
42. Teilhard, *The Phenomenon of Man*, p. 108.
43. G. G. Simpson, "On the Remarkable Testament," p. 204.
44. Teilhard, *The Phenomenon of Man*, p. 149.
45. *Ibid.*, pp. 36, 297–98.
46. *Ibid.*, p. 291.
47. Simpson, "On the Remarkable Testament," p. 204.
48. Teilhard, *The Phenomenon of Man*, p. 29.
49. Medawar, "Critical Notice," p. 106.
50. Simpson, "On the Remarkable Testament," p. 206.
51. See the description in Lukas and Lukas, *Teilhard*, pp. 234–235.
52. "Monitum," p. 526. For an interpretation, see G. Isaye, S. J., "Avertissement du 30 juin 1962 concernant les oeuvres du P. Teilhard de Chardin."
53. Teilhard, *Man's Place in Nature*, pp. 94–95.
54. *Ibid.*, p. 95.
55. Quoted in Claude Cuénot, *Teilhard de Chardin*, p. 247.
56. Teilhard, *Human Energy*, p. 134.
57. Teilhard, *The Phenomenon of Man*, p. 257.
58. Teilhard, *Human Energy*, p. 38.
59. Teilhard, *The Phenomenon of Man*, p. 279.
60. Teilhard, *Human Energy*, p. 127.
61. Huxley, in "Introduction," p. 21 and, on racial eugenics, Teilhard, *The Future of Man*, p. 234.
62. Teilhard, *The Future of Man*, p. 234.
63. Claude Cuénot, *Teilhard de Chardin*, p. 141.
64. He spoke of the Chinese on one occasion as "primitive as 'redskins' or American aborigines," and said they lived "in a spirit of complete utilitarianism, without any kind of idealism or hope." In letters in 1923 and 1924, while in China, Teilhard wrote, "The Chinese soul is naturally turbid, vague, unstable, elusive (even to itself), and also, as if by nature, materialistic, down to earth, and agnostic"; he also said, "So far, I have seen only races on their way to extinction, or primitives (the Chinese) who multiply without showing as yet any creativity. I firmly believe that it is our old Europe—and especially Paris—which is the promise of mankind today." Although Teilhard spent much of the time between 1923 and 1946 in China, according to his biographers he never learned a word of Chinese. See Lukas and Lukas, *Teilhard*, pp. 79, 119; Claude Cuénot, *Teilhard de Chardin*, pp. 49, 54.

65. Quoted in Robert Speaight, *The Life of Teilhard de Chardin*, p. 234.
66. Teilhard, *The Future of Man*, pp. 203–4.
67. Lukas and Lukas, *Teilhard*, pp. 237–38.
68. Teilhard, *The Future of Man*, pp. 140–48.
69. J. H. Woodger, *Biological Principles*; E. S. Russell, *The Interpretation of Development and Heredity*; Ludwig von Bertalanffy, *Kritische Theorie der Formbildung*; Ludwig von Bertalanffy, *Das Biologische Weltbild*.
70. Russell, *Interpretation of Development and Heredity*, p. 147.
71. Alex B. Novikoff, "The Concept of Integrative Levels and Biology."
72. For careful criticisms, see Ernest Nagel, *The Structure of Science*, pp. 428–46; and Kenneth F. Schaffner, "Antireductionism and Molecular Biology."
73. Loren R. Graham, *Science and Philosophy in the Soviet Union*.
74. Quoted in Sidney W. Fox, ed., *The Origins of Prebiological Systems and of Their Molecular Matrices*, pp. 53–55.
75. Herman J. Muller, "Lenin's Doctrines in Relation to Genetics," p. 464.
76. Loren R. Graham, "Why Study Soviet Science?"

12. Science and Values: Attempts at Historical Understanding

1. Stephen Toulmin, "How Can We Reconnect the Sciences with the Foundations of Ethics?" p. 47.
2. Max Black, "Scientific Neutrality," p. 58.
3. "Anthropology," *Encyclopedia Britannica* 1:973.
4. Stobaios, *Eclogarum Physicarum et Ethicarum*, A. Meineke, ed. (Lepizig: Teubner, 1860), 1:2, 3. I would like to thank Dr. Mark Pinson and Professor Čelica Milovanović-Barham for locating this passage and translating it from the Greek.
5. *Plutarch's Moralia* (Loeb Classical Library), vol. 9. Translation by E. L. Minar, F. H. Sandbach, and W. C. Helmbold. London: Heinemann, Cambridge: Harvard University Press, 1961), pp. 119, 121. I would like to thank Professor Čelica Milovanović-Barham and Dr. Mark Pinson for locating this passage for me.
6. A. Ia. Khinchin, "O vospitatel'nom effekte urokov matematiki." I am grateful to Dr. Mark Kuchment, Russian Research Center, Harvard University, for pointing out this passage to me.
7. I am grateful to Professor Ronald Giere of Indiana University for helpful discussions of this section.

BIBLIOGRAPHY

"An Act for the Development and Control of Atomic Energy," *United States Statutes at Large* (79th Cong., 2d Sess.), vol. 60, Washington, D.C., 1946.

Adams, Mark B. "The Founding of Population Genetics: Contributions of the Chetverikov School, 1924–1934." *Journal of the History of Biology* (1963), 1:23–39.

Adams, Mark B. "Towards a Synthesis: Population Concepts in Russian Biological Thought, 1925–1935." *Journal of the History of Biology* (1970), 3:107–29.

Akutsu, Tetsuzo. *Artificial Heart: Total Replacement and Partial Support.* Tokyo: Igaku Shoin, 1975.

Alexander, Samuel. *Space, Time, and Deity.* London: Macmillan, 1920.

Altman, Stuart R., ed. *Social Communication among Primates.* Chicago: University of Chicago Press, 1967.

Altner, Günter. *Weltanschauliche Hintergründe der Rassenlehre des Dritten Reiches: Zum Problem einer umfassenden Anthropologie.* Zurich: EVZ, 1968.

Ammon, Otto. *Die Gesellschaftsordnung und ihre natürlichen Grundlagen.* Jena: G. Fischer, 1895.

Arling, G. L. and H. F. Harlow. "Effects of Social Deprivation on Maternal Behavior of Rhesus Monkeys." In Thomas E. McGill, ed., *Readings in Animal Behavior.* 2d ed. New York: Holt, Rinehart, and Winston, 1973.

Baker, Keith Michael. *Condorcet: From Natural Philosophy to Social Mathematics.* Chicago: University of Chicago Press, 1975.

Ballantyne, J. W. *Manual of Antenatal Pathology and Hygiene.* Edinburgh: William Green, 1902.

Barker, Martin. "Racism: The New Inheritors." *Radical Philosophy* (1979), no. 21.

Barnett, S. A. *Instinct and Intelligence: Behavior of Animals and Man.* Englewood Cliffs, N.J.: Prentice-Hall, 1967.

Barthelemy-Madaule, Madeleine. *Bergson et Teilhard de Chardin.* Paris: Editions du Seuil, 1963.

"Bauman Amendment's Chances Down." *Science* (July 4, 1975), vol. 189.

Beach, Frank A. "The Snark was a Boojum." *The American Psychologist* (1950), 5:115–24.

Bennett, William and Joel Gurin. "Home Rule and the Gene." *Harvard Magazine* (October 1976) vol. 79, no. 2.

Berg, L. S. "Brachnost' rozhdaemost' i smertnost' v Leningrade za poslednie gody." *Priroda* (1924), nos. 7–12.

Bergson, H. "Controverse au sujet des temps fictifs et des temps réels dans la théorie d'Einstein: Réplique de M. André Metz; Deuxième réponse de M. Henri Bergson; Un dernier mot de M. A. Metz." *Revue de Philosophie* (July–August 1924), pp. 437–440.

Bergson, Henri. *Creative Evolution.* New York: Modern Library, 1944.

Bergson, H., "Les Temps fictifs et le Temps réel." *Revue de Philosophie* (May–June 1924), pp. 241–260.

Bergson, Henri. *The Two Sources of Morality and Religion.* R. Ashley Audra and Cloudesley Brereton, trans. New York: Henry Holt, 1935.

Bernstein, Jeremy. *Einstein.* New York: Viking Press, 1973.

von Bertalanffy, Ludwig. *Kritische Theorie der Formbildung.* Berlin: Gebrüder Borntraeger, 1928. English: *Modern Theories of Development,* J. H. Woodger, trans. Oxford: Oxford University Press, 1933.

von Bertalanffy, Ludwig. *Das biologische Weltbild.* Bern: A. Francke, 1949. English: *Problems of Life: An Evaluation of Modern Biological Thought.* New York: Wiley, 1952.

Beyerchen, Alan D. *Scientists Under Hitler: Politics and the Physics Community in the Third Reich.* New Haven: Yale University Press, 1977.

Black, Max. "Scientific Neutrality." *Encounter* (July 1978).

Black, Max. "Some Aversive Responses to a Would-Be Reinforcer." In Harvey Wheeler, ed., *Beyond the Punitive Society: Operant Conditioning, Social and Political Aspects* San Francisco: W. H. Freeman, 1973.

Bohr, N. "Analysis and Synthesis in Science." *International Encyclopedia of United Science*, vol. 1, no. 1. Chicago: University of Chicago Press, 1962.

Bohr, N. *Atomic Physics and Human Knowledge*. New York: Wiley, 1958.

Bohr, N. *Atomic Theory and the Description of Nature*. New York and Cambridge: Macmillan, 1934.

Bohr, N. "Atomic Theory and Mechanics." *Nature* (December 5, 1925), vol. 116.

Bohr, N. "Discussion with Einstein on Epistemological Problems in Atomic Physics." In P. A. Schilpp ed., *Albert Einstein: Philosopher-Scientist*, pp. 201–41. Evanston: Library of Living Philosophers, 1949.

Bohr, N. *Essays 1958–1962 on Atomic Physics and Human Knowledge*. New York: Interscience, 1963; Bungay: Richard Clay, 1963.

Bohr, N. "Light and Life." *Nature* (April 1, 1933), vol. 131.

Bohr, N. "Medical Research and Natural Philosophy." In *Papers on Medicine and the History of Medicine* (Acta Medica Scandinavica: Supplement 256), Copenhagen, 1952.

Bohr, N. "Natural Philosophy and Human Cultures." *Nature* (February 18, 1939), vol. 143.

Bohr, N. "On the Notions of Causality and Complementarity." *Science* (January 20, 1950), 3:312–19.

Bohr, N. "On the Notions of Causality and Complementarity." *Dialectica* (1948), vol. 2.

Bohr, N. "Physical Science and the Study of Religions." In *Studia Orientalia Ioanni Pedersen septuagenario* A.D. VII ID. NOV. ANNO MCMLIII. Einar Munksgaard, 1953.

Bohr, N. "The Quantum Postulate and the Recent Development of Atomic Theory." *Nature* (Supplement April 14, 1928), vol. 121.

Bohr, N. "Science and the Unity of Knowledge." In Lewis Leary, ed. *The Unity of Knowledge*. Garden City, N.Y.: Doubleday, 1955.

Bohr, N. "Über der Wirkung von Atomen bei Stössen." *Zeitschrift für Physik* (1925), no. 34.

Bohr, N. "Über die Anwendung der Quantentheorie auf den Atombau." *Zeitschrift für Physik* (1923), no. 13, pp. 142–57.

Bohr, N., H. A. Kramers, and J. C. Slater. "The Quantum Theory of Radiation." *Philosophical Magazine* (May 1924), 74:785–802.

Boni, Nell, Monique Russ, and Dan H. Laurence, eds. *A Bibliographical Checklist and Index to the Published Writings of Albert Einstein*. Paterson, N. J.: Pageant Books, 1960.

Brandenburg, Hans. *Goethe und Wir.* Weimar: Hermann Böhaus, 1937.

Bregman, David. *Mechanical Support of the Failing Heart.* Chicago: Year Book Medical Publishers, 1976.

de Broglie, Louis. "The Concepts of Contemporary Physics and Bergson's Ideas on Time and Motion." In P. A. Y. Gunter, ed., *Bergson and the Evolution of Physics.* Knoxville: University of Tennessee Press, 1969.

Bunak, V. V. "Evgenicheskie zametki." *Russkii evgenicheskii zhurnal* (1924) vol. 2.

Bunak, V. V. "Materialy dlia sravnitel'noi kharakteristiki sanitarnoi konstitutsii evreev." *Russkii evgenicheskii zhurnal* (1924), 2(2–3):142–52.

Bunak, V. V. "Novye dannye k voprosu o voine, kak biologicheskom faktore." *Russkii evgenicheskii zhurnal* (1923), vol. 1, no. 2.

Čapek, Milič. *Bergson and Modern Physics: A Reinterpretation and Reevaluation.* Boston Studies in the Philosophy of Science, vol. 7. New York: D. Reidel/Humanities Press, 1971.

Caplan, Arthur L., ed. *The Sociobiology Debate: Readings on the Ethical and Scientific Issues Concerning Sociobiology.* New York: Harper and Row, 1978.

Carossa, Hans. *Wirkungen Goethes in der Gegenwart.* Leipzig: Isel, 1938.

Challaye, Felicien. "Bergson vu par les Soviets." *Preuves* (1954), vol. 4, no. 44.

Chanyshev, A. N. *Filosofiia Anri Bergsona.* Moscow, 1960.

Chevalier, Jacques. *Henri Bergson.* Lilian A. Clare, trans. Freeport, N.Y.: Books for Libraries Press, 1970.

Chomsky, Noam. "The Case Against B. F. Skinner." *The New York Review of Books,* December 30, 1971.

Chomsky, Noam. *Reflections on Language.* New York: Pantheon Books, 1975.

Chulkov, N. P. "Genealogiia dekabristov Murav'evykh." *Russkii evgenicheskii zhurnal* (1927), 5(1):21–24.

Chulkov, N. "Rod grafov Tolstykh." *Russkii evgenicheskii zhurnal* (1923), vol. 1, no. 3–4.

Clagett, Marshall, ed. and comm. *Nicole Oresme and the Medieval Geometry of Qualities and Motions.* Madison: University of Wisconsin Press, 1968.

Clark, Ronald W. *Einstein: The Life and Times.* New York and Cleveland: World, 1971.

Clark, Ronald W. *The Huxleys*. London: Heinemann, 1968.

Cohen, Stanley F. "The Manipulation of Genes." *Scientific American* (July 1975), vol. 233.

Colton, Clark K. *A Review of the Development and Performance of Hemodialyzers*. Bethesda, Md.: National Institute of Arthritis and Metabolic Diseases, 1966.

Conrad-Martius, Hedwig. *Utopien der Menschenzüchtung: Der Sozialdarwinismus und seine Fölgen*. Munich: Kösel, 1955.

Copernicus, Nicolaus. *On the Revolution of the Heavenly Spheres*. A. M. Duncan, trans. New York: Barnes and Noble, 1976.

Crick, Francis. *Of Molecules and Men*. Seattle and London; University of Washington Press, 1966.

Crombie, A. C. *Medieval and Early Modern Science*, vol. 1. Cambridge: Harvard University Press, 1967.

Crowther, J. G. *British Scientists of the Twentieth Century*. London: Routledge and Kegan Paul, 1952.

Cuenot, Claude. *Teilhard de Chardin*. Baltimore: Helican, 1965.

Danilevskii, N. *Darvinizm, kriticheskoe issledovanie*. St. Petersburg, 1889.

Darwin, Charles. *The Descent of Man and Selection in Relation to Sex*. New York: D. Appleton, 1896.

Dedrick, Robert L., Kenneth B. Bischoff, and Edward F. Leonard, eds. *The Artificial Kidney*. New York: American Institute of Chemical Engineers, 1968.

De Vore, Irven. "Mother-Infant Relations in Free-Ranging Baboons." In Thomas E. McGill, ed., *Readings in Animal Behavior*. New York: Holt, Rinehart, and Winston, 1965.

De Vore, Irven and K. R. L. Hall. "Baboon Ecology." In Irven De Vore, ed., *Primate Behavior: Field Studies of Monkeys and Apes*. New York: Holt, Rinehart, and Winston, 1965.

De Vore, Irven, ed. *Primate Behavior: Field Studies of Monkeys and Apes*. New York: Holt, Rinehart, and Winston, 1965.

Dingle, Herbert. *The Sources of Eddington's Philosophy*. Cambridge: Cambridge University Press, 1954.

Djerassi, Carl. "Probabilities and Practicalities." *Bulletin of the Atomic Scientists* (December 1972), vol. 28, no. 10.

Dobzhansky, Theodosius. "The Teilhardian Synthesis." In his *The Biology of Ultimate Concern*, pp. 108–37. London: Rapp and Whiting, 1969.

Douglas, A. Vibert. *The Life of Arthur Stanley Eddington*. London: Thomas Nelson, 1956.

Draghiscesco, M. D. "L'influence de la philosophie de M. Bergson." *Le Mouvement Socialiste* (November 1911).

Dubinin, N. P. *Vechnoe dvizhenie*. Moscow, 1973.

Eddington, Sir Arthur. "The Decline of Determinism." *Annual Report of the Board of Regents of the Smithsonian Institution*. Washington, D.C., 1932. Reprinted from the *Mathematical Gazette* (May 1932), 16(218):141–57.

Eddington, A. S. *The Nature of the Physical World*. Cambridge: Cambridge University Press, 1928.

Eddington, A. S. *New Pathways in Science*. New York: Macmillan; Cambridge: Cambridge University Press, 1935.

Eddington, A. S. *The Philosophy of Physical Science*. Cambridge: Cambridge University Press, 1939.

Eddington, A. S. *Science and Religion*. London: Friends Home Service Committee, 1931.

Eddington, A. S. *Science and the Unseen World*. New York: Macmillan, 1929.

Edman, Irwin. "Foreword." In H. Bergson, *Creative Evolution*. New York: Modern Library, Random House, 1944.

Einstein, A. "Bemerkung zu der Arbeit von A. Friedmann, 'Über die Krummung des Raumes.'" *Zeitschrfit für Physik* (1922), 11.

Einstein, Albert. *Collected Works, 1901–1956*. New York: Readex Microprint, 1960.

Einstein, A. "Dialog über Einwände gegen die Relativitätstheorie." *Die Naturwissenschaften* (November 20, 1918).

Einstein, A. "Epilogue: A Socratic Dialogue." In Max Planck, *Where Is Science Going*? New York: Norton, 1932.

Einstein, A. "Folgerungen aus den Capillaritätserscheinungen." *Annalen der Physik*, 1901, ser. 4, vol. 4.

Einstein, Albert. "The Laws of Science and the Laws of Ethics." In Philipp Frank, *Relativity—A Richer Truth*. Boston: Beacon Press, 1950.

Einstein, A. "Notiz zu der Arbeit von A. Friedmann 'Uber die Krümmung des Raumes.'" *Zeitschrift für Physik* (1923), 16.

Einstein, Albert. "Religion and Science: Irreconcilable?" *The Christian Register* (June 1948).

Einstein, A. "Remarks on Bertrand Russell's Theory of Knowledge." In Paul A. Schilpp, ed., *The Philosophy of Bertrand Russell*. Evanston: Library of Living Philosophers, 1944.

Einstein, Albert. "Science and God: A Dialog." *Forum* (June 1930), vol. 83.

Einstein, Albert. "Science and Religion." In *First Conference on Science, Philosophy, and Religion.* New York, 1941.
Einstein, A. *Sobranie nauchnykh trudov*, vols. 1–4. Moscow, 1965–67.
Einstein, Albert. *The World as I See It.* New York: Civici Friede, 1934.
Einstein, Albert. "Why Socialism?" *Monthly Review* (May 1949), vol. 1.
"Encyclical Letter of His Holiness Pius XI by Divine Providence Pope. On Christian Marriage." National Catholic Welfare Conference, Washington, D.C., 1931.
Eisenberg, Leon. "The 'Human' Nature of Human Nature." In Arthur L. Caplan, ed., *The Sociobiology Debate.* New York: Harper and Row, 1978. Originally published in *Science* (April 14, 1972), 176:123–28.
Etkin, William. *Social Behavior from Fish to Man.* Chicago: Phoenix Books, University of Chicago Press, 1967.
Evans, Mary Alice and Howard Ensign Evans, *William Morton Wheeler: Biologist.* Cambridge: Harvard University Press, 1970.
Evans, Richard I. *Konrad Lorenz: The Man and His Ideas.* New York: Harcourt Brace Jovanovich, 1975.
"Evgenicheskaia sterilizatsiia v Germanii." *Russkii evgenicheskii zhurnal* (1925), vol. 3. no. 1.
"Evgenicheskie zametki." *Russkii evgenicheskii zhurnal* (1924), vol. 2, no. 1.
"Evgenika." *Bol'shaia Sovetskaia Entsiklopediia* (1931), vol. 23, cols. 812–819.
Fairweather, D. V. I. "Technique and Safety of Amniocentesis." In his edited work, *Amniotic Fluid: Research and Clinical Application.* Amsterdam: Excerpta Medica, 1973.
Farges, Albert. "L'erreur fondamentale de la philosophie nouvelle." *Revue thomist* (March–June 1909), vol. 17, pp. 182–97.
Ferris, Paul. "Nikolaas Tinbergen Seagull," *New York Times Magazine*, April 7, 1974.
Feyerabend, Paul K. "Dialectical Materialism and Quantum Theory." *Slavic Review* (September 4, 1966), vol. 25, no. 3.
Fields, Cheryl M. "Who Should Control Recombinant DNA?" *The Chronicle of Higher Education* (March 21, 1977), 14(4):1, 4–5.
Filipchenko, Iu. A. "Evgenika v shkole." *Russkii evgenicheskii zhurnal* (1925), vol. 3, no. 2.
Filipchenko, Iu. A. "Nashi vydaiushchiesia uchenye." *Izvestiia buro po evgenike* (1921), vol. 1.

Filipchenko, Iu. A. "Die russische rassenhygienische Literatur, 1921–1925." *Archiv für Rassen- und Gesellschaftsbiologie* (1925), vol. 3.

Finley, Sara C. "Genetic Counseling." *Pediatric Annals* (June 1978), vol. 7.

Fleming, Donald. "On Living in a Biological Revolution." *Atlantic Monthly* (February 1969).

Fleming, Donald. "Introduction" in Jacques Loeb, *The Mechanistic Conception of Life*. Cambridge: Harvard University Press, 1964.

Florinskii, V. M. *Usovershenstvovanie i vyrozhdenie chelovescheskogo roda*. St. Petersburg, 1866.

Fock, V. A. "Comments." *Slavic Review* (September 1966), vol. 25, no. 3.

Fock, V. A. "K diskussii po voprosam fiziki." *Pod znamenem marksizma* (1938), no. 1.

Fock, V. A. "Les principes mécaniques de Galilée et la theorie d'Einstein." *Atti del convegno sulla relatività generale: problemi dell'energie e onde gravitazionali*. Florence, 1952.

Fock, V. A. "Nekotorye primeneniia idei Lobachevskogo v mekhanike i fizike." In A. P. Kotel'nikov and F. A. Fock, *Nekotorye primeneniia idei Lobachevskogo v mekhanike i fizike*. Moscow-Leningrad, 1950.

Fock, V. A. "Ob interpretatsii kvantovoi mekhaniki." In P. N. Fedoseev et al., eds., *Filosofskie problemy sovremennogo estestvoznaniia*. Moscow, 1959.

Fock, V. A. "Osnovnye zakony fiziki v svete dialekticheskogo materializma." *Vestnik Leningradskogo Universiteta* (1949), no. 4.

Fock, V. A. "Otnositel'nosti teoriia." *Bol'shaia Sovetskaia Entsiklopediia*, vol. 31, pp. 405–11. 2d ed., 1955.

Fock, V. A. "Poniatiia odnorodnosti, kovariantnosti i otnositel'nosti." *Voprosy filosofii* (1955), no. 4.

Fock, V. A. "Protiv nevezhestvennoi kritiki sovremennykh fizicheskikh teorii." *Voprosy filosofii* (1953), no. 1, pp. 168–74.

Fock, V. A. "Sistema Kopernika i sistema Ptolemeia v svete obshchei teorii otnositel'nosti." In *Nikolai Kopernik*. Moscow, 1947.

Fock, V. A. "Sovremennaia teoriia prostranstva i vremeni." *Priroda* (1953), no. 12.

Fock, V. A. *The Theory of Space, Time, and Gravitation*. New York: Pergamon Press, 1959.

Fock, V. A. "Über philosophische Fragen der modernen Physik." *Deutsche Zeitschrift für Philosophie* (1955), no. 6.

Fock, V. A. "Velikii fizik sovremennosti" (K 60-letiiu A. Einshteina) *Leningradskii universitet* (March 23, 1939), no. 13.

Fok, Vladimir Aleksandrovich. Materialy k biobibliografii uchenykh SSSR (seriia fiziki, vyp. 7, AN SSSR). Moscow, 1956.

Forman, Paul. "Weimar Culture, Causality, and Quantum Theory, 1918–1927: Adaptation by German Physicists and Mathematicians to a Hostile Intellectual Environment." *Historical Studies in the Physical Sciences*, 3:1–115. Philadelphia: University of Pennsylvania Press, 1971.

Fox, Sidney W., ed. *The Origins of Prebiological Systems and of Their Molecular Matrices*. New York: Academic Press, 1965.

Frank, Philipp. *Einstein: His Life and Times*. New York: Knopf, 1947.

Friedmann, A. A. *Mir kak prostranstvo i vremiia*. Moscow, 1965 (reprinted from the 1923 edition).

Friedmann, A. A. "Über die Krümmung des Raumes." *Zeitschrift für Physik* (1922), no. 6, pp. 377–87.

Friedmann, Theodore and Richard Roblin. "Gene Therapy for Human Genetic Disease?" *Science* (March 3, 1972), vol. 175.

Fuchs, R. F. Review of A. Basler, *Einführung in die Rassen-und Gesellschafts-Physiologie*. In *Die Gesellschaft* (1926), 3.

Fuchs, Fritz and Lars L. Cederqvist. "Prenatal Diagnosis Based upon Amniotic Fluid Cells." In D. L. I. Fairweather, ed., *Amniotic Fluid: Research and Clinical Application*. Amsterdam: Excerpta Medica, 1973.

Fulton, James Street. "Bergson's Religious Interpretation of Evolution." *Rice Institute Pamphlet* (October 1956), vol. 43, no. 3.

Gallagher, Idella J. "Bergson on Closed and Open Morality." *New Scholasticism* (Winter 1968), vol. 42, no. 3.

Garaudy, Roger. *From Anathema to Dialogue: The Challenge of Marxist-Christian Cooperation*. London: Collins, 1967.

Gasman, Daniel. *The Scientific Origins of National Socialism: Social Darwinism in Ernst Haeckel and the German Monist League*. New York: American Elsevier, 1971.

Gelegentliches von Albert Einstein, Zum fünfzigsten Geburtstag 14 März 1929, dargebracht von der Soncino Gesellschaft der Freunde des judischen Buches zu Berlin. Berlin, 1929.

Giese, Annemarie. "Schriftverzeichnis von Werner Heisenberg," *Zeitschrift für Naturforschung* (1976), 23a:510–16.

Gillispie, Charles Coulston. *Genesis and Geology*. New York: Harper, 1959.

Gillouin, M. Rene. "L'influence de la philosophie de M. Bergson." *Le Mouvement Socialiste* (February 1912).

Gilson, Etienne. *The Philosopher and Theology*. Cécile Gilson, trans. New York: Random House, 1962.

Glick, Thomas, ed. *The Comparative Reception of Darwinism*. Austin: University of Texas Press, 1974.

Goodall, Jane. "Chimpanzees of the Gombe Stream Reserve." In Irven De Vore, ed., *Primate Behavior: Field Studies of Monkeys and Apes*. New York: Holt, Rinehart, and Winston, 1965.

Goodfield, June. *Playing God: Genetic Engineering and the Manipulation of Life*. New York: Random House, 1977.

Gorbunov, A. V. "Vliianie mirovoi voiny na dvizhenie naseleniia Evropy." *Russkii evgenicheskii zhurnal* (1922), vol. 1, no. 1.

Gould, Stephen Jay. "Biological Potential vs. Biological Determinism." *Natural History Magazine* (May 1976), 85:12–22.

Graham, Loren. "The Multiple Connections Between Science and Ethics." *Hastings Center Report* (June 1979), 9:35–40.

Graham, Loren R. "Quantum Mechanics and Dialectical Materialism" and "Reply." In *Slavic Review* (September 4, 1966), vol. 25, no. 3.

Graham, Loren R. *Science and Philosophy in the Soviet Union*. New York: Knopf, 1972.

Graham, Loren R. "Why Study Soviet Science? The Example of Genetic Engineering." In Linda L. Lubrano and Susan Solomon, eds., *The Social Context of Soviet Science*, pp. 204–40. Boulder, Colo.: Westview Press, 1979.

Gruber, Max. *Ursache und Bekämpfung des Geburtenrückgangs im deutschen Reiche*. Munich: J. F. Lehmann, 1914.

Gunter, P. A. Y., ed. *Bergson and the Evolution of Physics*. Knoxville: University of Tennessee Press, 1969.

Gurvich, K. "Ukazatel' literatury po voprosam evgeniki, nasledstvennosti i selektsii i sopredel'nykh oblastei, opublikovannoi na russkom iazike do I/1 1928 g." *Russkii evgenicheskii zhurnal* (1928), 6(2–3):121–43.

Haberer, Joseph. *Politics and the Community of Science*. New York: Van Nostrand Reinhold, 1969.

Hall, Everett W. *Modern Science and Human Values: A Study in the History of Ideas*. New York: Dell, 1956.

Hall, K. R. L. and Irven De Vore. "Baboon Social Behavior." In Irven De Vore, ed., *Primate Behavior: Field Studies of Monkeys and Apes*. New York: Holt, Rinehart, and Winston, 1965.

Hanna, Thomas, ed. *The Bergsonian Heritage*. New York: Columbia University Press, 1962.
Haraway, Donna. "The Biological Enterprise: Sex, Mind, and Profit from Human Engineering to Sociobiology." *Radical History Review* (Spring/Summer 1979), vol. 20.
Harlow, Harry F. "The Heterosexual Affectional System in Monkeys." *American Psychologist* (1962), 17:1–9.
Harlow, H. F., M. K. Harlow, R. O. Dodsworth, and G. L. Arling. "Maternal Behavior of Rhesus Monkeys Deprived of Mothering and Peer Associations in Infancy." *Proceedings of the American Philosophical Society*, (1966), 110:58–66.
Heberer, Gerhard and Franz Schwanitz, eds. *Hundert Jahre Evolutionsforschung: Das wissenschaftliche Vermächtnis Charles Darwin*. Stuttgart: G. Fischer, 1960.
Heisenberg, W. "Die Entwicklung der Quantentheorie, 1918–1928." *Die Naturwissenschaften* (1929), vol. 17.
Heisenberg, W. "Die gegenwärtigen Grundprobleme der Atomphysik." *Wandlungen in den Grundlagen der Naturwissenschaft*. 10th ed. Stuttgart: S. Hirzel, 1973.
Heisenberg, W. "Goethe és Newton szinelmélete a modern fizika megvilágitásában." *Matematikai és fizikai lapok* (1941), vol. 48.
Heisenberg, Werner. "Ideas of the Natural Philosophy of Ancient Times in Modern Physics." *Philosophic Problems of Nuclear Science*. London: Faber and Faber, 1952.
Heisenberg, W. "Kausalgesetz und Quantenmechanik." *Annalen der Philosophie und philosophischen Kritik* (1931), vol. 2, pp. 172–82.
Heisenberg, W. *Das Naturbild der heutigen Physik*. Hamburg: Rowohlt, 1955. English: *The Physicist's Conception of Nature*. New York: Harcourt, Brace, 1958.
Heisenberg, W. "Das Naturbild Goethes und die technisch-naturwissenschaftliche Welt." In Werner Heisenberg, *Zwei Vorträge*. Alexander von Humboldt-Stiftung, Mitteilungen Nr. 13 (June 1967).
Heisenberg, W. "On the Unity of the Scientific Outlook on Nature." *Philosophic Problems of Nuclear Science*. London: Faber and Faber, 1952.
Heisenberg, W., *Philosophic Problems of Nuclear Science*. London: Faber and Faber, 1952; New York: Pantheon, 1952.
Heisenberg, W. *The Physical Principles of the Quantum Theory*. Chicago: University of Chicago Press, 1930.
Heisenberg, W. *Physics and Beyond: Encounters and Conversations*.

Arnold J. Pomerans, tr. World Perspective Series, vol. 42. Ruth Nanda Anshen, ed. New York: Harper & Row, 1971.

Heisenberg, W. *Physik und Philosophie*. Stuttgart: Hirzel, 1959. English: *Physics and Philosophy*. New York: Harper and Row, 1958.

Heisenberg, W. "Die Plancksche Entdeckung und die philosophischen Grundfragen der Atomlehre." *Wandlungen in den Grundlagen der Naturwissenschaft*. Stuttgart, 1973.

Heisenberg, W. "Quantenmechanik." *Die Naturwissenschaften* (1926), vol. 14, pp. 989–94.

Heisenberg, W. *Schritte über Grenzen*. Munich: Piper, 1971. English: *Across the Frontiers*. New York: Harper and Row, 1974.

Heisenberg, W. "The Teachings of Goethe and Newton on Colour in the Light of Modern Physics." *Philosophic Problems of Nuclear Science*. London: Faber and Faber, 1952.

Heisenberg, W. *Der Teil und das Ganze: Gespräche in Umkreis der Atomphysik*. Munich: Piper, 1969. English: *Physics and Beyond*. New York: Harper and Row, 1971.

Heisenberg, W. "Über den anschaulichen Inhalt der quantentheoretischen Kinematik und Mechanik. *Zeitschrift für Physik* (1927), no. 43, pp. 172–98.

Heisenberg, W. "Über quantentheoretische Umdeutung kinematischer und mechanischer Beziehungen." *Zeitschrift für Physik* (1925), no. 33, pp. 879–983.

Heisenberg, W. "Wandlungen der Grundlagen der exakten Naturwissenschaft in der jüngsten Zeit." *Die Naturwissenschaften* (1934), vol. 22, pp. 669–75.

Heisenberg, W. *Wandlungen in den Grundlagen der Naturwissenschaft*. Leipzig: S. Hirzel, 1935.

Heisenberg, W. "Zum Artikel 'Deutsche und jüdische Physik.'" *Völkischer Beobachter*, Berlin (February 28, 1936).

Heisenberg, W. "Zur Geschichte der physikalischen Naturerklärung." *Berichte der sächsischen Akademie der Wissenschaften* (1933), vol. 85, pp. 29–40.

Hermann, Armin. "Johannes Stark." *The Dictionary of Scientific Biography*, vol. 12, pp. 613–16. New York: Scribner, 1975.

Holt, Niles R. "Monists and Nazis: A Question of Scientific Responsibility." *Hastings Center Report* (1975), vol. 5.

Holton, Gerald. "Mach, Einstein, and the Search for Reality." In his *Thematic Origins of Scientific Thought: Kepler to Einstein*, pp. 219–59. Cambridge: Harvard University Press, 1973.

"Human Genetic Engineering." *Medical World News* (May 11, 1973), pp. 45–57.

Hutton, Richard. *Bio-Revolution: DNA and the Ethics of Man-Made Life*. New York: New American Library, 1978.

Huxley, Sir Julian. "Introduction." In Pierre Teilhard de Chardin, *The Phenomenon of Man*, pp. 11–28. New York: Harper Torchbooks, 1961.

Huxley, T. H. In T. H. and Julian Huxley, *Evolution and Ethics, 1893–1943*. London: Pilot Press, 1947.

Iltis, Hugo. "Rassenwissenschaft und Rassenwahr." *Die Gesellschaft* (1927), 4:97–114.

Iltis, Hugo. "Der Schädelindex in Wissenschaft und Politik." *Die Gesellschaft* (1931), 8:549–62.

Infeld, Leopold. *Why I Left Canada: Reflections on Science and Politics*. Helen Infeld, trans.; edited and introduced by Lewis Pyenson, Montreal and London: McGill-Queen's University Press, 1978.

Isaye, G., S. J. "Avertissement du 30 juin 1962 concernant les oeuvres du P. Teilhard de Chardin." *Nouvelle Revue Theologique* (September-October 1962), 84(8):866–69.

"Iz otcheta o deiatel'nosti Russkogo Evgenicheskogo Obshchestva za 1923 g." *Russkii evgenicheskii zhurnal* (1924), vol. 3, no. 1.

Jacks, L. P. *Sir Arthur Eddington: Man of Science and Mystic*. Cambridge: Cambridge University Press, 1949.

Jacob, M. C. "The Church and the Formulation of the Newtonian World-View." *Journal of European Studies* (1971), vol. 1.

Jammer, Max. *The Conceptual Development of Quantum Mechanics*. New York: McGraw-Hill, 1966.

Jay, Phyllis. "The Common Langur of North India." In Irven De Vore, ed., *Primate Behavior: Field Studies of Monkeys and Apes*. New York: Holt, Rinehart, and Winston, 1965.

Jay, Phyllis. "Primate Field Studies and Human Evolution." In Phyllis C. Jay, ed., *Primates: Studies in Adaptation and Variability*. New York: Holt, Rinehart, and Winston, 1969.

Jaynes, Julian. "The Historical Origins of 'Ethology' and 'Comparative Psychology.'" *Animal Behaviour* (November 1969), vol. 17, no. 4a.

Johns Hopkins University: Charter, Extracts of Will, Officers, and By-Laws. Baltimore: John Murphy, 1874.

Johnson, Martin. *Time and Universe for Scientific Conscience*. Cambridge: Cambridge University Press, 1952.

Jonas, Hans. "Freedom of Scientific Inquiry and the Public Interest," *Hastings Center Report* (August 1976), vol. 6.

Joravsky, David. *The Lysenko Affair*. Cambridge, Mass.: Harvard University Press, 1970.

Joussain, Andre. "Bergsonisme et marxisme." *Ecrits de Paris* (April 1956), no. 137.

Kalikow, Theo J. "Konrad Lorenz's 'Brown Past': A Reply to Alec Nisbett," *Journal of the History of the Behavioral Sciences* (1978), vol. 14.

Kantorowicz, Miron. Review of Ernst Neumann, *Individual- Rassen- und Volkshygiene*. In *Die Gesellschaft* (1931), vol. 8.

Kantorowicz-Kroll, M. Review of Robert Sommer, *Familienforschung, Vererbungs- und Rassenlehre*. In *Die Gesellschaft* (1928), 5:92–94.

Kautsky, K. Review of Oda Olberg, *Die Entartung in ihrer Kulturbedingtheit: Bemerkungen und Anregungen*. In *Die Gesellschaft* (1926), vol. 3.

Kautsky, Karl. Review of A. Grotjähn, *Die Hygiene der menschlichen Fortpflanzung: Versuch einer praktischen Eugenik*. In *Die Gesellschaft* (1927), vol. 4.

Kautsky, Karl. "Unterbrechung der Schwangerschaft: Individualistische oder sozialistische Lösung?" *Die Gesellschaft*. (1925), 2:354–66.

Khinchin, A. Ia. "O vospitatel'nom effekte urokov matematiki." *Matematicheskoe prosveshchenie* (1961), no. 6.

Kilmister, C. W. *Men of Physics: Sir Arthur Eddington*. Oxford/New York: Pergamon Press, 1966.

Kilmister, C. W. and B. O. J. Topper, *Eddington's Statistical Theory*. Oxford: Oxford University Press, 1962.

Kindermann, Heinz. *Das Goethebild des 20 Jahrhunderts*. Darmstadt: Wissenschaftliche Buchgesellschaft, 1966.

Kinkade, Kathleen. "A Walden Two Experiment." *Psychology Today* (January 1973 and February 1973).

Klickstein, Herbert S. "A Cumulative Review of Bibliographies of the Published Writings of Albert Einstein." *Journal of the Albert Einstein Medical Center* (July 1962).

Kline, George. "Darwinism and the Russian Orthodox Church." In Ernest J. Simmons, ed., *Continuity and Change in Russian and Soviet Thought*. Cambridge, Mass.: Harvard University Press, 1955.

Kobzarev, I. Iu. "Otnositel'nosti teoriia." *Bol'shaia Sovetskaia Entsiklopediia*, vol. 18, pp. 623–28. 3d ed., 1974.

Koch, Franz. *Goethe und die Juden*. Hamburg: Hanseatische Verlagsanstalt, 1937.

Kolff, W. J. *Artificial Organs*. New York: Wiley, 1976.

Kol'tsov, N. K. "Evgenicheskie s"ezdy v Milane v Sentiabre 1924 goda." *Russkii evgenicheskii zhurnal* (1925), 3(1):73–77.

Kol'tsov, N. K. Review of E. Baur, E. Fisher, and F. Lenz, *Grundriss der menschlichen Erblichkeitslehre und Rassenhygiene*. In *Russkii evgenicheskii zhurnal* (1924), vol. 2, no. 2–3.

Kol'tsov, N. K. "Noveishie popytki dokazat' nasledstvennost' blagopriobretennykh priznakov." *Russkii evgenicheskii zhurnal* (1924), 2(1):159–67.

Kol'tsov, N. K. "Die rassenhygienische Bewegung in Russland." *Archiv für Rassen- und Gesellschaftsbiologie* (1925), vol. 17.

Kol'tsov, N. K. "Rodoslovnye nashikh vydvizhentsev." *Russkii evgenicheskii zhurnal* (1926), 4(3–4):103–43.

Kol'tsov, N. K. "Vliianie kul'tury na otbor v chelovechestve," *Russkii evgenicheskii zhurnal* (1924), 2(1):3–19.

Kuhn, Thomas. *Structure of Scientific Revolutions*. Chicago: University of Chicago Press, 1962.

Kuznetsov, B. *Einstein*. Moscow, 1965.

Lederberg, Joshua. "DNA Splicing: Will Fear Rob Us of Its Benefits?" *Prism* (November 1975), vol. 3.

Legendre, M. Maurice. "L'influence de la philosophie de M. Bergson." *Le Mouvement Socialiste* (July-August 1911).

Lenin, V. I. *Materialism and Empirio-Criticism: Critical Comments on a Reactionary Philosophy*. Moscow, 1952.

Lenz, Fritz. "Die Stellung des Nationalsozialismus zur Rassenhygiene." *Archiv für Rassen- und Gesellschaftsbiologie* (1931), vol. 23.

Lenz, Fritz. "Der Fall Kammerer und seine Umfilmung durch Lunatscharsky." *Archiv für Rassen- und Gesellschaftsbiologie* (1929), 21:311–18.

Lenz, Fritz. "Notizen: Günthurs Berufung nach Jena." *Archiv für Rassen- und Gesellschaftsbiologie* (1931), vol. 23.

Lenz, Fritz. Review of L. F. Claus, *Die nordische Seele*. Review of Alfons Fischer, *Grundriss der sozialen Hygiene*. Review of Kurt Gerlach, *Fernstenliebe*. In *Archiv für Rassen- und Gesellschaftsbiologie* (1925), vol. 6.

Lenz, Fritz. Review of Alfred Grotjähn, *Leitsätze zur sozialen und generativen Hygiene.* In *Archiv für Rassen- und Gesellschaftsbiologie* (1923), vol. 15.

Lenz, Fritz. Review of Friederich Hertz, *Rasse und Kultur.* In *Archiv für Rassen- und Gesellschaftsbiologie* (1926), vol. 18.

Lenz, Fritz. Review of Hermann Paull, *Wir und das kommende Geschlecht.* In *Archiv für Rassen- und Gesellschaftsbiologie* (1924), vol. 15.

Lenz, Fritz. Review of W. Schallmayer, "Einführung in die Rassehygiene," in W. Weichardt, ed., *Fortsetzung des Jahresberichts über die Ergebnisse der Immunitätsforschung.* In *Archiv für Rassen- und Gesellschaftsbiologie* (1918), vol. 13.

Lenz, Fritz. Review of Wilhelm Schmidt, *Rasse und Volk.* In *Archiv für Rassen- und Gesellschaftsbiologie* (1928), vol. 21.

Lepin, T. K., Ia. Ia. Luss, and Iu. A. Filipchenko. "Deistvitel'nye chleny byvsh, imperat., nyne Rossiiskoi, Akademii za poslednie 80 l. (1864–1924 gg.)." *Izvestiia buro po evgenike* (1925), vol. 3.

Leppman, Wolfgang. *The German Image of Goethe.* Oxford: Clarendon Press, 1961.

Le Roy, Edouard. *Une Philosophie Nouvelle: Henri Bergson.* Paris: Librairie Felix Alcan, 1912.

Levien, Max. "Stimmen aus dem deutschen Urwalde (Zwei neue Apostel des Rassenhasses)." *Unter dem Banner des Marxismus* (1928), no. 2, pp. 150–95.

Lippmann, Walter. "Most Dangerous Man in the World." *Saturday Review of Literature* (August 23, 1947), vol. 30, no. 34.

Liubimov, S. V. "Predki grafa S. Iu. Vitte." *Russkii evgenicheskii zhurnal* (1928), vol. 6, no. 4.

Liublinskii, P. I. "Sovremennoe sostoianie evgenicheskogo dvizheniia." *Russkii evgenicheskii zhurnal* (1926), vol. 4, no. 2.

Lockard, Robert B. "Reflections on the Fall of Comparative Psychology: Is There a Message for Us All?" *American Psychologist* (February 1971), 26(2):168–79.

Loeb, J. "Assimilation and Heredity." *Monist* (July 1898).

Loeb, J. *The Dynamics of Living Matter.* New York: Columbia University Press, 1906.

Loeb, J. *The Mechanistic Conception of Life.* Chicago: University of Chicago Press, 1912.

Loeb, J. *The Mechanistic Conception of Life.* Donald Fleming, ed. Cambridge: Harvard University Press, 1964.

Loeb, J. "La nature chimique de la vie." *Revue philosophique* (1921), no. 11–12.

Loeb, J. *The Organism as a Whole: From a Physicochemical Viewpoint.* New York and London: Putnam, 1916.

Loeb, J. "Die Sehstörungen nach Verletzung der Grosshirnrinde." *Pflüger's Archiv für die gesamte Physiologie des Menschen und der Tiere* (1884), 34:67–172.

Lorenz, K. *Civilized Man's Eight Deadly Sins.* New York: Harcourt Brace Jovanovich, 1974.

Lorenz, K. "Betrachtungen über das Erkennen der arteigenen Triebhandlungen bei Vögeln." *Journal für Ornithologie* (1932), 80:50–98.

Lorenz, K. "Companions as Factors in the Bird's Environment." In *Studies in Animal and Human Behavior,* vol. 1. Cambridge: Harvard University Press, 1970.

Lorenz, Konrad. "Durch Domestikation verursachte Störungen arteigenen Verhaltens." *Zeitschrift für angewandte Psychologie und Charackterkunde* (1940), vol. 59.

Lorenz, K. "Instinctive Behavior Patterns in Birds," in *Studies in Animal and Human Behavior,* vol. 1. Cambridge: Harvard University Press, 1970.

Lorenz, Konrad. *On Aggression.* New York: Harcourt, Brace, and World, 1966.

Lorenz, Konrad. "Systematik und Entwicklungsgedanke im Unterricht." *Der Biologe* (January-February, 1940), vol. 9.

Longmore, Donald. *Spare-Part Surgery: The Surgical Practice of the Future.* Garden City, N.Y.: Doubleday, 1968.

Lovejoy, Arthur Oncken. "The Practical Tendencies of Bergsonism." *International Journal of Ethics* (July 1913), vol. 53, no. 4.

Lukas, Mary and Ellen Lukas. *Teilhard.* Garden City, N.Y.: Doubleday, 1977.

Maritain, J. *Bergsonian Philosophy and Thomism.* M. L. Anderson, trans. New York: Philosophical Library, 1955.

Maritain, J. *Bergsonian Philosophy and Thomism.* New York: Greenwood Press, 1968.

"Materialen zur Anthropologie der Ukraine." *Archiv für Rassen- und Gesellschaftsbiologie* (1931), vol. 23.

Medawar, P. B. "Critical Notice." *Mind* (January 1961), vol. 70, no. 277.

Metz, A. "Le Temps d'Einstein et la Philosophie." *Revue de Philosophie* (January–February 1924), pp. 56–88.

Metz, Andre. "Einstein's Time and Philosophy." In P. A. Y. Gunter, ed. *Bergson and the Evolution of Physics*. Knoxville: University of Tennessee Press, 1969.

Meyer-Abich, Klaus Michael. *Korrespondenz, Individualität und Komplementarität*. Wiesbaden: Franz Steiner, 1965.

"Monitum." *Acta Apostolicae Sedis* (June 30, 1962), vol. 54.

Monod, Jacques. *Chance and Necessity: An Essay on the Natural Philosophy of Modern Biology*. Austryn Wainhouse, trans. New York: Knopf, 1971.

Montgomery, William. "Germany." In Thomas Glick, ed., *The Comparative Reception of Darwinism*. Austin: University of Texas Press, 1974.

Morgan, C. Lloyd. *Emergent Evolution*. London: Williams and Norgate, 1923.

Morison, Samuel Eliot. *Three Centuries of Harvard, 1636–1936*. Cambridge: Harvard University Press, 1964.

Morris, Desmond. *The Naked Ape: A Zoologist's Study of the Human Animal*. New York: McGraw Hill, 1967.

Muller, Herman J. "Lenin's Doctrines in Relation to Genetics." In Loren R. Graham, *Science and Philosophy in the Soviet Union*. New York: Knopf, 1972.

Myers, G. H. and V. Parsonnet. *Engineering in the Heart and Blood Vessels*. New York: Wiley-Interscience, 1969.

Nagel, Ernest. *The Structure of Science: Problems in the Logic of Scientific Explanation*. New York: Harcourt Brace and World, 1961.

Nelidov, Iu. A. "O potomstve barona Petra Pavlovicha Shafirova." *Russkii evgenicheskii zhurnal* (1925), vol. 3, no. 1.

Nelkin, Dorothy. "The Science-Textbook Controversies." *Scientific American* (April 1976), 234:33–39.

Nelkin, Dorothy. *Science Textbook Controversies*. Cambridge: MIT Press, 1977.

Niels Bohr Memorial Session recorded in *Physics Today* (October 1963), 16:21–62.

Nisbett, Alec. *Konrad Lorenz*. New York: Harcourt Brace Jovanovich, 1976.

Nisbett, Alex. "Reply." *Journal of the History of the Behavioral Sciences* (1978), vol. 14.

Novikoff, Alex B. "The Concept of Integrative Levels and Biology." *Science* (March 2, 1945), 101:209–15.

"NSF: Congress Takes Hard Look at Behavioral Science Course." *Science* (May 2, 1975), vol. 188.

"O deiatel'nosti Russkogo Evgenicheskogo Obshchestva za 1921 god." *Russkii evgenicheskii zhurnal* (1922), vol. 1, no. 1.

"Obsuzhdenie Norvezhskoi evgenicheskoi programmy na zasedaniiakh Leningradskogo Otdeleniia R. E. O." *Russkii evgenicheskii zhurnal* (1925), vol. 3, no. 2.

Olberg, Oda. Review of Stavros Zurukzoglu, *Biologische Probleme der Rassenhygiene und die Kulturvölker.* In *Die Gesellschaft* (1926), vol. 3.

Olberg, Oda. Review of von Bohr-Pinnow, *Die Zukunft der menschlichen Rasse: Grundlagen und Forderungen der Vererbunglehre.* In *Die Gesellschaft* (1926), vol. 3.

Osterhout, W. J. V. "Biographical Memoir of Jacques Loeb 1859–1924." In *Biographical Memoirs*, vol. 13, no. 4. Washington, D.C.: National Academy of Sciences, 1930.

Peel, J. D. Y. *Herbert Spencer: The Evolution of a Sociologist.* New York: Basic Books, 1971.

Petersen, Aage, "The Philosophy of Neils Bohr." *Bulletin of the Atomic Scientists* (September 1963), vol. 19, no. 7.

Pickens, Donald K. *Eugenics and the Progressives.* Nashville: Vanderbilt University Press, 1968.

Planck, Max. *The New Science.* Greenwich: Meridian Books, 1959.

Planck, Max. *Das Weltbild der neuen Physik.* Leipzig: J. A. Barth, 1929.

Plekhanov, G. V. *Izbrannye filosofskie proizvedeniia*, vol. 3. Moscow, 1957.

Ploetz, Alfred. *Die Tüchtigkeit unserer Rasse und der Schutz der Schwachen: Ein Versuch über Rassenhygiene und ihr Verhältnis zu den humanen Idealen, besonders zum Socialismus.* Berlin: S. Fischer, 1895.

Politzer, Georges. *La fin d'une parade philosophique: Le bergsonisme.* Utrecht: Jean-Jacques Pouvert, 1967. First published in 1929 under the pseudonym François Arouet.

Powledge, Tabitha M. "From Experimental Procedure to Accepted Practice." *Hastings Center Report* (February 1976), vol. 6.

Powledge, Tabitha M. and John Fletcher. "Guidelines for the Ethical, Social, and Legal Issues in Prenatal Diagnosis." *New England Journal of Medicine* (January 25, 1979), vol. 300.

"Programma prakticheskoi evgenicheskoi politiki," *Russkii evgenicheskii zhurnal* (1927), vol. 5, no. 1.

Raabe, August. *Goethes Sendung im Dritten Reich.* Bonn: Ludwig Rührscheid, 1934.

Ramsey, Paul. "Genetic Engineering." *Bulletin of the Atomic Scientists* (December 1972), vol. 28, no. 10.

Rauschenberger, Walter. *Goethes Abstammung und Rassenmerkmale*. Leipzig: Robert Noske, 1934.

Redding, J. S. *Life Support: The Essentials*. Philadelphia: Lippincott, 1977.

"Rerun of German Genetic Therapy." *Medical World News* (August 25, 1975), pp. 30–31.

Ritchie, A. D. *Reflections on the Philosophy of Sir Arthur Eddington*. Cambridge: Cambridge University Press, 1948.

Rogers, Carl R. and B. F. Skinner. "Some Issues Concerning the Control of Human Behavior." *Science* (November 30, 1956), vol. 124, no. 3231.

Rogers, James Allen. "Charles Darwin and Russian Scientists." *The Russian Review* (1960), vol. 19.

Rogers, James Allen. "Russia: Social Sciences." In Thomas Glick, ed., *The Comparative Reception of Darwinism*. Austin: University of Texas Press, 1974.

Rosenfeld, Leon. "Bohr, Niels Henrik David." *Dictionary of Scientific Biography*, 2:239–54. New York: Scribner, 1970.

Rosenfeld, L. "Niels Bohr." *Nuclear Physics* (1963), vol. 41.

Rosenfeld, Leon. *Niels Bohr: An Essay Dedicated to Him on the Occasion of His Sixtieth Birthday*. Amsterdam: North-Holland, 1961.

Roszak, Theodore. *Where the Wasteland Ends: Politics and Transcendence in Postindustrial Society*. Garden City, N.Y.: Doubleday, 1972.

Rozental, S., ed. *Niels Bohr, His Life and Work as Seen by His Friends and Colleagues*. Amsterdam: North-Holland, 1967.

"Rukovodiashchie polozheniia nemetskogo obshchestva rasovoi gigieny." *Russkii evgenicheskii zhurnal* (1923), vol. 1, no. 3–4.

Russell, Bertrand, *The Philosophy of Bergson*. Cambridge: Bowes and Bowes, 1914.

Russell, E. S., *The Interpretation of Development and Heredity: A Study in Biological Method*. Oxford: Oxford University Press, 1930.

Sahlins, Marshall D. *The Use and Abuse of Biology*. Ann Arbor: University of Michigan Press, 1976.

Saller, K., *Die Rassenlehre des Nationalsozialismus in Wissenschaft und Propaganda*. Darmstadt: Progress, 1961.

Saunders, Peter and Philip Rhodes. "The Origin and Circulation of the Amniotic Fluid." In D.V.I. Fairweather, ed., *Amniotic Fluid: Research and Clinical Application*. Amsterdam: Excerpta Medica, 1973.

Schadewaldt, Wolfgang, Bernhard Schweitzer, and Johannes Rtroux. "An die Leser der 'Antike.'" *Die Antike* (1937), vol. 13, no. 2.

Schaffner, Kenneth F. "Antireductionism and Molecular Biology." *Science* (August 11, 1967), vol. 157:644–47.

Schaller, George B. "Behavioral Comparisons of the Apes." In Irven De Vore, ed., *Primate Behavior: Field Studies of Monkeys and Apes*. New York: Holt, Rinehart, and Winston, 1965.

Schaller, George B. "The Behavior of the Mountain Gorilla." In Irven De Vore, ed., *Primate Behavior: Field Studies of Monkeys and Apes*. New York: Holt, Rinehart, and Winston, 1965.

Schaller, George B., *The Mountain Gorilla*. Chicago: University of Chicago Press, 1963.

Schallmayer, Wilhelm. *Vererbung und Auslese im Lebenslauf der Völker: Eine staatswissenschaftliche Studie auf Grund der neueren Biologie*. Jena: G. Fischer, 1903.

von Schirach, Baldur, ed. *Goethe an Uns: Ewige Gedanken des grossen Deutschen*. Berlin: Zentralverlag der NSDAP, 1943.

Schmeck, Harold M., Jr. *The Semi-Artificial Man: A Dawning Revolution in Medicine*. New York: Walker, 1965.

Senghor, Leopold. *Pierre Teilhard de Chardin und die afrikanische Politik*. Cologne: M. DuMont Schauberg, 1968.

Serebrovskii, A. S. "Genealogiia roda Aksakovykh." *Russkii evgenicheskii zhurnal* (1922), vol. 1, no. 1.

Serebrovskii, A. S. "O zadachakh i putiakh antropogenetiki." *Russkii evgenicheskii zhurnal* (1923), 1(2):107–16.

Serebrovskii, A. S. "Teoriia nasledstvennosti Morgana i Mendelia i marksisty." *Pod znamenem marksizma* (1926), no. 3.

Sertillanges, A. G. *Henri Bergson et le Catholicisme*. Paris: Flammarion, 1941.

Shankland, R. S. "Conversations with Albert Einstein." *American Journal of Physics* (1963), vol. 31.

Sherman, May. *Artificial Organs: Kidney, Lung, Heart*. Bethesda, Md.: National Institutes of Health, 1965.

Shmidt, O. Iu. *Izbrannye trudy*. 4 vols. Moscow, 1959–1960.

Shmidt, O. Iu. *Zhizn' i deiatel'nost*. Moscow, 1959.

Simpson, George Gaylord. "On the remarkable testament of the Jesuit paleontologist Pierre Teilhard de Chardin." *Scientific American* (April 1960), 202(4):201–7.

Sinsheimer, Robert. "Comments." *Hastings Center Report* (August 1976), vol. 6.

Skinner, B. F. *Beyond Freedom and Dignity*. New York: Knopf, 1971.

Skinner, B. F. *Walden II*. New York: Macmillan, 1976.

Slater, Noel B. *Development and Meaning of Eddington's Fundamental Theory*. Cambridge: Cambridge University Press, 1957.

Slepkov, Vasilii. "Nasledstvennost' i otbor u cheloveka (Po povodu teoreticheskikh predposylok evgeniki)." *Pod znamenem marksizma* (1925), no. 4, pp. 102–22.

Sloan, Douglas. "The Teaching of Ethics in the American Undergraduate Curriculum, 1876–1976." In his edited *Education and Values*. New York: Teachers College Press, 1980.

Smith, Alice Kimball. *A Peril and a Hope: The Scientists' Movement in America: 1945–47*. Chicago: University of Chicago Press, 1965.

Smuts, Jan Christiaan, *Holism and Evolution*. London: Macmillan, 1926.

Sorel, Georges. "L'Evolution creatrice." *Le Mouvement Socialiste* (October 15, 1907), pp. 257–82.

Sorel, Georges. "Sur L'evolution créatrice." *L'indépendant* (May 1, 1911).

"Sovremennoe sostoianie voprosa o sterilizatsii v Shvetsii." *Russkii evgenicheskii zhurnal* (1925), vol. 3, no. 1.

Speaight, Robert. *The Life of Teilhard de Chardin*. New York: Harper and Row, 1967.

Stark, Johannes. *Nationalsozialismus und Wissenschaft*. Munich: Zentralverlag der NSDAP, 1934.

Stent, Gunther. *The Molecular Biology of Bacterial Viruses*. San Francisco: W. H. Freeman, 1963.

Szende, Paul. "Bergson, der Metaphysiker der Gegenrevolution." *Die Gesellschaft: Internationale Revue für Sozialismus und Politik* (December 1930), 7(12):542–68.

Teilhard de Chardin, Pierre. *The Future of Man*. New York: Harper and Row, 1964.

Teilhard de Chardin, Pierre. *Human Energy*. New York: Harcourt Brace Jovanovich, 1969.

Teilhard de Chardin, P., *Man's Place in Nature*. New York: Harper and Row, 1966.

Thibaudet, Alfred. *Le Bergsonisme*, vol. 2. Paris: Editions de la Nouvelle Revue Francaise, 1923.

Thompson, Jerry N. "Prenatal Detection of Heritable Metabolic Disorders." *Pediatric Annals* (June 1978), vol. 7.

Timiriazev, K. A. *Charlz Darvin i ego uchenie*. 3d ed. Moscow, 1894.

Tinbergen, N. "Ethology." In his *The Animal in Its World*, vol. 2. Cambridge: Harvard University Press, 1973.

Tonquédec, Joseph de. "Comment interpréter l'ordre du monde?"

Études par des Pères de la Compagnie de Jésus (March 5, 1908), vol. 114, pp. 577–97.

Tonquédec, Joseph de. "M. Bergson est-il moniste?" *Études par des Pères de la Compagnie de Jésus* (February 20, 1912), vol. 130, pp. 506–16.

Toulmin, Stephen. "Can Science and Ethics Be Reconnected?" *Hastings Center Report* (June 1979), 9:27–34.

Toulmin, Stephen. "How Can We Reconnect the Sciences with the Foundations of Ethics?" In H. Tristram Engelhardt Jr. and Daniel Callahan, eds., *Knowing and Valuing: The Search for Common Roots*. Hastings-on-Hudson, N.Y.: Hastings Center, 1980.

Toynbee, Arnold, "Great Expectations." In Harvey Wheeler, ed., *Beyond the Punitive Society: Operant Conditioning: Social and Political Aspects*. San Francisco: W.H. Freeman, 1973.

Trueblood, D. Elton. *The People Called Quakers*. New York: Harper and Row, 1966.

van Lawick-Goodall, Jane. "A Preliminary Report on Expressive Movements and Communication in the Gombe Stream Chimpanzees." In Phyllis C. Jay, ed., *Primates: Studies in Adaptation and Variability*. New York: Holt, Rinehart, and Winston, 1969.

Vaihinger, Hans. *The Philosophy "As-If."* C. K. Ogden, trans. New York: Harcourt, Brace, and World, 1924.

Vavilov, S. I., *Sobranie sochineniia*. 4 vols. Moscow, 1954–56.

Vecoli, Rudolph. "Sterilization: A Progressive Measure?" *Wisconsin Magazine of History* (1960), vol. 43.

Vermel', S. S. "Prestupnost' evreev." *Russkii evgenicheskii zhurnal* (1924), vol. 2, no. 2–3.

Verrall, Richard. "Sociobiology: The Instincts in Our Genes." *Spearhead* (March 1979), no. 127.

Veselov, M. G., P. L. Kapitsa, and M. A. Leontovich. "Pamiati Vladimira Aleksandrovicha Foka." *Uspekhi fizicheskikh nauk* (October 1975), vol. 117, no. 2.

Vietroff, J. "L'influence de la philosophie de M. Bergson." *Le Mouvement Socialiste* (January 1912).

Volotskoi, M. V. "K istorii evgenicheskogo dvizheniia." *Russkii evgenicheskii zhurnal* (1924), vol. 2, no. 1.

Volotskoi, N. M. "Spornye voprosy evgeniki." *Vestnik kommunisticheskoi akademii* (1927), no. 20, pp. 224–25.

Vucinich, Alexander. "Russia: Biological Sciences." In Thomas Glick, ed., *The Comparative Reception of Darwinism*. Austin: University of Texas Press, 1974.

Waldman, Mark. *Goethe and the Jews: A Challenge to Hitlerism*. New York: Putnam, 1934.

Warshofsky, Fred. *The Rebuilt Man: The Story of Spare-Parts Surgery*. New York: Thomas Y. Crowell, 1965.

Washburn, S. L. and D. A. Hamburg. "Aggressive Behavior in Old World Monkeys and Apes." In Phyllis C. Jay ed., *Primates: Studies in Adaptation and Variability*. New York: Holt, Rinehart, and Winston, 1969.

Washburn, Sherwood L. and David A. Hamburg. "The Implications of Primate Research." In Irven De Vore, ed., *Primate Behavior: Field Studies of Monkeys and Apes*. New York: Holt, Rinehart, and Winston, 1965.

Washburn, Sherwood L. and David Hamburg. "The Study of Primate Behavior." In Irven De Vore, ed., *Primate Behavior: Field Studies of Monkeys and Apes*. New York: Holt, Rinehart, and Winston, 1965.

Washburn, S. L. and Phyllis C. Jay, eds. *Perspectives on Human Evolution*. New York: Holt, Rinehart, and Winston, 1968.

Washburn, S. L. and C. S. Lancaster. "The Evolution of Hunting." In S. L. Washburn and Phyllis C. Jay, eds., *Perspectives on Human Evolution*. New York: Holt, Rinehart, and Winston, 1968.

Wasmuth, Carl E. "Euthanasia." *The Encyclopedia Americana*, vol. 10. New York: Americana, 1978.

von Weizsäcker, C. F. *Zum Weltbild der Physik*. Stuttgart: Hirzel, 1960.

Westman, Robert S. "The Melanchthon Circle, Rheticus, and the Wittenberg Interpretation of the Copernican Theory." *Isis* (June 1975).

Wheeler, Harvey, ed. *Beyond the Punitive Society: Operant Conditioning, Social and Political Aspects*. San Francisco: W. H. Freeman, 1973.

Whittaker, E. T. *Eddington's Principle in the Philosophy of Science*. Cambridge: Cambridge University Press, 1951.

Whittaker, Edmund. *From Euclid to Eddington*. Cambridge: Cambridge University Press, 1949.

Wildiers, N. M. *An Introduction to Teilhard de Chardin*. New York: Harper and Row, 1968.

Wilson, E. O. *Sociobiology: The New Synthesis*. Cambridge: Harvard University Press, 1975.

Wilson, E. O. "The Superorganism Concept and Beyond." In

L'effet de Groupe chez les Animaux. Paris: Colloques internationaux du Centre National de la Recherche Scientifique, 1967.

Wittgenstein, L. *Tractatus Logico-Philosophicus.* Milan and Rome: Bocca, 1954.

Woodger, J. H. *Biological Principles: A Critical Study.* London: Kegan Paul, Trench, Trubner, 1929.

Yakhot, Jehoshua. "The Theory of Relativity and Soviet Philosophy." *Crossroads* (Israel Research Institute of Contemporary Society), Autumn 1978, pp. 92–118.

Yolton, J. W. *The Philosophy of Science of A. S. Eddington.* The Hague: Martinus Nijhoff, 1960.

Zavadovskii, B. M., Darvinizm i lamarkizm i problema nasledovaniia priobretennykh priznakov," *Pod znamenem marksizma* (1925), 10–11:79–114.

Zolotarev, V. "Dekabristy." *Russkii evgenicheskii zhurnal* (1928), 6:178–97.

INDEX